T0143211

Predicting the Weather

Predicting the Weather

Victorians and the Science of Meteorology

KATHARINE ANDERSON

The University of Chicago Press

CHICAGO AND LONDON

KATHARINE ANDERSON is associate professor
in the science and society program at York University.

The University of Chicago Press, Chicago 60637
The University of Chicago Press, Ltd., London

Printed in the United States of America

14 13 12 11 10 09 08 07 06 05   1 2 3 4 5

ISBN: 0-226-01968-3 (cloth)

Library of Congress Cataloging-in-Publication Data

Anderson, Katharine.
Predicting the weather : Victorians and the science of meteorology / Katharine Anderson.
p. cm.
Includes bibliographical references and index.
ISBN: 0-226-01968-3 (cloth : alk. paper)
1. Meteorology—Great Britain—History—19th century.
2. Weather forecasting—Great Britain—History—19th century.
3. Great Britain—History—Victoria, 1837-1901. I. Title.
QC857.G77A53 2005
551.5'0941—dc22   2004020178

⊗ The paper used in this publication meets
the minimum requirements of the American National
Standard for Information Sciences—Permanence of Paper
for Printed Library Materials, ANSI Z39.48-1992.

FOR WILLIAM, DOMINIQUE, & HOPE

Contents

Acknowledgments

I AM grateful for the opportunity to record here my debts to many individuals and institutions whose support made this book possible. It emerged from a dissertation on the British meteorological department, supervised by Harold Perkin, Bill Heyck, and Joel Mokyr at Northwestern University. Their insights and encouragement made all the difference. Two generous fellowships bookend the evolution of that initial project into this book: the Social Sciences and Humanities Research Council of Canada funded a postdoctoral fellowship in 1995–96; and a fellowship from the Faculty of Arts at York University in 2001 allowed me to complete the writing. During these years, I benefited from the collegiality and sharp questions of faculty and students during extended visits to the History and Philosophy of Science Department at Cambridge University, the History of Science Department at Harvard, and the History Department at Princeton University. The wide intellectual curiosity and good cheer of my own colleagues in the Division of Humanities at York University always made me glad to return, even though it meant leaving the reading rooms of the Cambridge, Widener, or Firestone libraries behind.

The encouragement of Susan Abrams at the University of Chicago Press was as critical to me as it was to many other historians of science. After her final illness intervened, Alan Thomas and then Catherine Rice applied their formidable energy and expertise to the manuscript. The book also profited greatly from the meticulous copyediting work of Erin Milnes. Several individuals deserve particular thanks for having read the manuscript in whole or in part: Timothy Alborn, Graeme Burnett, James Fleming, Jan Golinski, Eileen Groth, Bernard Lightman, Anne Secord, Jim Secord, Simon Schaffer,

Alison Winter, and, last but not least, the two anonymous reviewers for the press. They rescued me from many errors and infelicities but of course bear no responsibility for those that remain. I am doubly grateful that Jim Burton not only shared his detailed knowledge of and enthusiasm for the history of the Meteorological Office, but insisted that some of our conversations should take place in Yorkshire rather than central London. In Edinburgh, Margery Roy and Mary Bruck introduced me to the Scottish Meteorological Society and the Piazzi Smyth archives respectively. I owe special thanks to the resource-sharing staff at Scott Library, the librarians of the Royal Edinburgh Observatory, Mick Wood and Maurice Crewe at the Meteorological Office library and archives, Jane Insley at the Science Museum, Peter Hingley at the Royal Astronomical Society, and Derek Barlow, who indexed the Meteorological Office papers now held at the Public Record Office. I have the great good fortune to possess an aunt, Margaret Beer, who not only read my work with apparently genuine interest but also often provided a bed three minutes' walk from the Public Record Office.

Most of all, I have depended on the love and support of family and friends—especially Rosemary and Peter Anderson, Carol and Ernie Trischuk, Sarah Anderson, Jennifer Schmidt, Anna Beer, and of course my own family: William, Nicky, and Hope. Like all the rest, but more openly, Nicky has spent the last two years wondering whether this book would ever be done and I am glad finally to be able to answer her.

INTRODUCTION

A Science of the Weather

WEATHER, like death and taxes, has an inevitable quality. Its phenomena are rhythmic, like the seasons or droughts and floods, or fleeting, like the clouds. In either case, weather seems universal and timeless. Yet of course our ideas about the weather have a history. In this sense—quite separate from the question of actual climate change—Victorians had different weather. They encountered its phenomena armed with views of nature and science that we do not share. Above all, to Victorians the weather seemed a puzzle that would be solved. In this era, natural philosophers explored invisible forces like electricity and magnetism and developed new concepts of heat and energy. They built global networks to collect physical data, stretched telegraphic cables across the oceans, and with camera and spectroscope penetrated the chemistry of the sun and planets. Surrounded by such impressive new theories and new technologies, many Victorians assumed that the problems of meteorology were ones of scale and complexity: challenging, but given enough observation and coordination, ultimately tractable. Like other kinds of physical knowledge, the weather would become mathematical and certain. This book explores that expectation, because it tells us not only about the history of meteorology, but also about the nature of science in public life.

For weather prediction was a sharply modern science. Its development hinged on a dramatic new form of communication, the telegraph, which for the first time allowed weather data to travel faster than the winds themselves. The electric telegraph was developed as a working instrument between 1836 and 1844 and spread rapidly in both Europe and the United States. By the early 1850s, the first Channel cable had been laid and the European capitals were linked by telegraphy; within twenty years, ocean cables had extended these links globally.[1] The existence of these networks of wire and cable transformed conditions for the study of the weather. Within particular nations, telegraphy connected observers with a central office; internationally, the daily exchange of data between European observatories made it possible to track atmospheric change across a wide area with regularity and immediacy. Moreover, because the telegraph was intimately connected with commercial development of railways and newspapers, its role in the science perhaps more subtly encouraged expectations about the dissemination of practical knowledge as well. In short, the telegraph introduced a new dynamic to meteorology.

But the science of weather was also modern in another sense. It called attention to the new scale of scientific activities, which required coordination and centralization that put the work beyond any one individual's abilities and resources. This element of scale raised new and urgent questions about public funding of science. In 1854, a British government office to collect meteorological data was formed, the precursor of the present Meteorological Office. Its first director built a telegraphic network for warning the public about storms and began to publish regular "forecasts" in the newspapers. Then, in 1865, he committed suicide, an act of private despair that was widely interpreted as a response to the pressures of his public office. The country's most elite scientific organization, the Royal Society, adopted the management of government meteorology and promptly canceled forecasting. These events lifted meteorological science into the clouds of controversy. What was proper scientific prediction, and what was it for? What merits distinguished observatory science from popular knowledge about weather conditions? Whose interests should guide scientific research? In short, what sort of knowledge, for what ends, and under whose leadership? An examination of the development of meteorology, then, especially the problem of predicting the weather, provides a revealing way to analyze science and culture.

The chief questions of this book involve scientific practices and the public worlds in which those practices developed in Britain from about 1840 to 1880. In 1859 Britain was in fact the third European country, after France and Belgium, to create a telegraphic network for weather forecasting, and all three were preceded by an American project, based at the Smithsonian Institution

1. For a good summary of telegraphy in Britain in this period, see "Telegraph" in the *Encyclopaedia Britannica*; see also Headrick, *The Invisible Weapon*; and Marvin, *When Old Technologies Were New*. For a different account of the impact of telegraphy in the physical sciences, see Hunt, "Doing Science in a Global Empire."

under Joseph Henry. Why then did British efforts prove to be the most energetic and ambitious of the European initiatives? There was one obvious reason it should not: British meteorology labored under a considerable geographical handicap, especially obvious when its situation was compared to the large landmasses of the North American continent. Without the means of collecting data from the empty spaces of the Atlantic that stretched west of Ireland, British weather telegraphy was often painfully limited. Yet this obstacle was offset by some other practical considerations. Above all, the nation was economically, technologically, and politically well placed to see meteorology's implications for maritime trade. Britain had the largest merchant marine in the world. The annual figures from 1868, at the height of disputes over government forecasting efforts, listed some twenty-two thousand vessels (more than five million tons) and two hundred thousand men involved in regular shipping, coastal and international. Significantly, the vast majority (more than 92 percent) of this tonnage was under sail. Of the new vessels built in 1868, most were still sailing vessels: 879 as compared to 232 steamers. This was the era of the clippers, fast sailing ships initially designed for the China tea trade where the first cargo of the season brought in huge premiums. The discovery of gold in California in 1848 and then in Australia a few years later increased the value of rapid sea passages. Not until the development of the Suez Canal (1869) and the steam turbine (1884) did sail give way to steam. Before then, a science of the winds had an obvious national interest.

The meteorological controversies displayed to great effect the tensions surrounding scientific authority, popular knowledge, and government in Victorian society. Meteorology was exactly fitted to reveal the ambitions of elite science. Scientists invested the construction of a government-sponsored science with great significance: an immature field, meteorology would be guided and developed by a combination of intellect and efficient administration. This vision explicitly rejected another characteristic of the science: the strength of popular traditions of weather knowledge, including the lively traditional medium for their expression, the almanacs. By the 1870s, there were dozens of periodicals and popular texts in which meteorological fact and opinion were exchanged. Their pages presented the contests among different interests in science and provided the terrain on which individuals and scientific institutions staked their claims to authority and expertise. In parallel with these literary debates, meteorological practices also highlighted the exchange and comparison of scientific knowledge. The central office, with its fanned network of observers and its public announcements, could not help but underline the relationships between different interests and different kinds of observation.

Study of the weather linked the seaman watching storm signals, or the reader of newspaper weather charts, with the council of the Royal Society and the standardized instruments of national observatories. The exchanges between such individuals, networks, and institutions made visible the processes of fashioning facts into science, and the standards and value of these processes became part of the debates.

The story of weather prediction accordingly connects two kinds of narratives about nineteenth-century science. One is the institutionalization of science: the growth of specialized societies and journals, the assertive presence of men of science in cultural life after mid-century, the movements for scientific education and research funding. The other is the strength of popular interest in science—evidenced by the sales of general books on natural history or astronomy, or the coverage of scientific debates in general periodicals. These stories often seem to fit together as an account of leadership by a cultural elite, harnessing the popular enthusiasm for science. But in fact meteorology shows a much less stable picture. Meteorology emerged in this period as a physical science, a science of precision instruments, increasingly associated with elite observatories and statistical manipulation of data—all the ingredients of high scientific authority. Despite this authority, the scientific men of the Meteorological Office could not control popular expectations, nor could they decisively separate scientific *observation* of the weather from unscientific *experience* of the weather. However, even to phrase the differences this way is misleading, suggesting a territory of a modern kind in which the policing of boundaries between lay and expert interests was simply unsuccessful. Instead, there was a much more ambitious project under way. Meteorology displayed the characteristically Victorian emphasis on the integration of culture. No one had really agreed to differ, and all assumed that a whole vision of nature, once discovered, would address all interests and communities.

This account reflects the recent transformation of historical studies of nineteenth-century science. The picture of the development of a professional and largely secular scientific community, based intellectually on the achievements of Faraday, Maxwell, Darwin, or Kelvin and based socially and politically on the influence of national societies, educational reforms, and the leadership of a metropolitan elite, is no longer so clear. Knowledge-making and intellectual life now seems much more heterogeneous, spread across a diversity of settings and social groups, and the scientific professionalization of the metropolitan leaders seems correspondingly less secure. As a result of this changing picture, tracing the engagement of scientific theories and practices with Victorian cultural life has become a steadily richer enterprise. Attention to popular science, publishing conventions, visual culture, instruments, or gender have given new perspectives on the character and values of Victorian natural philosophy.[2] This study of weather prediction is indebted to these histories in terms of the questions it asks, and the wide range of sources it consults.

The focus of this book, on weather forecasting in mid-century, is designed to explore the place of science in Victorian society, but the book also makes a claim about the character of meteorology more broadly. Meteorology is a science of observation and of arguments about observation; it is hard to grasp its development as a discipline by an exclusive focus on rival theories about atmospheric change. The historical action is in ideas about instruments or how statistics are used, about the ways that popular weather prophets communicated with the public, how networks of observers were built, and how government officials interacted with scientists. This claim deserves some attention because it contains a general and a particular argument. Why is a focus on practices the way to study nineteenth-century meteorology? A general answer to that question invokes a style of history: a highlighting of the web of material objects, like instruments, texts, or buildings, within which intellectual developments take their shape, and an analysis of how science intersects with other social and cultural concerns. But a second answer is particular and explains why meteorology is a rich subject for the history of science. Meteorology, and especially forecasting, provides a way of exploring critical contemporary debates on the relationship of speculation and observation. In this period, implicit and explicit arguments about method—about the relative worth of speculation or exact observation—were a constant feature of meteorology. Meteorologists developed a distinction between theory (the physics of the atmosphere) and observation practices (weather forecasting) and, moreover, readily discussed the implications of that distinction. Such self-conscious methodological debate in meteorology readily exposed the mixture of epistemological, political, and social concerns that shaped the science.

One of the guiding assumptions of this book, therefore, is that a study of the social context of weather forecasting illuminates the history of the development of scientific disciplines, and vice versa. Over the course of the century, meteorology came to represent a test case of the merits of speculation or disciplined, Baconian observation. The close connections between accounts

2. For guides to this extensive literature, see the bibliographies in Lightman, ed., *Victorian Science in Context*.

of scientific method and social relationships are obvious in even a brief summary of the attitudes of prominent natural philosophers toward theory and observation in the study of the weather.

The criticism of theory in Britain, for instance, was especially noteworthy in contrast to the heated debates that took place in the United States in the 1830s and 1840s, when rivals in Boston and Philadelphia argued about how and whether storms rotated, whether atmospheric electricity was a cause or effect of rain and winds, and the causes of rain.[3] John Frederick William Herschel, one of the leading men of science in the first half of the century, certainly followed the American exchanges and thought they showed how "theoretical speculations" might afford fruitful "collisions of intellect." But he also noted, in an article on meteorology for the *Encyclopaedia Britannica,* that meteorology was a science of induction rather than deduction. "If there be one part of dynamical science more abstruse and unapproachable than another, it is the doctrine of the propagation of motion in fluids, and especially elastic fluids like the air."[4] Surfaces are not uniform, and land and seas, mountain and plain, differently affected the motion of bodies of air passing over them. The local heat from sun, moderated by variable clouds, combined with more regular variation of temperatures according to latitude, and with the heat moved around by evaporation and rain, to create a bewildering assembly of dynamic forces. He concluded it was "one of the most marked peculiarities of this our science," that when we trace the forces of the atmosphere "as exhibited in a specified time and place, theory affords us no assistance."[5] Herschel's conclusion, that meteorology was a "science of detail" in which "laborious and continued observation" was the key, continued to dominate approaches to the subject in the nineteenth century.[6] By 1850, the president of the British Meteorological Society, Samuel Charles Whitbread, was making an almost ritualized statement of method as he exhorted its members to "observe, record and collate, if we would arrive at first causes" and insisted that "above all things" the society's object must be the collection of "strict scientific data

3. Feldman, "Late Enlightenment Meteorology"; Frisinger, *The History of Meteorology*; Garber, "Thermodynamics and Meteorology"; Hildebrandsson and Teisserenc de Bort, *Les bases de la météorologie dynamique*; Khrgian, *Meteorology*; Kutzbach, *The Thermal Theory of Cyclones*; Middleton, *A History of the Theories of Rain*; Shaw, *Manual of Meteorology*.

4. Herschel, *Meteorology*, 3. Herschel's article was widely known and was reprinted as a separate text in 1861.

5. Ibid., 5.

6. Ibid.

by accurate instrumental means."[7] Meteorology, according to these influential accounts, was a science of observation.

Yet, there were advocates of deductive reasoning who seized on meteorology. Writing to a colleague in 1831, the Cambridge natural philosopher William Whewell announced that meteorology was his "new pet," a ripe field of investigation.[8] In the next few years Whewell continued to press for "general laws" as well as systematic observation. He spurred on his Edinburgh colleague James Forbes in 1833 with a list of observational and theoretical questions about clouds, the formation of rain, storm theories, and upper air currents.[9] Whewell's list insisted that observation serve theoretical inquiry, and Forbes, in response, agreed that meteorology most of all needed its Newton. Half a century later, in 1882, the physicist John Tyndall was busy with another blow for theory. "Meteorology," he wrote in an article on thermodynamics, "seems to me to abound in facts which it has hitherto been incompetent to explain." He had no tolerance for the meteorologists who claimed that difference between laboratory investigation and the open conditions of atmosphere meant that theoretical work was of little use. "Experiments," noted Tyndall severely, "are necessary to rescue their science from empiricism."[10] By the end of the century, meteorologists increasingly applied thermodynamics to the study of weather, but the lead was taken by continental scientists like Heinrich Hertz and Hermann von Helmholtz—a reflection of the continuing British preoccupation with defining the significance of observations.

This emphasis on observation enacted understandings about who did meteorology. The tension between observation and speculation blended with accounts of the familiarity of the science, indeed its domesticity. The weather provided a notably accessible field of study. Meteorologists could work "[i]n the secluded parts of a crowded city, in the smallest room, amid the busiest haunts of man," as one journal noted.[11] Observation required only methodical habits and precision, and no more than an hour a day. Meteorology gained a reputation as a science "within the grasp of every mind,"[12] where disciplined observation without specialized knowledge could go a long way. "There is no science in which so much may be accomplished by private observers," urged

7. Whitbread, "President's Address," 2.

8. William Whewell to Rev. R. Jones, July 23, 1831, in Todhunter, *William Whewell*, 2:124.

9. William Whewell to Edward Forbes, June 10, 1833, in Todhunter, *William Whewell*, 2:165–67.

10. Tyndall, "Action of Free Molecules on Radiant Heat," 344–45.

11. Webster, *A Meteorological Journal for October* 1841, 5.

12. Scott, "A Word about the Weather."

David Brewster, another of the founding members of the British Association for the Advancement of Science.[13] "If each does what he can, and contributes his mite, much may be accomplished," announced the Reverend Charles Clouston, pursuing his own meteorological observations in a remote area of Scotland.[14] With the emergence of newspaper weather forecasts and weather maps, Victorians could see science still more readily as a daily practice accessible to all. Meteorology seemed to preserve a special role for the observers scattered far from the centers of British science. Weather prediction offered benefits that potentially affected everyone, but especially those at the periphery in a literal as well as figurative sense—the coasts and ports of Britain.

How did climate—weather over the long term—fit into this picture? Although ideas about climate are not a focus of this account, they did bear on the identification of meteorology as a science that addressed urgent, contemporary inquiries about public health, sanitation, and even imperial power. Climate was traditionally a question of nature and place, looking backward to the Hippocratic tradition of the study of "airs, waters, places." The development of ideas about climate in the nineteenth century, accordingly, often emerged within accounts of geology, paleontology, medicine, and travel, rather than accounts of the dynamics of the atmosphere.[15] But the study of climate nevertheless represented an important background to debates about weather prediction. It provided an obvious context in which to articulate the national value of scientific knowledge. The General Register Office, for instance, was founded in 1838 to gather statistics of mortality and, as a related activity, collected and published comprehensive weather information.[16] Particular epidemics, too, stimulated special interest in meteorology, following the miasmatic theory that contagion spread through the atmosphere. In the wake of the cholera epidemic of 1853–54, London's General Board of Health

13. [Brewster], "The Weather and Its Prognostics," 204.

14. Clouston, *Popular Weather Prognostics of Scotland,* 2.

15. For the tradition of climate and locality, see Jankovic, *Reading the Skies*; White, *Natural History of Selbourne*; Howard, *The Climate of London*; for the development of ideas about climate, see Bowler, *The Fontana History of the Environmental Sciences,* 220–29; Feldman, "The Ancient Climate"; Fleming, *Historical Perspectives on Climate Change*; Greene, *Geology in the Nineteenth Century*; Hamlin, "James Geikie, James Croll, and the Eventful Ice Age"; Le Roy Ladurie, *Times of Feast, Times of Famine*; Ospovat, "Lyell's Theory of Climate." Examples of discussions of climate, health, and travel are Abercromby, *Seas and Skies in Many Latitudes*; Clark, *The Influence of Climate in the Prevention and Cure of Chronic Diseases*; Kevan, "Quests for Cures"; Sargent, *Hippocratic Heritage*; Thomson, *A Change of Air.*

16. Szreter, "The General Register Office and the Public Health Movement"; Burton, "Meteorology and the Public Health Movement."

commissioned meteorological analysis of this and past cholera years from the meteorologist at Greenwich Observatory.[17] Finally, the Victorian experience with tropical climates was deeply influential. Over the late eighteenth and nineteenth centuries, the British built new accounts of bodies and climate. Tropical climates like those of India became interpreted as fundamentally hostile, and acclimatization impossible: conclusions that reflected ideas about government and empire as well as airs and places.[18] Victorian accounts of climate, then, although in most ways separate from the controversies over weather prediction, critically reinforced the perception of meteorology as a public concern.

In this way, the study of the weather promised to bring things together—the hour and the season, the particular fact and the general law, the local record and the central office, practical applications and theoretical advance, the provincial observer and the elite scientist, the nation and the empire. At the same time, this promise of philosophical and social coherence made the question of establishing the proper relationship of these several elements correspondingly important. This concern, under various guises, provides a continuous thread in the history of Victorian meteorology and explains its importance for the study of science as both a set of philosophical questions and social practices.

The path of this book is both thematic and chronological. It begins by tracing ideas about prediction and prophecy in diverse sciences and in other fields of inquiry, such as history, law, and theology. For Victorians, the meanings of prediction and prophecy were formed and reformed against other convictions about precision, national identity, economic progress, and religion. The first chapter argues that men of science built a spectrum of knowledge that connected the most prestigious and impressive kinds of prediction, in astronomy, with other lesser investigations in order to represent science as the highest form of reason. But those arguments about a spectrum of knowledge could turn into a liability, since meteorology could as easily appear to slide toward the unreason of astrology or religious prophecy as aspire to astronomy. The central place of prediction in accounts of a cultural hierarchy of knowledge made the struggle with partial and inexact forecasts in meteorology a critical matter. These debates about prediction provide a framework for

17. Glaisher, *Report on the Meteorology of London.*

18. Livingstone, *The Geographical Tradition*, 221–28; Livingstone, "Tropical Climate and Moral Hygiene"; Harrison, *Climates and Constitutions*; Mathe-Shires, "Imperial Nightmares."

understanding meteorological controversies as they unfolded from the 1840s to the 1880s.

The next two chapters consider different settings in which the science of meteorology developed around mid-century and analyze the implications of these settings for the organization of the science and its claims to cultural authority. Chapter 2 concentrates on the 1840s, a decade before the establishment of the Meteorological Department, and considers an energetic circle of popular weather prophets who shaped discussions about meteorology. Their influence leads to questions about publicity, reputation, and the changing circulation of scientific knowledge. Integrating popular weather prophets into the history of meteorology requires attention to the role of the almanacs, as a traditional but revitalized form of publication in this period. The story of the prophets and their almanacs shows how the marketplace, popular culture, and scientific authority took shape in British meteorology. Meteorologists and meteorological organizations from the 1840s on, whether they explicitly acknowledged it or not, worked in awareness of the lively popular tradition of weather prophecy and its distinctive form, the almanac.

Chapter 3 considers the development of meteorology as a model of collective science, partly in response to the disorderly world of popular weather prophecy. This model, establishing a careful hierarchy of observers and leaders, was a critical influence on the Meteorological Department from 1854 to 1867. As an observation network and as a government enterprise, meteorology represented the collective organization of science working in official or semi-official bodies, interacting with public audiences. All these forms of collective enterprise raised questions about personal responsibility, public interests, and the proper scale of government activity. A public scientific office like the Meteorological Department in particular exposed the sharp split between popular interest in forecasting and the distaste of many men of science for utilitarian goals in science.

The fourth chapter shifts attention to methods of observation in the science: what was being measured, and why. These questions emerge within a narrative of the history of the Meteorological Office under the reforming management of the Royal Society. By the 1860s and 1870s, increasingly uncomfortable with the public nature of their enterprise, the Royal Society sought to dismiss forecasting and confine meteorology to the observatory. But this quickly led to discussions about the merits of precision, statistical knowledge, and determinism. When the order of quantitative evidence failed to satisfactorily control the disorder of atmospheric phenomena, meteorology became a battleground for scientific and religious authorities to debate the limits of natural law.

Further exploration of ideas about precision brings us to the visual tools and products of meteorology, discussed in chapter 5. Meteorologists experimented eagerly with maps, charts, and cloud cameras, especially from the 1860s on, as new technologies of printing and photography emerged. By analyzing the development of visual methods in relationship to ideas about weather wisdom, I examine forms of knowledge that challenged those of instrument and number. Maps built bridges to a model of knowledge known as weather wisdom, a term for the popular, apparently intuitive insight of sailors and shepherds that seemed to succeed where government weather forecasting failed. Because they embodied the intersection of popular and conventionally scientific approaches to the study of the weather, maps and their counterparts offered a different kind of history of precision.

Finally, chapter 6 returns to the theme of control and national identity that emerged in ideas about scientific prediction and examines the international dimensions of Victorian forecasting. This dimension involved not so much a comparison of British work with that of other European countries or of the United States, but rather a shift in focus to the empire. In the 1870s, meteorological science moved away from the vagaries of the British weather to India's intense but regular tropical conditions. By connecting rainfall, sunspots, and famine, meteorologists revived dramatic claims about the promise of forecasting. Their research on the cyclical patterns of rainfall and the management of the British Empire suggested how science and the state became mutually reinforcing models of rational order. Amid political debate over the future of the empire in the 1870s, meteorology provided a natural foundation for British imperialism. India, in turn, provided a new field in which to trace atmospheric laws and prove the relevance of science to the modern state.

This account of Victorian meteorology, then, will consider elements as diverse as institutions and popular almanacs, statistics and cloud photography, and will feature a cast of characters that includes astrologers, seamen, government administrators, and presidents of the Royal Society. Looking at the experience of the observer at either end of the century will summarize the broad direction of change over this period.

One exemplary observer, the spinster daughter of a baronet, Caroline Molesworth began her record in 1823 when she moved from London to Cobham, Surrey. Her observations continued for the next forty-four years, on a plan derived from her reading of meteorological works by Thomas Forster, Luke Howard, and John Frederic Daniell. Eventually she took a daily record of nineteen columns. She had an "ordinary Mahogany pedestal barometer" in her hall, made before 1810 by the London instrument maker George Adams, and a thermometer of the same vintage and maker hanging nearby. The barometer was compared with a standard instrument only once, in 1859. A rain gauge on a design by Howard was sunk into her north lawn about thirty feet from the house, and a post with outdoor thermometers stood in the same spot. Another thermometer was placed on the south grounds of the house, and she also had an "unreliable" weather vane and a "Tagliabue's storm glass." (This last was a mixture of camphor, potassium nitrate, and ammonium chloride in alcohol and water, in which particles apparently crystallized more or less strongly under different weather conditions; it was described by her biographer dismissively as "scarcely more than a scientific toy.") She also noted the appearance of clouds and made animal and plant observation, recording migratory birds, the first flowering of shrubs, and the budding and leaf fall of selected trees. Her journals were published in 1880, several years after her death, and dedicated to the Royal Meteorological Society, though with some apology for their old-fashioned quality.[19]

A century later, Lewis Fry Richardson took a post at Eskdalemuir Observatory in 1913, in a remote site in Dumfriesshire in the Scottish lowlands. The demands of a solitary, persevering routine remained, but Richardson's Cambridge education in physical science, the instruments, and his attitude to observing were quite different. Napier Shaw, the director of the Meteorological Office, who appointed Richardson, delivered a letter of advice that apologized for "the routine of tabulation etc" that Richardson would face, even noting a colleague's recent suggestion that "the best thing for meteorology would be for everybody to stop observing for 5 years." He urged Richardson nevertheless to take advantage of the moving panorama of his array of self-recording instruments, which would provide "a continuous representation of observations of magnetism, meteorology and seismology like the pictures of a kinematograph." Though the solitude gave Richardson this opportunity for visual contemplation, Shaw warned that "co-operative research was essential. You will find that you can do nothing without using somebody else's observations as well as your own."[20] Eskdalemuir was part of a British network of observatories, and meteorology was a collective science of mechanized observations and university-trained observers, linked to national and international networks.

19. Ormerod, ed., *The Cobham Journals*, viii–xiii. Tagliabue was a London instrument maker (Turner, *Nineteenth Century Scientific Instruments*, 231–32).

20. Quoted in Ashford, *Prophet or Professor?* 47–48.

Although the experience of observers like Molesworth and Richardson were quite different, there were also marked continuities. Meteorology wrestled throughout this period with the public nature of its science, the challenges of organizing observation, and the problem of converting facts to philosophical advances. Writing a historical account of his discipline in the 1920s, Shaw considered the meteorology of the previous century a failure. His explanations for that failure drew attention to particular historical conditions. Telegraphy and the weather map gave the science dramatic new tools, Shaw pointed out, while the era of world trade by sail gave it urgency. By the 1880s, the former had lost their glamour and the latter had given way to steam travel. Public funding for meteorology in Britain was stationary for about a quarter century; it ballooned again only during the First World War in response to the needs of aviation. Meteorology, Shaw suggested, might have fared better had not "meteorologists been compelled to issue a series of 50,000 sets of forecasts, only more or less correct." But, he continued, this was to wish the impossible. "The universal desire for information about future weather opens the main artery of communication between the science and public, and is the chief vindication for public funds; and when once a scientific subject enjoys public money it is difficult to persuade anyone that it is not provided for in all particulars."[21]

This dismissal of Victorian forecasting emerged at the moment when the methods of Vilhelm Bjerknes were transforming the science. A Norwegian physical scientist, Bjerknes developed the equations for hydro- and thermodynamics of the atmosphere that for the first time offered a solid mathematical foundation for meteorology. Out of Bjerknes's researches came a reconceptualization of atmospheric circulation and new models of cold and warm fronts as the origin of cyclonic motion. Yet the triumphs of meteorology were somehow equivocal. The equations, even as brilliantly condensed by the former Eskdalemuir observer Richardson, required too much time to solve to become useful in weather prediction until the advent of computers, a century after the foundation of the Meteorological Office.[22] In the nineteenth century, the forecasting controversies petered out, rather than arriving at some clear resolution. This inconclusive ending may be the key to the history of weather forecasting. Modern science, with its computers, satellites, and chaos theory, has not so much answered the questions of the nineteenth century as it has displaced them. As the philosopher Karl Popper has famously said, the essence of physical

21. Shaw, *Manual of Meteorology*, 1:1–2, 9.

22. Friedman, *Appropriating the Weather*.

determinism was the belief that "*all clouds are clocks*—even the most cloudy of clouds." Twentieth-century science assembled the reverse conclusion: that "all clocks are clouds."[23] However full our knowledge, a limited ability to predict is built into the scientific theories that we now use to explain atmospheric change. In that way, the ambitions of meteorologists, trying to unlock the mechanisms of the weather, have receded into the past, and an account of their efforts seems to call for something like sympathy. This is the condescension of history, a special hazard in the history of science, in which each successive generation's knowledge can seem fuller and more justified. But the study of meteorology can also restore a fundamental sense of historical change, capturing the difference of past experience with familiar phenomena. "What apparently can be less subject to rule and law, even to a proverb, than the changeful wind, and the treacherous wave?" its president cheerfully asked the assembled men of science of the British Association for the Advancement of Science in 1854, as he introduced the new Meteorological Department. He anticipated that his question set the scene for scientific triumphs of meteorology in the near future.[24] In fact, that future receded before him.

23. Popper, "Of Clouds and Clocks," 210, 213, original emphasis.

24. Dudley Ryder, Earl of Harrowby, "Presidential Address," *Report of the British Association for the Advancement of Science*, 1854, lix.

Prediction, Prophecy, and Scientific Culture

POISED between divination, opinion, and calculation, weather prediction was a controversial enterprise. When he began to send statements about the coming weather to the newspapers in August 1861, the first director of the Meteorological Department of the Board of Trade adopted, as a new term, the word "forecast." "Prophecies or predictions they are not," Robert Fitzroy wrote. "The term forecast is strictly applicable to such an opinion as is the result of a scientific combination and calculation."[1] Eighteen years later, Robert Scott, Fitzroy's successor, turned to a metaphor of contagion made newly potent by Louis Pasteur's researches, when he warned that "the smallest lurking germ" of prophecy endangered meteorology.[2] The anxiety was deserved: Fitzroy and his successors quickly became known as the "government Zadkiel," a reference to Victorian Britain's most notorious astrologer.[3] As a *Punch* satire pointed out, Zadkiel made the same kind of prophecies as "them voorcasts, what a' calls 'Weather Predictions'"—but he supplied them for a whole year in advance and for all England at once. "Meteorology? Yaa! What's that to the Voices o' the Stars?"[4]

The complex identity of prediction and prophecy in Victorian culture shaped these different responses to the project of forecasting the weather. Weather prediction may be the single most important feature of the history of

1. Fitzroy, *Weather Book*, 171.

2. [Scott], "The Weather and Its Prediction," 495.

3. *Times*, January 17, 1883.

4. "Meteorology," *Punch*, May 17, 1879, 221; cf. Wynter, "The Clerk of the Weather," 202.

meteorology, because it was a signpost for the development of the discipline and showed how it was embedded in a wider culture. Claims about prediction engaged a set of scientific, historical, and religious ideas about the future, the past, and the human ability to discover both. What was the relationship of the calculations of an eclipse, to a prediction about geological strata, to an astrological prophecy of the illness or death of a sovereign, or to the interpretation of contemporary politics as fulfillments of Biblical prophecy? This range of ideas was particularly significant in a period when scientific men sought to extend their authority, proposing scientific method as the foundation of all reliable knowledge. Prediction was a demonstration of the powers of science. "There is no more convincing proof of the soundness of knowledge than that it confers the gift of foresight," William Stanley Jevons summarized in his *Principles of Science* (1874).[5] Because prediction underpinned the extension of scientific reasoning to all studies and all questions, clarifying the nature of proper forms of prediction was critical. In the words of John Stuart Mill, "A complete logic of the sciences would be also a complete logic of practical business and common life."[6] Scientific predictions that failed were challenges to the eminence of scientific thought, and to the social order that sound knowledge upheld.

The standards of science, then, supported much more than inquiry into the natural world; they were the scaffolding for rational human relations. Prediction, moreover, seemed to make that framework visible; that is, it involved a social form of demonstration as well as a logical one. Prediction took scientific reasoning into the public world. It allowed, or required, men of science to display their wares and impress their audience, and it linked science to a public presentation of its value. As a critical element in debates about scientific reasoning, prediction seemed especially important in investigations of phenomena that were beyond direct experience. Such investigations proliferated in the scientific world of the nineteenth century, in historical sciences like geology or paleontology, in studies of psychology and perception, and in a science of social behavior. Subjects as diverse as gold prospecting, mesmeric clairvoyance, the history of Roman Britain, legal treatments of circumstantial evidence, and theological discussions of prophecy shared a common thread. They all sought to assess the status of knowledge of matters beyond direct experience.[7] Like these, weather prediction raised broad questions about styles of knowledge and the justifications of scientific method.

5. Jevons, *The Principles of Science*, 536–38.

6. Mill, *Collected Works*, 7:284.

7. Welby, *Predictions Realized*; cf. Beer, "The Reader's Wager"; Winter, *Mesmerized*, 122–25, 143–46.

The scope of this analysis can be traced in an anonymous work first published in 1865. *Modern Prophecies* paid extended tribute to "the development of a science of prophesying" and reflected the significance that ideas about prediction had assumed in the mid-nineteenth century. "We are compelled by an absolute necessity to *forecast* the future, since all the activities of man are a preparation for it," the author wrote, "and in this consists our life." The book outlined a series of forecasts divided into three categories of religion, politics, and free trade. The Bible would cease to become the standard of authority, and Catholicism would grow while the Church of England declined. The British Empire would increase mightily, Europe would form itself into a confederation of Europe, and the United States would acquire a new role, in which it "will reject the policy of non-interference, and join in the intrigues, policy and wars of Europe." The factory system would transform "the conditions of labour in all parts of the world" while the newspapers, "a sort of whispering gallery which echoes every prediction," would progress from their informal "parliamentary" role into a formal wing of government. All professions would be opened, even to women, the Church of England would be separated from the State, and the government would lose its postal monopoly.[8] None of these predictions were particularly shocking. As their author pointed out, they were the accumulated stock of conversations over his lifetime. The interest of this account lies not with the predictions themselves but with the framework within which they were set.

For the author, the list was an opportunity to define the importance of prediction. Just as chemists can "foretell accurately the results of certain experiments," just as "the gathering of storms is known before they exist" and "the astronomer predicts eclipses," human affairs may develop into a science of the future tense. Precision of this sort is a distant goal, the writer acknowledged. But even in the present state of knowledge, he argued, prophecy was an immensely powerful force. Although prophecy was increasingly repudiated in religious doctrine, it still summarized the imagined identity of every nation, including that of Britain. "A thousand years of honour lie before the people . . . She shall be greater than Rome, as Cromwell said; and the instructress of nations, as Milton prophesied." Such visions of times to come, noted the author, now governed commerce as well. "The prophecies of the future are eagerly discounted upon mercantile exchanges, and the prices of securities form new and most sensitive barometers of the state of nations." This prophetic picture was built on a natural account of constant change. Referring to geology and

8. Lupton, *Modern Prophecies*, 4–5, 33–36.

to Darwin's theory of natural selection, the author suggested that "a sublime and harmonious principle of growth pervades the universe." This principle extends to "all the institutions of man." The most important feature of the age was the gradual extension of our knowledge of the past and the future through an understanding of sciences as varied as astronomy, geology, meteorology, and mental physiology. Our lives are lived either in the past or future.[9]

The author's identity was revealed when the work was republished in 1887 by his son, Harry Lupton, a resident of Chesterfield, in northeast England. Arthur Lupton, who had died in 1867, was a graduate of the University of London. He had been, according to his son, "devoted to intellectual pursuits," especially "the study of history, theology and metaphysics, and the more popular branches of science."[10] As was obvious from his book, Lupton was a Liberal committed to free trade. Writing a generation later, the son endorsed the father's characterization of prediction as the spirit of the age. Extending his father's remarks on commerce, Harry wrote an account of stock trading in Liverpool. In this city, trading on what would now be called futures had emerged in the 1840s and 1850s in cotton, a direct consequence of steam navigation and the establishment of Samuel Cunard's mail lines across the Atlantic. Cunard could bring news of crops and even transport samples more than six weeks faster than the sailing ships crossing the same distance, so that merchants could sell cargoes "to arrive" in advance of the actual supply. This type of market expanded in the years of the cotton famine of the 1860s, when the American Civil War interfered with supply of cotton. The pattern of such speculation was well in place by the time of a further revolution in communication that took place just about the time of Arthur Lupton's death—the completion of the Atlantic cable in 1866.[11] Men like the Luptons, then, familiar with mid-century Liverpool, had a ringside seat for the emergence of these forms of trading; by 1887, Lupton could present this activity as an ordinary matter of fact. "By far the greater part of transactions [in the stock market] are purely speculative, and every post carries thousands of predictions all over the country as to the rise or fall of stocks." If we consider this transformation of commerce, Harry Lupton insisted, his father's idea of a science of prophesying, in which we could weigh and calculate even the most difficult matters, was not then so visionary. To him, the parallels with another telegraphic novelty were

9. Ibid., 4–5, 8–9.

10. Harry Lupton, "Preface," in Lupton, *Modern Prophecies*, 2nd ed., 2.

11. Rees, *Britain's Commodity Markets*; Ellison, *The Cotton Trade of Great Britain*; Milne, *Trade and Traders*.

obvious. "The daily forecast of the weather now published in the papers" was simply "a remarkable instance" of the prophetic spirit of his times. Under these conditions, he concluded, it is difficult to say "where 'experimenting' ends and 'prophesying' begins."[12] Science was rewriting old categories of knowledge, and weather forecasts were both experiment and prophecy.

The vision of prediction, ranging from astronomy to stock trading, that emerged in the Luptons' text was not unusual. But their confident account made the contests among these activities invisible, ignoring the controversy that surrounded such a comprehensive extension of scientific authority. This controversy, as later chapters will reveal, found a particular focus in weather forecasting, but it can be seen more generally in the debates about the place of prediction in scientific knowledge. The status of prediction was critical to whole-hearted advocates, like the Luptons, but it was similarly important to the critics of modern scientific culture. Defining prediction was a way to separate scientific naturalism from troublesome forms of supernaturalism, like astrology or religion. How scientific men treated questions of prediction became then a kind of barometer of their claims to intellectual, moral, and cultural leadership.

The Ideal of Prediction

In the course of the nineteenth century, as the Luptons' work made clear, striking new events, disciplines, and practices established contemporaries' expectations of scientific work. In addition to weather forecasting, these included the discovery of new planets, gold in Australia, and statistical regularities in human behavior. But the chief stories of the predictive power of science came from astronomy, precisely founded on the twin pillars of Newton's celestial mechanics and impressive new technologies, from giant telescopes to the photo-registration of observations. "The age of mere wonder" at eclipses, in John Herschel's view, was past; it had been replaced by more enlightened appreciation of the calculations and certainty involved. "A page of 'lunar distances' from the *Nautical Almanac*," wrote Herschel, "is worth all the eclipses that have ever happened for inspiring this necessary confidence in the conclusions of science."[13] At the center of any discussion of astronomy and prediction lay the discovery of Neptune, the powerful culmination of exact science. In

12. Harry Lupton, "Preface," in Lupton, *Modern Prophecies*, 2nd ed., 1–2. Cf. an account of thinking about the future drawn from early science fiction: Clarke, *The Pattern of Expectation*.

13. Herschel, *Preliminary Discourse*, 27.

the summer of 1846, at a time when dedicated observers with powerful new telescopes were regularly bringing new planetary bodies to light, the presence of a planet, which became called Neptune, was predicted by the analysis of irregularities in the orbit of Uranus by both John Couch Adams of Cambridge Observatory and Urbain LeVerrier in Paris. It was immediately verified by observations taken at Berlin in September. Priority disputes aside,[14] the incident exemplified scientific reasoning about unseen phenomena. One popular account described LeVerrier "advanc[ing] step by step along the intricate maze of his researches, . . . cheerfully executing calculations which diffused are almost appalling to contemplate; until a bright flood of light is finally over all his labours, and the distant member of our system, which human eye has not yet seen, discloses itself to his purely intellectual scrutiny with all the certainty of demonstrative reasoning."[15] Prediction was a demonstration of the power of this pure intellectual scrutiny.

Dramatic, well-publicized international expeditions to view eclipses followed. In 1851, for a July total eclipse of the sun, George Airy, the Astronomer Royal, directed an extended series of observations in Sweden and Norway, and, nine years later, an even more elaborate expedition was mounted in northern Spain. These travels confirmed for the Victorian world the prestige of astronomical science.[16] In parallel, physical astronomy, the study of constituent matter of the stars and planets using the spectroscope, began to venture predictions that went further and further into the future. By the mid 1860s, with the development of theories of entropy, these predictions extended, as John Stuart Mill put it, "even to the dissipation of the sun's heat and the heaping up of the solar system into one dead mass of congelation."[17] Astronomy had reached as far as the end of time.

During the same era, the authority of astronomy also underwrote the most novel, controversial forms of scientific prediction: a statistics of society. Starting in the 1830s and 1840s, social science developed on the strength of its use of a technique for measuring error in astronomical observations. Looking at large numbers, statisticians found startling regularities, even in the most disorderly and arbitrary events, such as crimes or—in one renowned example—the number of badly addressed letters that languished unclaimed in post offices. These regularities allowed for an extraordinary kind of prediction, such as

14. Chapman, "Private Research and Public Duty"; Standage, *The Neptune File*.

15. Grant, *History of Physical Astronomy*, 197.

16. Pang, "Social Event of the Season."

17. John Stuart Mill to William George Ward, February 14, 1867. Mill, *Collected Works*, 16:1241.

that in the mortality tables described by William Farr, director of the General Register Office, founded to collect population data in 1838. "It would have been considered a rash prediction in a matter so uncertain as human life to pretend to assert that 9000 of the children born in 1841 would be alive in 1921," he explained in his 1843 annual report. "Such announcement would have been received with as much incredulity as Halley's prediction of a return of a comet." But just like Halley, the statistician can rely upon "the constancy of nature; hence [Halley] ventured from an observation of parts of the comet's course to calculate the time in which the whole would be described." Predictions just as remarkable, concluded Farr, are now possible elsewhere. "Although we little know the labours, the privations, the happiness or misery, the calms or tempests, which are prepared for the next generation of Englishmen, we entertain little doubt that about 9000 of 100,000 of them will be found alive at the distant Census in 1921."[18] For Farr and his contemporaries, statistics exemplified the new and striking powers of scientific reasoning.

Even when the kind of precision to which Farr lay claim was not the goal, nevertheless the possibility of prediction was seen as the badge of the scientific character for this new and controversial form of scientific knowledge. Mill, whose *System of Logic* took great pains to lay down the grounds for a "science of human nature," was typical in putting astronomy and prediction at the heart of his arguments. "All phenomena of society are phenomena of human nature," he wrote, and are therefore subject to fixed laws. Prediction of the "history of society" will never be quite the same as the prediction of "celestial appearances," but this "difference of certainty" is due to the inadequacy of our data and not to any limitations to the law-like character of social phenomena themselves. In astronomy the "data . . . are as certain as the laws themselves"; in social science "the multitude of causes is so great as to defy our limited powers of calculation." In this respect, as Mill had noted earlier, the study of society is just like the study of weather or the tides. In these cases, we do not succeed in making exact predictions of rain or sunshine, "yet no one doubts that the [meteorological] phenomena depend on laws, and that these must be derivative laws resulting from known ultimate laws, those of heat, electricity, vaporization and elastic fluids." Meteorology as a science, then, faces the same problems of partial data that social science does, since for both "the data requisite for applying [the science's] principles to particular instances would rarely be procurable." But weather, tidal, and social phenomena do

18. Quoted in Eyler, *Victorian Social Medicine*, 73.

occur according to fixed laws and hence possess the possibility of prediction, in the same way as astronomy.[19]

Mill wrote most of his *System of Logic* between 1837 and 1843, greatly stimulated by his readings of the philosophy and histories of science published by John F. W. Herschel and William Whewell.[20] As he made clear in his commitment to the extension of its methods and principles to the study of society, the exciting feature of science in the modern era was its ability to lay down new avenues of knowledge. This ability was equally impressive in another field, that of geology. In the mid-1840s, Roderick Murchison had established his reputation as a geologist by identifying the oldest fossil-bearing strata, the Silurian. He had confirmed his discovery by tracing the Paleozoic rocks into northern Russia, hitherto unexplored by geologists, on a field excursion of 1842. Collecting observations from distant Australia, Murchison embarked on comparisons between European and antipodal formations. In 1844, in an address to the Royal Geological Society, he noted that all evidence suggested that the Australian central range was Silurian rock, thrown up in the Permian period. This kind of rock, he suggested, would hold gold, accumulating near the surface in quartz veins. A few years later, Murchison encouraged unemployed tin miners in Cornwall to emigrate to the Australia to seek gold. When the first of the strikes in Australia was made in 1851, Murchison took full credit for the prediction. His prediction of gold had been "*worked out* mentally . . . [before] the diggers discovered it," he told Whewell proudly.[21] In 1854, his account of his geological work, *Siluria*, promoted his theories, including his prognostications for gold and other minerals. Lectures featuring his maps and diagrams were offered for departing Australian emigrants at the London School of Mines. When Murchison accepted the position of director general of the Geological Survey of the United Kingdom (an organization that he was to push increasingly into imperial researches), the members of Parliament reportedly broke into cheers. The future of British geology seemed secure in the hands of a man who had demonstrated the national value of science.[22]

Murchison's remarkable prediction suggested also then how scientific reasoning was linked to national identity, history, and modernity. The son of a Scottish landowner, Murchison was always acutely aware of the relationship

19. Mill, *Collected Works*, 8:844–85, 877–78.

20. Robson, "Textual Introduction," xlvix–cviii.

21. Roderick Murchison to William Whewell, February 19, 1852, quoted in Stafford, *Scientist of Empire*, 39, original emphasis.

22. Stafford, *Scientist of Empire*, 31–39.

between his geological researches, economic development, and national glory. In 1846, he was the one of the first scientific men in the nineteenth century to be knighted for his work, a reward he energetically sought by bringing his international honors before government figures of the day. At the geological section of the British Association for the Advancement of Science in 1848, he presented geology as the science of the industrial age, and Britain as leaders of both. His ability to predict thus linked past, present, and future. Knowledge of geological actions of the distant past would provide control over the destiny of Britain.

But the past could be claimed in other ways. In the 1850s and 1860s, George Airy, the Astronomer Royal, published several studies on Julius Caesar's invasion of Britain in 54 and 55 BC, which brought evidence about natural phenomena to bear on this disputed antiquarian question. Lively debates surrounded both the identity of the port from which Caesar departed and at which he first landed to face the opposing Britons. To resolve these questions, inquirers turned to different sorts of evidence. There were the commentaries of Caesar himself, and other writings of near contemporaries, whose words had to be translated, weighed, and compared. There was the linguistic analysis of surviving place names: Was the Roman Port Itius, from which Caesar and his two legions departed, near the small but ancient village of Isques, now above the port called Boulogne? Was Romaney Marsh on the English coast so-called after the invasion of the Romans? Finally, there was evidence from natural phenomena, such as the measurement of tides and changes to the shoreline and harbors over long centuries from alluvial deposits. Using this last type of evidence to deal with the controversy had absorbed an earlier English astronomer, Edmund Halley. He was first to draw attention to the description of the tides in the documentary accounts and use them to decide upon a location for the landing. Airy, a century later, resurrected the interest of his fellow astronomer and emerged with a novel conclusion. Caesar had departed from the mouth of the Somme, and not Boulogne, Calais, or Wissant, and had landed at Pevensey, not Hythe or Deal.[23]

Although Airy certainly did not shy away from literary evidence, he built his case with data on tidal effects and coastal shelves, gathered by modern scientific authorities. He contrasted these authorities with those of his opponents, sharply dismissing, for example, the witnesses of another scholar, Thomas Lewin. Lewin had cited a railway official's estimate on the distance from Boulogne to Folkestone or humble French curés with anecdotes about

23. Airy, *Essays on the Invasion of Britain by Julius Caesar.*

local history whose ignorance of scholarly controversies, Lewin claimed, made them "the most unprejudiced witnesses."[24] But natural fact, Airy felt, clearly trumped Lewin's appeal to local experience. And when Lewin dismissed Airy on the grounds that the Somme was far further than the thirty miles' distance indicated by Caesar's description, Airy countered that, before the modern age, all such estimates of distance were unreliable. Only the triangulation measurements of 1787 resolved the insoluble question of Britain's distance from the Continent. Caesar's were mere "eye observations," and his literary testimonial could not outweigh modern evidence.[25] Using these exchanges, we can distinguish two steps to the Astronomer Royal's approach to these antiquarian questions. First, he tied together the historical sciences, such as geology, with those that relied most heavily on the precision of mathematical and Newtonian sciences, such as tidal theory or triangulation methods to determine distance. Second, he insisted that these natural facts must frame literary evidence, which unlike the natural evidence, required no comparable special expertise. With this dual approach, Airy claimed to resolve long-standing disputes about human history. He firmly snubbed a critic who called his "guess" "unlucky." Lucky or unlucky, his theories were not guesses; they were the "result of careful thought" and "scientific observations."[26]

In private letters, Airy chose to portray the Caesar's landing work as nothing more than the mental sport of an overworked Astronomer Royal, which had justified some pleasant excursions to the Sussex coast. Certainly the interest in Roman Britain was not in itself unusual—antiquarian and archaeological associations burgeoned in the 1830s and 1840s.[27] It was also a patriotic exercise: Boulogne, favored by Airy's opponent, Lewin, was generally the choice of French scholars on the disputed question, and it was also the port selected by Napoleon I for his planned invasion at the beginning of the nineteenth century. The Somme estuary, on the other hand, as Airy pointed out in his arguments, had launched the invasion force of William the Conqueror in 1066. Airy's theory thus linked Caesar and William, architects of British destiny, against modern French designs, making the studies a modern parallel to the Romans' military conquests. Airy was extending the domain of scientific approaches and attitudes to questions that seemed beyond its range. Airy published his

24. Lewin, *Invasion of Britain by Julius Caesar*, 15–21.

25. Lewin, *Invasion of Britain by Julius Caesar*, 2nd ed., x–xi; Airy, *Essays on the Invasion of Britain by Julius Caesar*, 24.

26. Airy, *Essays on the Invasion of Britain by Julius Caesar*, 24.

27. Westherall, "The Growth of Archeaological Societies."

collected writings on Caesar's landing privately in 1865 and distributed the book to a list of colleagues that reads like a who's who of Victorian science: Whewell, of course, and John Herschel, the leading gentleman of science in Britain, Michael Faraday, the famous electrical experimenter at the Royal Institution, Adam Sedgwick, the Cambridge geologist, Edward Forbes in Edinburgh, George Peacock, leader of mathematical reforms in Cambridge, and Col. Edward Sabine, dedicated to the global science of terrestrial magnetism.[28] Airy's hobby was an assertion of the power of science. The head of British astronomy was claiming judgment over historical questions that lay at the heart of Britain's imperial identity, its link to Roman glories.

In what sense was this exercise a demonstration of prediction? It is first helpful to note as an aside that the study of Roman glories connected Airy's investigation to the most famous prediction of the Victorian era: Thomas Babington Macaulay's brief description of a Maori tourist of the distant future contemplating the ruins of London, thrown off in an essay on Rome and the Papacy. This vision seized the imagination of contemporaries to an unprecedented degree, becoming a symbol of national destiny (fig. 1.1).[29] More directly, however, just as Murchison had made geological predictions about the remote land of Australia, Airy's project announced that another sort of inaccessible knowledge should become the domain of men of science. In his historical and philosophical writings on scientific method, Whewell explicitly identified this key analogy between prediction of past and future events. In *The Philosophy of the Inductive Sciences* (1840), Whewell defined prediction as a claim built on past events, looking forward to events that have not yet happened. But he also emphasized that it would be impossible to distinguish reasoning about the future from reasoning about the past. If a scientific explanation is true, it is true of "all particular instances" wherever or whenever they are. "That these cases belong to past or to future times, that they have or have not already occurred, makes no difference in the applicability of the rule to the case. Because the rule prevails, it includes all cases; and will determine them all, if we can only calculate its real consequences."[30] Murchison's marshalling of geological evidence led to a prediction of a future event—the discovery of gold by human diggers—but, equally, it could be considered a "prediction" about the past—gold had been deposited in those quartz veins near the present surface

28. See the distribution list in Airy Papers, RGO 6/466/168.

29. Macaulay, *Essays and Lays of Ancient Rome*, 548. Woodward, *In Ruins*; Dingley, "Ruins of the Future." On Victorians and history, see Burrows, *A Liberal Descent*.

30. Whewell, *The Philosophy of the Inductive Sciences*, 2:62–63.

of the land in a distant age. This was no different from Airy's argument that Julius Caesar had landed at Pevensey in 54 BC. Both the geological and the human past were beyond direct interrogation, but the scientific treatment of the evidence would resolve the mystery.

Some years later, Thomas Huxley made the same point with the invention of a new term, "retrospective prophecy," in an 1880 essay on "The Method of Zadig." Zadig, in a fable written by Voltaire, was a Babylonian philosopher skilled in reading small natural clues, which allowed him to deduce precise knowledge of events. Canvassed by soldiers hunting for the queen's spaniel and horse, which had strayed from a royal hunt, Zadig described the animals in minute detail, but claimed not to have laid eyes on them. The suspicious soldiers dragged the philosopher to court, where he was fined for "saying he had seen that which he had not seen." Zadig explained that faint tracks left by the missing animals had allowed him to arrive at his description. In Huxley's inimitable retelling of the tale, he noted that while the philosopher's "profound and subtle discernment" won him a refund, it was swallowed in legal fees, and for all that some of the judges were admiring, others still wanted to burn him as a sorcerer.

Huxley probably encountered Zadig in his readings on paleontology. Long before Huxley's essay, the French naturalist Georges Cuvier had first brought the Zadig story into scientific literature; later, Louis Agassiz's essay on classification adopted its implications when he wrote on "prophetic types" in biological development, which "now appear like a prophecy . . . exhibiting in a striking manner the antecedent consideration of every step in the gradation."[31] So the fable perfectly suited Huxley's zealous promotion of natural knowledge. In his analysis, Huxley, like Whewell, denied any distinction between retrospective prediction and foretelling: "it is obvious that the essence of the prophetic operation does not lie in its backward or forward relation to the course of time, but in the fact that it is the apprehension of that which lies out of the sphere of immediate knowledge." Unlike predictions about the future, we can never test retrospective prophecies by experience, Huxley admitted. But we proceed with confidence in our methods, and in the uniformity of nature. "All that can be said is, that the prospective prophecies of the astronomer are always verified; and that, inasmuch as his retrospective prophecies are the result of following backwards the very same method as that which invariably leads to verified results when worked forward, there is as much reason for placing full confidence in one as the other." Working prediction backward "is

31. Agassiz, *Essay on Classification*, 166–74.

FIGURE 1.1 Gustave Doré's 1872 rendering of the most famous prediction of the nineteenth century, a New Zealander sketching the ruins of London. Briefly mentioned by Macaulay in an essay on the history of Rome, this image took on a life of its own. Despite the gloom and mystery, this perhaps should not be interpreted chiefly as a melancholy vision: contemplating the imperial Roman past and the imperial British future both provided ways to dwell on the nation's significance. Macaulay's image suggests then the ways that prediction and prophecy shaped national identity. (Gustave Doré, *The New Zealander*, in Jerrold, *London: A Pilgrimage* [New York: Dover, 1970], opp. 188.)

therefore a legitimate function of astronomical science; and if it is legitimate for one science it is legitimate for all," Huxley concluded.[32] Both the past and the future belonged to science.

32. Huxley, "On the Method of Zadig," 135, 136, 140. Cf. Shortland and Somerville, "Thomas Henry Huxley, H. G. Wells and the Method of Zadig"; Block, "T. H. Huxley's Rhetoric and the Popularization of Victorian Scientific Ideas."

Like Airy in his researches on Roman Britain, Huxley's essay placed questions of historical accuracy in a different and inferior category of knowledge. He began the essay by emphasizing that the "conspicuous merits" of Zadig's first "biographer," Voltaire, did not include "strict historical accuracy." Instead, what concerns us, said Huxley, are Zadig's "conceptions"—the "light . . . that shows the way"—even though in this reading we "reduce Zadig himself to the shadowy condition of a solar myth." History or myth were irrelevant. "What [Zadig] was like in the flesh, indeed whether he existed at all, are matters of no great consequence."[33] To be really valuable, historical investigation must be transmuted into natural knowledge. This was the whole point of the introduction to Huxley's essay, which from that point on turned to fossils. As the rest of the essay made clear, it was paleontological experts like Huxley who can take "matters of no great consequence" and turn them into genuine knowledge. Unraveling all the evidence for the apparently insignificant fossil Belemnites by the end of his essay, Huxley could indeed demonstrate how things existed "in the flesh," describing the Belemnites's internal physiology and even the color of its ink. The story of Zadig, with its parable of myth, history, religion, and prophecy, was there to set the scene for the triumph of the scientific method.

The contrast between historical and scientific forms of testimony can be appreciated further if we add a third, contemporary model of knowledge, that of the evaluation of evidence in a legal setting. In both the antiquarian researches of George Airy into Caesar's landing, and the presentation of retrospective prophecy in Thomas Huxley's "Method of Zadig," we can see how comparisons with legal standards of evidence informed their accounts. In Airy's case, his chief opponent, Lewin, was a lawyer who claimed that his professional training in juggling insufficient evidence had produced his conclusion about Caesar in 54 BC. When he rejected Lewin's theories, Airy rejected this claim of expert judgment. Huxley's hero, Zadig, exposed courts and judges as a sham, although he remained, ironically, within their power. Yet the legal model of examining relationships among a complex series of facts provoked obvious comparison with the scientific method. In a recent study of circumstantial evidence and literary narrative, Alexander Welsh has shown how the treatment of evidence and inference to things unseen developed together in law and natural religion from the seventeenth century. Sciences such as geology and paleontology were indebted to the analogical reasoning

33. Huxley, "On the Method of Zadig," 939.

of natural religion, and hence to legal scholarship.[34] Comparisons to the law surfaced repeatedly in the 1860s and 1870s, partly as a result of changes in the legal profession and renewed interest in legal reform,[35] but partly also as a response to scientific controversy. In both situations, debates about the evaluation of evidence reflected concern with the methods of science. In the case of legal reforms, the interested parties were anxious to ground their claims on principles of sound knowledge and therefore sought to relate their practices to scientific ones. In the latter case, scientific controversies like spiritualism, for instance, prompted renewed debate about testimony and reliable evidence.

In a long and fascinating article published anonymously in the *Westminster Review* in 1865, the barrister Sheldon Amos used circumstantial evidence and the criminal trial as a focus for a general discussion of evidence: "the error and uncertainty that ever and again seems to mock the most cautious and deliberate inquiry." Through an examination of the rules of evidence, the article sought to establish "the accuracy and precision" of the legal usage of circumstantial evidence. The scope of the subject was made clear at the beginning of the article. Most of our understanding of anything was circumstantial, thus a good acquaintance with the standards involved in evaluating evidence was important to everyone. The "main source of our knowledge," argued Amos, comes from neither personal belief, nor trusted hearsay, but is sewn out of inferences drawn from both. Any inquiry into evidence thus reached far beyond its "merely local or professional significance," he continued, and its "methods and logical position" need to be coordinated with those in "other branches of science and art."[36]

Amos, a recent graduate of Cambridge, was a frequent contributor to the *Westminster Review*, then at the height of its reputation as an organ of liberal intellectuals. He was a keen student of the law and was elected in 1869 to the chair of jurisprudence at University College, London. One of his principal concerns was the parallel of legal and scientific methods. He had published two treatises on the subject, *A Systematic View of the Science of Jurisprudence* in 1872, and *Science of Law* in 1874, both of which emphasized the connections between science and law to an extent that a contemporary biographer found excessive.[37] Yet, as Amos insisted, the underlying rationale for all attention

34. Welsh, *Strong Representations*, 10–18, 152–84; cf. Schramm, *Testimony and Advocacy*.

35. Manchester, *Modern Legal History*; Twining, *Theories of Evidence*; Shapiro, *"Beyond Reasonable Doubt."*

36. [Amos], "Circumstantial Evidence," 160, 161, 162.

37. *Dictionary of National Biography*.

to circumstantial evidence was the knowledge that all things are connected. "Every physical and psychological fact throughout the earth is indissolubly joined with an indefinite number of others." In that sense, defining a proper standard of evidence was critical to all human activity. The development of modern society, moreover, made such expertise vitally important. "As things are now, the reading public are being flooded with facts pouring in upon them in an unintermittent stream, relating to an indefinite number of subjects, and possessing every variable degree of value."[38]

Of course legal decisions were not predictions. But they were judgments about testimony—the drawing of conclusions from observations. Moreover, it is worth noting that one of the chief features of the legal analogy was an emphasis on the consequences of correct judgment, rather than the evidence or testimony itself. Making a decision based on the evidence in the legal sense was a potentially monumental activity—most dramatically, as in murder trials, for instance, it meant the difference between life and death. On a more mundane level, the legal analogy points to the significance of the claims about scientific method, and their reach into practical affairs. In this sense, judgment *was* comparable to prediction. In the words of Whewell, "To trace order and law in that which has been observed, may be considered as interpreting what nature has written down for us . . . But to predict what has not been observed, is to attempt ourselves to use the legislative phrases of nature."[39] Whewell's suggestive phrase indicated a view of natural law far removed from the blind amoral operation of cause and effect. Legislation, after all, was a set of rules of conduct, shaping the organization of society in some kind of mirror of the moral values that guided a member of that society as an individual. But in this context, "legislative" more directly referred to law and legal processes—the making of decisions. In this sense a "legislative" view of nature was an applied view, relevant to the practical demands of applied sciences like meteorology. In a similar fashion, Mill distinguished in the *System of Logic* between the insights of a judge or advocate as he accumulates evidence and the rigorous inquiry required when the time comes for evaluating such evidence and arriving at judgment. If the former can proceed with native wit and experience, the latter stage demands the "sagacity" of the trained, scientific inquirer.[40] This reference to "legislative phrases of nature" then marked Whewell's acknowledgment that prediction had a public face and practical implications.

38. [Amos], "Circumstantial Evidence," 160, 164.

39. Whewell, *The Philosophy of the Inductive Sciences*, 2:64–65.

40. Mill, *Collected Works*, 7:285.

Huxley's Zadig displayed a parable about modern scientific practices, the technique of reasoning and standards of evidence that natural philosophers followed. But it was also a parable about the character of scientific and religious authorities. The status of religious knowledge was a dominant thread in this pattern of ideas about prophecy and prediction. As one collection of prophecies noted, "much of the interest of the Predictions . . . may be secular; but it is impossible to read them without feeling the holier influence of their mysteries in prompting thoughts of a much higher nature."[41] William Fowler, a lawyer and Liberal M.P., made the same point when he brought Airy's discussion of Caesar into an essay on natural law and miracles. It was "excessively absurd," he commented, "that a man who believes in the invasion of Britain by Caesar on testimony should reject all testimony as to the resurrection of our Lord."[42] Victorian debates about "testimony" and prediction were concerned with the order of nature. Could that order be interrupted by supernatural events or miracles, or did events unfold always under the infallible pressure of natural law? Prophecy raised the same questions as miracles about the weight of natural law: indeed, prophecies were conventionally considered as a specially conclusive form of miracles, grounded in "the uncontested verdict of history" that did not, for instance, raise the problem of credulous observers.[43] However, prophecy was even more important than miracles to the contests between religious and scientific authorities for a different reason. Prophecy linked questions about reasoning and evidence to debate over the character and leadership of society. In theological terms, prophecy expressed the connection between religious and secular authority; this, along with its implications for understanding natural law, made it particularly evocative during an era when religious and scientific leaders jostled for influence.

In the range of Victorian ideas about scientific prediction, religious prophecy enters on two different levels. In the first place, prophecy suggested dramatic beliefs about the governing connections between the natural and super-

41. Welby, *Predictions Realized*, vii.

42. Fowler, *Mozley and Tyndall on Miracles*, 20–21. On Fowler, see Stenton, ed., *Who's Who of British Members of Parliament*, 1:148.

43. On prophecy as demonstration, see Harrison, "Prophecy," 244; and Force, "Hume and Johnson on Prophecy and Miracles." On Victorian treatment of miracles, see Cannon, "The Problem of Miracles in the 1830s"; Chadwick, *The Victorian Church*, 2:40–112; Powell, *The Order of Nature*; Chalmers, *The Evidence and Authority of the Christian Revelation*; and Mozley, *Eight Lectures on Miracles*.

natural worlds. These beliefs epitomized the kind of charismatic religious enthusiasm that was anathema to a scientific understanding. On a less dramatic but more influential level, prophecy emerged as a way to criticize the character of scientific authority and the combination of intellectual ambition and moral failure that religious critics found in scientific culture. To appreciate both these levels, however, it is necessary to understand that religious prophecy in the Victorian period was itself both notorious and very ordinary.

Given the range of denominations and theological positions in Victorian Britain, any simple account is bound to distort some aspects of its history. In general terms, prophecy connected the Old Testament with the figure of Jesus Christ in the New Testament, making Christ the historical fulfillment of ancient predictions about a messiah. This proof of Christianity opened then the search for further confirmations of future stages of Christian history in events of the past and present, based on further interpretation of the texts of the Old and New Testament.[44] We know most about the charismatic forms of the study of prophecy and its associations with anticipations of the apocalypse, or the Second Coming, which drew from the book of Revelations and hence some of the most memorable imagery of the Bible, with descriptions of seven headed beasts, lakes of fire, sword-bearing angels on white horses, the whore of Babylon, and the destruction of Jerusalem. Combining the drama of Revelations with the bulk of the books of the Old Testament prophets, prophecy was simply a commonplace of Christian experience, the subject of texts regularly encountered by a church- or chapel-goer.[45]

Among those deeply engaged with the interpretation of prophecy, of course, opinions varied enormously. Millenarians, awaiting the Second Coming of the Messiah, were split into camps, some seeing most of the prophetic signs as fulfilled gradually over the recorded Christian era and others arguing that these signs would be compressed into a single generation. Postmillenarians saw Christ appearing at the conclusion of a thousand years of godly government, whereas premillenarians looked for Christ's appearance as its inauguration. Such divides had critical social and political implications: postmillenarians, like the evangelical seventh earl of Shaftesbury, worked to create the conditions that would culminate in a divine social order, whereas premillenarians, like

44. Brooke, *Science and Religion*, esp. 226–74; Clements, "The Study of the Old Testament"; Reardon, *Religious Thought in the Victorian Age*; Rogerson, *Old Testament Criticism in the Nineteenth Century*; Storr, *The Development of English Theology in the Nineteenth Century*.

45. Harrison, *The Second Coming*; Oliver, *Prophets and Millenialists*; Taylor, *Natural History of Enthusiasm*; Wilks, ed., *Prophecy and Eschatology*; "On Divination and on Prophecy."

Thomas Chalmers, perhaps the most influential evangelical minister of the 1830s, often emphasized their detachment from the material world.[46] The premillenarians produced the notorious contemporary figures, like Joanna Southcott and Edward Irving. Southcott, a Methodist farmer's daughter from Devonshire, attracted large crowds across the country in the early years of the nineteenth century to hear her visions of the millennium and her prophecies that she would bear a new Messiah. Irving was a Scottish Presbyterian who established an impressive and fashionable following after moving to London in 1825. In 1830, "manifestations" began to occur in Irving's Regent Square church, as some of his followers began speaking in tongues, a sign that the last days had arrived. Alarmed by the resulting disorder, leaders in the Scottish church expelled Irving in 1833, and he died a year later in Glasgow.[47]

These histories remained vivid to later generations, and figures like Southcott and Irving helped establish the reputation of prophecy in Victorian times as a spiritually and politically alarming enthusiasm. In the second half of the nineteenth century, the study of prophecy often seemed to be gloomily conservative, a world away from the political radicalism of Southcott's followers. This perspective is well illustrated by Rev. John Charles Ryle, a leading evangelical in the Church of England, who became the first Bishop of Liverpool in 1880. In 1867, amid debates over extensions of the franchise, Ryle doubted "whether there ever was a time in the history of our country, when the horizon on all sides, both political and ecclesiastical, was so thoroughly black and lowering. Happy is he who has learned to expect little from Statesmen or from Bishops, and to look steadily for Christ's appearance."[48] As Ryle's example suggests, the significance of prophecy cannot be restricted to the consideration of spectacular figures on the radical fringe.

One critical place in which to trace ideas about prophecy was the study of history and geography, as we can read in the subtitle of Alexander Keith's *Evidence of the Truth of the Christian Religion Derived from the Literal Fulfillment of Prophecy Particularly as Illustrated by the History of the Jews and by the Discoveries of Recent Travellers*. First published in 1830, *Evidence* quickly became a standard work. Chalmers called Keith, a Presbyterian who joined the Free Church of Scotland in the split of 1843, "a household word throughout the

46. Brown, *Chalmers and the Godly Commonwealth*; Hilton, "Chalmers as Political Economist."

47. Gilley, "Edward Irving: Prophet of the Millennium"; Flegg, *Gathered under the Apostles*; Oliphant, *The Life of Edward Irving*.

48. Lyon, *Politicians in the Pulpit*; Ryle, *Coming Events and Present Duties*, xii–xiii; Cornish, *The English Church*, 2: 214.

land," and the work was in its thirty-eighth printing by 1861.[49] Its argument was straightforward. According to Keith, if an index to the history of the world can be found in the Bible, and if we can further confirm this history in the ruins of Middle Eastern civilizations, we have here "a more than human testimony to the truth of Christianity." Keith continued that "the fulfillment of the prediction is thus inscribed as upon a public monument, which every man who visits the countries in questions may behold with his own eyes; and is expressed in a language so universally intelligible, that every man may be said to read it in his own tongue."[50] For Victorians, prophecy was a familiar subject, whether encountered in liturgy or theology, history or travel.

Under the influence of historical criticism of the Bible, while discussions of prophecy continued to raise questions about testimony, evidence, and natural law, they turned increasingly to the significance of the person of the prophet. Two cases exemplify this transition. First, a striking exchange by two giants of Victorian debate, the philosopher John Stuart Mill and William George Ward, a leading Catholic intellectual, showed how both recognized that prophecy offered key grounds for debate about the sphere of natural knowledge. Second, an essay of the noted liberal theologian William Robertson Smith on prophecy illustrated the same recognition in quite a different way. Smith acknowledged the comparison between prediction and prophecy but refused to grant it, arguing instead that the knowledge offered by religion was of a different order from that offered by science. In taking this position, he rejected both Ward's challenges to scientific method and a historicist interpretation of prophecy.

Ward's article, titled "Science, Prayer, Free Will and Miracles," was published in the *Dublin Review* for April 1867. The critical section of the essay focused on problems of foreknowledge or prediction; in this section Ward explicitly challenged Mill as a representative of modern scientific philosophy.[51] Ward argued that scientific thinkers, holding that events had only natural antecedents, denied the possibility of providential interference in natural law, and hence were ultimately determinists, who eliminated the possibility of free will. Ward identified the root of this overwhelming structure as "*the abstract power of indefinite prediction*," or the belief that, given sufficiently detailed

49. Chalmers, "On Prophecy," 134.

50. Keith, *Evidence of the Truth of the Christian Religion*, 13, 365. For Keith, see *Dictionary of National Biography*.

51. In a letter, Ward told Mill he wrote "against you." William Ward to John Stuart Mill, February 7, 1867, quoted in Ward, *William George Ward*, 282–83. A more complete history of this exchange would trace it back to the reception of Mill's *System of Logic* and forward to a further discussion of free will between the two in 1872 and 1873.

information, we could trace cause and effect in the present and therefore "infallibly predict the whole series of future phenomena."[52] While no living scientific figure claimed to have mastered it, Ward insisted that indefinite prediction summarized the goal or ideal of science. Yet Ward denied that there had in fact been any advance in this direction, despite all the progress in the sciences. Indefinite prediction extended only to "cosmical phenomena" like eclipses and stumbled hopelessly everywhere else. Ward insisted, then, that a clear-eyed view of induction would show scientific knowledge was in fact tentative. As a Catholic philosopher, he turned this skepticism to religious account, arguing that our fundamental absence of certainty allowed for a belief in providential interference with natural law. Ward's article showed how central the idea of prediction had become as a definition of advanced scientific thought. It summarized a simple opposition, with on one hand the belief in spiritual power and direction (epitomized by God working in nature and history, respectively, through miracles and prophecy) and on the other a scientific materialism in which events worked indefinitely forward and backward, and human history played no evident special role.

In a private reply to Ward, Mill agreed that God's absence could not be proven, while disagreeing that this negative evidence required us to believe. Yet—the important point for us here—Mill also accepted Ward's characterization of prediction as the key. At present human knowledge is limited "to effects that depend on a very small number of causes" as in astronomy, but this reflected only the present state of knowledge and more complex events are nonetheless "*abstractedly* capable of calculation."[53] That nature is something about which we can make accurate predictions, to the degree we can understand its complexity, was for Mill the fundamental point to uphold, just as for Ward it was the fundamental point to resist.

Smith treated prophecy differently, while sharing Ward and Mill's sense that it was a distinguishing element of both science and liberal theology. Smith was a minister in the Free Church of Scotland, appointed in 1870 to the chair of Old Testament studies in the Free Church College of Aberdeen. He was also, however, one of relatively few in his position with a close acquaintance with scientific work. At university he won a mathematical scholarship, and summers passed studying in Bonn and Göttingen introduced him to German science as well as German theology. On his return in 1868 to finish his theological course

52. [Ward], "Science, Prayer, Free Will, and Miracles," 274, original emphasis.

53. John Stuart Mill to William Ward, February 14, 1867, quoted in Ward, *William George Ward*, 292–93, original emphasis.

at Edinburgh, he worked for two years as an assistant to Peter Guthrie Tait at the time that Tait was setting up his teaching laboratory in experimental physical sciences. Familiar then with scientific practice, he had read widely on scientific method; early essays dealt with the conduction of electricity, Hegel, calculus, and the work of John Stuart Mill. At Aberdeen, he began to publish and teach historical criticism of the Bible, but he soon surpassed the bounds of tolerance of the conservative Free Church. In 1875, he was denounced for heresy after publication of an article for the *Encyclopaedia Britannica* on the Old Testament and, after proceedings that dragged out for five years, dismissed from his college position. He spent the remainder of his career in Cambridge, editing the ninth edition of the *Encyclopaedia Britannica* and publishing his theological studies.[54]

One of his most successful works was *The Prophets of Israel*, originally presented as lectures to audiences in Edinburgh and Glasgow in 1881–82; yet an earlier essay, written in 1876 as he faced proceedings for heresy, most explicitly revealed the contrast Smith marked between science and religion. In the tradition of historical scholarship, Smith was concerned with establishing prophecy in the context of the ancient world. Prophecy is, he began, "a fact of history"; the prophets were "the children of their country and their age"; and the consciously "supernatural" aspects of prophecy can be understood in terms of the beliefs and demands of an earlier state of society. Yet he dismissed the idea that the prophets of the Old Testament could be examined in an empirical fashion for their predictive accuracy, as the theologian Abraham Kuenen had done. "No prophecy," he asserted, "can be mechanically divided into a part which has only historical value and a part which is ideal truth." This step in his argument proceeded then to a description of the inductive method, setting out how a scientific analysis of prophecy would proceed: it would describe, arrange, and then question the facts. He then concluded that this process was irrelevant. "If such a rigorous investigation could really be applied in all strictness and carried through to a successful conclusion, there is no doubt it would prove most valuable. It may, however, be doubted whether in the nature of things any purely religious fact falls within the range of strict induction. The facts of religion are transcendental . . . and rise quite above the region of phenomena."[55] For Smith, the focus on prediction took historical scholarship

54. Booth, *William Robertson Smith*; Rogerson, *The Bible and Criticism in Victorian Britain*, 113–29; Smith, *Lectures and Essays*.

55. W. R. Smith, "On Prophecy," 348–50.

of the Bible to the same boggy ground of biblical literalism. Although historical and literal readings of the Bible adopted different standards of evidence, both were mistakenly preoccupied with identifying fulfillment of prophecy.

Smith's separation of prophecy from prediction sought to turn prophecy away from concerns with exact meanings that exacerbated questions of evidence like those that pitted Ward against Mill. This shift of the terms incorporated another contemporary theological approach, which made the focus not prophecy but the prophet. In this view, prophecy was not a kind of magic, but rather an attribute of leadership. Foreknowledge in the days of the prophets of Israel conferred moral authority, enabling the prophet to serve as a counterweight to the temporal powers of the ancient biblical world. Robert Payne Smith, a conservative Anglican theologian appointed professor of divinity at Oxford in 1865, warned in 1869 that prediction was not the principal meaning of prophecy. We can have, he argued, "too complete an identification of prophecy with the foretelling of future events."[56] Rather the prophecies should be seen as a "pledge" of the supernatural that showed God's involvement in the world. Thomas Arnold similarly argued that prophecy was not anticipated history, but morality. "Prophecy fixes our attention on principles, on good and evil, on truth and falsehood."[57] One of the leaders of the Broad Church rational theology of the period, Rowland Williams, emphasized the identity of the prophet in his *Lampeter Theology* of 1856. Prophecy meant not only foretelling, but "preaching, or telling forth" and hence prophecy was not about "actual happenings in the future . . . but the moral vision and understanding which came from a perfect communion with God."[58] Prophecy, then, was not just about evidence, but about leadership, or how to understand the active role of those who held special knowledge.

Church leaders argued that the proper understanding of prophets showed such figures had recognized social and political functions, as a voice of moral leadership that could provide a counterpoise to temporal powers. Prophecy in ancient Israel, as in modern Britain, connected knowledge of the future with cultural and moral leadership in the present. As Smith put it in 1899, "the careers of the Prophets were contemporary with the development of Hebrew society from an agricultural to a commercial condition, and with the rise of the city. The social evils, therefore, with which the Prophets deal, are those

56. R. Payne Smith, *Prophecy*, 44.

57. Arnold, *Sermons*, 1:372.

58. Williams, *Lampeter Theology*, 42.

still urgent among ourselves."[59] As constructed by the liberal theology of mid-century, the prophet was a moral and spiritual guide in an era of social and political upheaval. In his personal authority he demonstrated the force of the divine in secular concerns.

The "modern prophet," then, was one of the rhetorical terrains on which the contests between religious and scientific authority were played out. In 1877, *Fraser's Magazine* outlined this contest for the role of the prophet in an anonymous article, titled simply "Modern Prophets." The article was an analysis of the moral and social implications of scientific thought in the 1870s. The author, "Unus de multis," began by describing his rebellion against the dogmatism of religion. "In my youth," he began, "the chief object of my intellectual and moral animosity was the Theologian." Here, he had first thought, lay the "main evil-doer" in the world, "poisoning every source of knowledge, vitiating every social relation." Shades of doctrine did not concern him: there was "not a pin to choose between one false dogmatism and another" and he raged against all the "old fantasies and falsities," longing to find the "Rightful Authority" to which he could "submit . . . without shadow of scruple or reserve." Yet, in maturity he found that the men of science were rapidly turning into dogmatists themselves, assaulting the benevolent impulses of religious faith itself rather than simply the "theologic dogmatism" of the churches. A mechanistic, evolutionary view of the universe was undermining "deepest notions of right and wrong"—and even more objectionable than the content of these views was their tone, their "pugnacity and paradox." "Unus de multis" singled out Huxley's description of the evolution of living protoplasm as an example of "retrospective expectation," which required a mental journey through time that was as absolute and as "entirely superhuman" as the religious beliefs Huxley delighted in attacking.[60] In short, the modern prophets—the "Prophet Scientists" like Professors Huxley and Tyndall—may be as dangerous as the old. (It should be noted here that, in Smith's distinction between false and true prophets, he labeled the former "the professional prophets," who handled subjects by formula and convention.)[61] By characterizing the professional man of science as a modern prophet, the article showed both the appeal of the prophetic figure for Victorians, the prophet's mantle of authority, and at the same time, its distasteful associations with enthusiasm, dogmatism, and false claims to knowledge.

59. Smith, *Modern Criticism* (1899), quoted in Cheyne, *The Transforming of the Kirk*, 136.

60. William Allingham [Unus de multis, pseud.], "Modern Prophets," 273, 280–81.

61. Smith, "Prophet," 817.

Conclusion

In science, prediction summarized what the uniformity of nature meant for the reach of the scientific observer. The natural philosopher saw into the past and the future and dominated all territories of knowledge. Prediction embodied knowledge that dismissed other authorities and could uniquely meet the demands of modern society. The debates over prediction and prophecy sought to identify those with reliable knowledge who therefore could provide sound leadership. A knowledge of the past or the future represented a form of control that had far-reaching importance for national identity. Victorians set themselves and their society consciously within a historical stream, and predictive knowledge marked out their claims on history.

Because prediction pointed to a kind of knowledge that reached beyond direct access of the investigator, it formed the connection between a wide range of phenomena and investigations. Some of these were established as scientific subjects, like the prediction of planetary movements, others were in the process of being constructed as such, like theories of the deposition and arrangement of geological strata, or the study of psychological states, as in mesmerism. But prediction also established a spectrum along which comparisons could occur to forms of knowledge that were not conventionally scientific. Divination and prophecy, the interpretation of legal testimony, trading in futures, and historical documents also dealt with indirect evidence and knowledge that was remote in time or place from the inquirer. As scientific culture increasingly claimed jurisdiction over all forms of knowledge and reasoning, the existence of this spectrum was a strength. It upheld the authority of science, insisting that the standards of natural knowledge, like the impressive and exact calculations of astronomy, applied to all fields of inquiry. Such standards, however, could also be a liability. When the understanding of natural phenomena was not wholly secure, the spectrum linked scientific work to spurious or supernatural knowledge. And it exposed scientific men to a critique based on the idea of false prophets, marked by shallow charisma and filled with intellectual and spiritual arrogance.

The problems of weather forecasting in Victorian Britain are hard to interpret without a consideration of this spectrum of ideas about prediction, prophecy, and prophets. Debates over method and epistemology led directly to consideration of the character of the scientific community and its leadership in modern society. This escalation was especially obvious in questions about prediction because of their public quality, offering demonstrations of truth that would compel the less informed or less inspired. Because weather prediction so often and so publicly failed, it could not stand as a model of scientific reasoning. When the Meteorological Department began to issue daily weather predictions in 1862, Lord Wrottesley wrote incredulously to Airy at Greenwich. Wrottesley, a prominent figure in the British Association for the Advancement of Science, had been involved in the pressure to establish a government meteorological office some years earlier. But he was appalled at the idea of weather prediction. He could not believe reports that both Airy and Herschel had "favourable opinions" of these "prophecies," which, said Wrottesley, are disguised under the terms "forecasts" and "warnings."[62] In answer to Wrottesley, Airy wrote a supportive, but cautious account of Fitzroy's experiment:

> On the whole, I am favourable, though not unmixedly. I think that in important instances . . . [Fitzroy] has been right and done good service. In some he has been wrong. But this I think surely, that he is going to work the right way . . . Also I think he is doing right in predicting boldly:—I believe this is peculiarly a case of "nothing ventured, nothing gained" and that a prophet, who cannot face the anticipation and bear the reproach of sometimes going wrong will have no chance of afterwards going steadily right.[63]

Weather forecasting, as Airy's reply made clear, was about risk, and Airy was somewhat unusual among his scientific colleagues in thinking it a risk worth pursuing. In 1846 the astronomer François Arago had famously remarked that any sort of weather prediction put one's reputation at stake. "However science may advance, worthy philosophers who care for their reputation will never venture to predict the weather."[64] Two decades later, Heinrich Dove, at the Prussian Meteorological Institute in Berlin, agreed, worried about the glamour of prediction. "Advances in meteorology," he wrote to the Royal Society in 1865, "are based on long-continued labours: we seem now to want to take it by storm; this may dazzle the public, but the results need control if they are to be recognized as really such."[65] To understand these risks to scientific authority and personal reputation, we need to turn to the notorious world of astrologers, almanac writers and popular weather prophets.

62. Lord Wrottesley to George Airy, June 17, 1862, Airy Papers, RGO 6/703.

63. George Airy to Lord Wrottesley, June 18, 1862, Airy Papers, RGO 6/703.

64. "Jamais, quels que puissent être les progrés des sciences, les savants de bonne foi et soucieux de leur réputation ne se hasarderont à prédire les temps." These remarks were quoted widely in meteorological literature, e.g., [Scott], "The Weather and Its Prediction," 489.

65. Heinrich Dove to Edward Sabine, June 12, 1865, quoted in "Correspondence between the Board of Trade and the Royal Society," 318.

CHAPTER TWO

Weather Prophets and the Victorian Almanac

IN AN 1849 meeting of the British Association for the Advancement of Science (BAAS), John Ball, an Irish naturalist and later a liberal Member of Parliament, commented on the exciting potential of the electric telegraph in the science of meteorology. "The ordinary rate [of atmospheric disturbances] does not seem to exceed twenty miles per hour"; he noted, "so that with a circle of stations extending about 500 miles in each direction, we should . . . be enabled to calculate the state of the weather twenty-four hours in advance."[1] Two years earlier, at the Smithsonian Institution in Washington, Joseph Henry's first outline of a meteorological network referred to the same possibility; from 1856, Henry collected observations and posted the data daily on a large weather map in the Smithsonian lobby. Henry made the step from reports to forecasting when he began issuing occasional forecasting announcements to the press the following year.[2] In France, Urbain LeVerrier promoted the collection of data by telegraph from 1860, although he waited three years for his suggestions to receive funding. Buys Ballot in the Netherlands had a small circuit of observers and warnings in place by 1860.[3] The British Meteorological Department began to collect data by telegraph in the fall of the same year

1. Ball, "Transactions of the Section: Weather Telegraphy," *Report of the British Association for the Advancement of Science [BAAS]*, 1872, 12–13.

2. Fleming, *Meteorology in America*, 141–47; on Henry, see *Dictionary of Scientific Biography*; Bruce, *The Launching of Modern American Science*, 187–201.

3. Lamotte and Lantier, *Urbain LeVerrier*, 140–43; Fleming, "Meteorological Observing Systems before 1870."

and issued its first warnings and general forecasts in 1861.[4] These opening forays into government meteorological services spread and developed over the following two decades, until most countries in Europe had some sort of weather reporting network (see table 2.1 for a survey of the most important of these).

But none of these official bodies or institutions was the first to publish weather predictions. That distinction belongs to the almanacs. Almanac publishing defined popular meteorology. From the 1830s, as controls over the press lifted, almanacs became one of the most ubiquitous forms of popular literature in Britain. Adapting their traditional associations with prophecy, both natural and political, they became the vehicle for debate over meteorological theories and the modern possibilities of weather prediction. The answer to the question of who predicted the weather in the nineteenth century, then, must begin with weather prophets and almanacs and not with institutions and telegraphy.

Popular prophecy was much more than a shadowy presence in the background of the discipline of meteorology.[5] The central fact of Victorian weather forecasting was that comparisons between government and popular weather prophets were constant, inevitable, and, from the point of view of scientific authorities, troubling. The world of popular weather prophecy calls attention to the way that meteorologists of all classes worked in a public theater with large and expectant audiences. A novelty as a government science, weather prediction was an established practice elsewhere. Splitting popular from elite knowledge, sound from unsound science, therefore, seems problematic as a way of understanding the impact of nineteenth-century meteorology. Weather prediction of all stamps appeared in the most public of settings, in newspapers and periodicals that reached a wide readership.

The effort that went into establishing sound scientific authorities on the weather was similar to other Victorian contests for control over natural knowledge. In the decade before the development of official centers for weather prediction, meteorology was one among many lively areas of inquiry into natural forces. Theories of the weather shared the stage with scientific demonstrations and instrumental exhibitions in the Victorian market for "Scientific Amusement."[6] Humphry Davy, Michael Faraday, and their successors expertly

4. Fitzroy, *Weather Book*, 168–206, 341–66; Burton, "Robert FitzRoy."

5. Symons, "History of the English Meteorological Societies"; Walker, "The Meteorological Societies of London" and "Pen Portraits of Presidents."

6. Altick, *Shows of London*; Cooter and Pumfrey, "Separate Spheres and Public Places"; Desmond, *The Politics of Evolution*; Morus, *Frankenstein's Children*; Morus, Schaffer, and Secord, "Scientific

crafted the fashionable lectures of the Royal Institution in London at the same time that other, less exclusive presentations of scientific novelties filled the assembly halls and theaters of the metropolis. New and controversial sciences like mesmerism challenged the medical profession with phenomena that could not be confined to their surgical theaters or hospital wards. Dozens of periodicals gave accounts of scientific ideas, from the *Philosophical Magazine* to the *Athenaeum* to *Zadkiel's Magazine or Record and Review of Astrology, Phrenology, Mesmerism and Other Sciences*.[7] Given the range of phenomena that early Victorian science explored, and the diversity of settings in which these explorations occurred, defining a definite, orthodox approach to many sorts of inquiries was impossible. Mesmerism is a key example: in the winter of 1837–38, a series of public experimental displays at University College London, run by John Elliotson, professor of practical medicine, demonstrated the exciting but ambiguous phenomena that observers uncertainly related to concepts of mental *and* physical forces. In the same year, 1837, Andrew Crosse, a gentleman experimentalist, discovered mites when he passed weak electrical currents through crystals, and his tentative communication of the experiment to fellow electrical experimenters in London triggered a controversy about spontaneous generation. Labeling such inquiries as marginal, fraudulent, or quackery captures neither the possibilities nor the anxieties of the era.[8]

After mid-century, some of the uncertainty over how to identify a scientific authority, and hence to manage the boundaries between genuine knowledge and fraud, disappeared. The scientific community more easily controlled education, accreditation, and publishing venues. But the debates over the public role of science did not disappear. Indeed, the more tightly managed scientific world in some ways faced its public roles more directly than before because of an increasingly literate, urban, and democratic culture. For these reasons, it can be misleading to emphasize the fragmentation of common cultural context for science as a new generation of natural philosophers defined disciplines, built specialized observatories, and developed powerful accounts of heat, energy,

London"; Ophir and Shapin, "The Place of Knowledge"; Outram, "New Spaces in Natural History"; Secord, "Science in the Pub"; Winter, *Mesmerized*. For a meteorological example, see Day, *Meterology* [*sic*] *as Applied to Practical Science*.

7. Brock, "The Development of Commercial Science Journals"; Henson et al., eds., *Culture and Science in the Nineteenth Century Media*; Shattock and Wolff, *Victorian Periodical Press*; Sheets-Pyenson, "Popular Science Periodicals in Paris and London"; James, *Print and the People*; Vincent, *Literacy and Popular Culture*; Anderson, *The Printed Image*.

8. Winter, "The Construction of Orthodoxies and Heterodoxies" and *Mesmerized*; Secord, "Extraordinary Experiment."

TABLE 2.1 Comparisons of meteorological services and institutions

Austria-Hungary
A central institute for meteorology was established in Vienna in 1848, and an extensive network of paid observers was developed. The institute was led by a series of respected directors in K. Kriel (1851–62), K. Jelinek (1863–76), and J. Hann (1877–97). The Austrian Meteorological Society was founded in 1863, and its journal became the prestigious *Meteorologische Zeitschrift*.

Belgium
Meteorological observations were taken at the Royal Observatory in Brussels from its founding in 1831, directed by Alphonse Quetelet. Its small network of observers was not supervised, and the international meteorological committee in the 1870s considered it one of the weakest organizations in Europe.

England
The Greenwich Meteorological and Magnetical Department was founded in 1840 and headed by James Glaisher under George Airy. Kew Observatory was founded in 1842, and the Meteorological Department of the Board of Trade in 1854, under Robert Fitzroy. The latter became the British Meteorological Office in 1867, with Robert Scott as director until 1900. A telegraphic network was established in 1860; the first warnings and forecasts were issued in 1861. The British (later Royal) Meteorological Society was formed in 1850; it was a successor to the short-lived Meteorological Society of London, founded in 1823.

France
In France, meteorological services were divided between the national observatory, Paris, and the Ministry of Marine. The Ministry of Marine received warnings from the British service from their inception in 1861. Under Urbain LeVerrier, director of Imperial Observatory in Paris 1854–70 and 1873–77, a national service began in 1863, with a daily bulletin containing forecasts, but it was suspended due to personnel conflicts and concerns about the value of general forecasts in 1865–66. Storm warnings resumed in 1866. From 1876, a French agricultural meteorological service also began observations and telegraphic reports. Finally, in 1878, after LeVerrier's death, a separate central meteorological office was established, directed by Emile Mascart. A private meteorological society began in 1852.

German states: Prussia, Bavaria, Hamburg
The Meteorological Institute was founded in 1847 in Berlin and directed by Heinrich Dove from 1849 to 1879. It began to issue forecasts in 1875. A marine service, the Norddeustche (later Deutsche) Seewarte, under G. Neumayer, based on the Hamburg network (eleven stations in 1877) issued storm warnings beginning in 1868. The Munich Observatory under Johann von Lamont was also highly regarded.

TABLE 2.1 (continued)

Italy
The Central Meteorological Office in Rome, an office to collect data rather than an observing center, was founded in 1863 as a department of the Ministry of Agriculture and Commerce. At Florence, the Ministry of Marine with the assistance of the Royal Observatory developed a telegraphic weather service in 1866, under the leadership of Giovanni Donati. Another Italian center for meteorology was the Montcalieri Observatory near Turin. Starting in 1859 the director Francesco Denza published a monthly bulletin of observations from Italian and international sources. Denza's network grew into the Italian Meteorological Association (founded 1880).

Netherlands
A central meteorological institute was founded at Utrecht in 1854 and eventually encompassed a network of thirty-seven stations. The director, Buys Ballot, published his account of the law relating wind direction and air pressure in 1857 and was the first in Europe to issue storm warnings in 1860, but he remained opposed to general forecasting.

Russia
The central physical observatory was located in St Petersburg, heading 104 stations in 1874 across the largest geographical area of any network. Observations standards became more reliable after the appointment of Heinrich Wild, 1868–95. Telegraphic storm warnings began in 1874.

Scotland
The Scottish Meteorological Society was founded in 1860. Though a private organization, it was very active in practical meteorology, and by the 1870s, it had a valuable network of more than 100 stations. In the 1870s and 1880s, its secretary, Alexander Buchan, often attended international meetings as a British representative alongside those from the Meteorological Office in London.

United States
The United States meteorological services, with its military observers, strictly trained and inspected, telegraphing reports several times a day, was by the 1870s the standard by which all other organizations were judged. Joseph Henry had founded a weather reporting network in 1849; its work shifted to the Weather Service of the U.S. Army Signal Office in 1870. Cleveland Abbe directed the office from 1871 to 1915.

Sources: Compiled from *Report on Weather Telegraphy and Storm Warnings*; *Report of the Treasury Committee*, 897–908 (Parliamentary Papers); Davis, "Weather Forecasting"; Fleming, *Meteorology in America* and "Meteorological Observing Systems before 1870"; Hellmann, "Die Organisation des meteorologischen Dienstes"; Hildebrandsson and Teisserenc de Bort, *Les bases de la météorologie dynamique*; Khrgian, *Meteorology*, 97–137.

or evolution. Even though boundaries between popular and elite forms of knowledge appear to have been more solid in the second half of the century, the interaction between different audiences for sciences was as extensive as ever.

Beyond a portrayal of this energetic exchange, the story of the weather prophets shows how much the possibilities and challenges of Victorian science were embedded in periodical literature. The relations between popular and elite science were negotiated and renegotiated in a constant flood of print as the material and social conditions of print were transformed in the nineteenth century. Steam printing, new reproduction techniques for illustrations, expanding literacy, cheap postal service, changes to the taxation of paper—all these dramatically affected who, what, and how much was published and read. Historians know something about the far-reaching consequences of these changes for the scientific world. Analyzing the periodical press after 1859 has provided a way to trace the assimilation and transmutation of Darwinian theory in Victorian culture. Particular publishing projects have been examined, like the Bridgewater treatises in the 1830s, an ambitious summary of modern scientific thought framed within the natural theology tradition, or the International Scientific Series in the 1870s and 1880s, a transatlantic venture in which eminent scientists wrote surveys of their field. The reading history of Robert Chambers's sensational *Vestiges of the Natural History of Creation* has shown how debates over authorship, reading, and reviewing practices all shaped Victorian scientific culture. And maps of the flourishing genre of popular science books and science journalism, associated with figures like Margaret Gatty, Richard Proctor, and Mary Agnes Clerke, have begun to appear.[9] The transformations of print culture that can be seen in these studies were especially important for a science like meteorology because it was so closely associated with a mass readership. Popular meteorology shows how knowledge, identity, and cultural leadership intersected on the page.

"A Few Wild Ideas"? Lunar and Planetary Influences on the Weather

Lunar influences on the weather had been a common subject of investigation from the eighteenth century onward. Stormy weather was thought to accompany the changes of the moon; wind was thought to be stronger when clouds

9. MacLeod, "Evolutionism, Internationalism, and Commercial Enterprise"; Ellegard, *Darwin and the General Reader*; Topham, "Science and Popular Education in the 1830s" and "Beyond the 'Common Context'"; Secord, *Victorian Sensation*; Lightman, "'The Voices of Nature.'"

cover the moon; and a month's weather could be predicted by extrapolating from the weather experienced during the moon's first quarter. It was also speculated that the effects of the influence of the moon on the atmosphere was stronger over large bodies of water, which explained both why belief in the moon's powers was so prevalent among seamen and why its effects were difficult to trace in the land-based observatories of elite science.[10] Giuseppe Toaldo, a Padua astronomer, argued in 1777 that atmosphere pressure varied according to the moon's position relative to the earth. In France, at the beginning of the nineteenth century, Jean-Baptiste Lamarck frequently published speculations on the influence of the moon on the atmosphere; and it was these speculations that prompted Pierre-Simon Laplace to call in 1814 for more rigorous statistical investigation in meteorology.[11] In Britain, Luke Howard turned to the subject of lunar influence in his detailed study of London weather.[12] Howard, a successful chemical manufacturer, was one of the most well-known meteorologists in the first half of the century, renowned for his application of the classification methods of Carl Linnaeus to cloud shapes. In *Barometrographica* he concluded that the fluctuations of the barometer showed periodicity, which he linked speculatively to the moon's gravitational effect on the atmosphere, and he analyzed an eighteen-year cycle of weather, based on the moon's relative position to the sun and earth. At mid-century, James Glaisher, meteorologist at Greenwich Observatory, took up Laplace's call for statistical analysis, correlating wind force, rain, and lunar period, and concluded that there was no evidence that the moon influenced precipitation, and only slight evidence regarding wind.[13] J. Park Harrison, a member of the British Meteorological Society, challenged Glaisher with data correlating the moon's phases and terrestrial temperature. His data, which he sent to the French Academy of Sciences and to the Royal Society, suggested that the moon at certain times dispersed cloud in the atmosphere, which affected temperatures.[14] Such studies, never quite resolving the case for or against lunar influence, reflected not

10. A. S. [Andrew Steinmetz?], "The Moon's Influence on the Weather," *English Mechanic*, July 19, 1867, 308–9; for other summaries of lunarist theories, see Browne, *The Moon and the Weather*; Howard, *Barometrographica*; and Saxby, *Foretelling the Weather*.

11. Sheynin, "History of the Statistical Method in Meteorology," 56–62.

12. Hamblyn, *The Invention of Clouds*; Howard, *Barometrographica*, *The Climate of London*, *A Cycle of 18 Years*, *Essay on the Modifications of Clouds*, *Papers on Meteorology*; Scott, ed., *Luke Howard*.

13. Glaisher, "The Influence of the Moon on the Direction of the Wind" and "The Influence of the Moon on Rainfall."

14. Harrison, "Lunar Influence on Temperature."

simply the state of data and statistical methods, but the persistent popular *and* scientific interest in the question.

Lunar theories of the weather linked ideas of atmospheric influence with electrical and fluid theories that had a wide appeal across disparate areas of scientific investigation. Some lunarists indeed did not distinguish between gravitational forces and electrical ones: subtle fluids or influences-at-a-distance blended together in their explanations, as they did in other nineteenth-century sciences. The appeal of these ideas was typically far-reaching. In 1862, Robert Fitzroy lowered his stock with his scientific mentors, John Herschel and George Airy, when he published his speculations on what he called the lunisolar effects.[15] The subject of electrical influences on the atmosphere elicited an extravagant style that made colleagues wince: "This all-pervading agency—is so intimately engaged in every atmospheric change, opposition, movement, or combination, that it should never be left out of mind. Imponderable, intangible, ubiquitous—nay, materially, almost omnipotent, this [is the] most marvellous of all the elements of our wonderful world."[16] This was just the sort of rhetoric that Herschel condemned. In a careful section on atmospheric electricity in his own account of meteorology, Herschel insisted, with emphasis, that "*as a cause of winds, or any atmospheric movements not merely molecular, we attribute to it* [electricity] *no importance whatever*."[17] Yet electricity had a plausible appeal within an atmospheric theory based, as Fitzroy's was, on the idea of friction between colliding wind currents. Although he sometimes insisted that he saw electricity as effect rather than cause, he elsewhere argued that polar and equatorial atmospheric currents might be differently charged, that the "electrical tension" was strongest in the polar or northern wind currents, and that the pressure of the surrounding electric ether concentrated the atmosphere and helped cause rotating wind currents, which led to storms.[18] In his detailed analysis of a destructive storm of 1859, presented to the BAAS in 1860, he stressed the electrical and magnetic disturbances that had accompanied the

15. Fitzroy, *Weather Book*, 244–56. Robert Fitzroy to George Airy, December 24, 1862, Airy Papers, RGO 6/703/27–28; id., January 15, 1863, Airy Papers, RGO 6/703/25–26; id., May 7, 1863, Airy Papers, RGO 6/703/31–32; id., June 11, 1863, Airy Papers, RGO 6/703/33–34; Robert Fitzroy to John Herschel, April 21 and 25, 1862, J. F. W. Herschel Papers, HS 7/259–60.

16. Fitzroy, *Weather Book*, 97, 451–58. Cf. his correspondence on theoretical questions with C. Marie-Davy of the meteorological section in the Observatoire Impériale, Paris (Met. Office Papers, PRO BJ7/792–99).

17. Herschel, *Meteorology*, 132.

18. Fitzroy, *Weather Book*, 453.

storm.[19] Although Fitzroy denied he was a "lunarist," it is easy to see how some of his readers found the denial ill founded.

By the 1860s and 1870s, lunar influence on the weather was a classic example of popular superstition. In 1875, urging the authorities at the Meteorological Office to pursue rigorous methods of physical inquiry, the physicist William Thomson (Lord Kelvin) called for the "weight of scientific investigation" to crush lunarists.[20] Yet it had long proved difficult to dismiss lunar influence as popular error. Throughout the century, Sir William Herschel and his son John were both widely associated with lunar meteorological theories. In the 1830s, the gardening encyclopedia of J. C. Loudon cited the elder Herschel as the authority for its account of lunar influences, and the attribution circulated as well in popular calendars like that shown in figure 2.1.[21] Based on a theory of "the attraction of the Sun and Moon" and "confirmed by the experience of many years" the inner levels of this circular table gave the dates and time at which the moon entered into new, full, or the first or last quarter. Following that point outward, one could read off the probable weather grouped into two seasons, summer (i.e., spring and summer), and winter (fall and winter). John Herschel, William's equally prestigious son, reinforced the established link between the scientific name of Herschel and lunar theories more than once. In 1863, he commented in the family weekly *Good Words* on the full moon's ability to bring a serene and calm night—"a tendency of which we have assured ourselves by long-continued and registered observation." Although he went on to denounce the popular lunar tables "falsely ascribed" to his father, Herschel found it difficult to shake the public's belief that he, too, was a lunarist.[22]

Herschel's experience showed that the distinction between prophecy and science centered as much on how it was circulated as what claims it contained. In the fall of 1860, rumors circulated in the west of England that Herschel had predicted heavy floods and cold weather for the coming months. Although Herschel wrote to energetically contradict the reports, his response outlined an interesting ambivalence to both the charge that he was a weather prophet and to speculations about natural causes that could shape the weather. "I do plead guilty," he wrote, "to having formed an opinion, from some remarkable phenomena exhibited by the sun last year . . . that this year would prove to

19. Fitzroy, "The *Royal Charter* Storm," *Report of the BAAS*, 1860, 13–14.

20. William Thomson to Richard Strachey, December 14, 1875, Add. MSS 60631:84–87, British Library.

21. Loudon, *Encyclopedia of Gardening*, 511.

22. Herschel, "The Weather and Weather Prophets," 59; Browne, *The Moon and the Weather*, 72–90.

FIGURE 2.1 The theory of lunar influence on the atmosphere was treated by many men of science as the height of vulgar superstition, to be eradicated by disciplined statistical investigations. Yet the astronomer William Herschel and his son, John F. W. Herschel, were both identified in the popular mind with lunar theories, with some reason, since both men published brief speculations about the moon's effect on the atmosphere. Based on a "Philosophical consideration of the attraction of the Sun and Moon," this chart of 1815 claimed it would "without trouble suggest to the Observer what kind of Weather will most probably follow." A reader, with a separate lunar almanac giving times of the lunar phases at hand, looked from the center of the picture outward. When the moon entered into the first or last quarter between noon and two o'clock, for instance, the prediction was "Very Rainy" in summer months, and "Snow or Rain" in the winter months. ("Herschel Table of the Weather" [London: R. and E. Williamson, 1815]. Reproduced by permission of the Bodleian Library, University of Oxford. John Johnson Collection: Calendars Box 1.)

be a rainy one." Solar activity, then, he suggested, could provide long-term indications of future weather. "I am disposed to regard the meteorology of the last twelve months as more pregnant with instruction than that of any equal lapse of time on record." Yet Herschel insisted he was not a prophet for two different reasons. First, he had not specified any time or location, "a thing which I consider at present quite beyond the power of any meteorologist." And second, Herschel was no prophet, he insisted, because he "expressed that opinion [about the character of the coming year] in private conversation, among friends, . . . assuredly never in such a way as I could suppose would come to be public." Herschel learned quickly that a philosopher of his eminence had difficulties keeping opinions private and "received many letters . . . some informing me that I stand charged with predicting the most dreadful storm ever known in the memory of man and asking me when it would take place." He was asked by another gentleman correspondent to explain to the *Times* how "a thick sheet of ice" had become "interposed between the earth and the sun, thereby causing this cold summer." Herschel of course disclaimed such wild notions; instead his view, he argued, was the opinion of "an observant person, connecting many scattered indications and some very remarkable and unusual phenomena with speculations on their possible or probable consequence." Nevertheless, the exchange revealed some of the grounds on which popular meteorologists could justly connect their claims to those circulating in elite circles.[23]

Ideas about the influence of the moon on the earth's atmosphere were not obscure and shared much with other contested natural phenomena, especially in the 1830s and 1840s. But one critical fact set lunar or planetary influence apart from other controversial scientific phenomena. An inquiry into the nature of these influences was inevitably associated with astrological practices, and in Victorian Britain, astrology was not simply intellectually dubious, it was technically illegal. An 1824 act designed to punish vagrancy had placed all species of fortune-tellers in its catch-pool of "idle and Disorderly persons, and Rogues and Vagabonds." The practitioners of Victorian and Edwardian astrology, as Patrick Curry has noted, therefore faced "real danger of prosecution and imprisonment"—something that never threatened mesmerists or supporters of spontaneous generation.[24] The effect of the criminalization of astrology can

23. [Herschel], "Sir John Herschel and the Weather," *Liverpool Journal*, September 15, 1860, 3.

24. This provision was not repealed until 1989. Howe, *Urania's Children*, 37–38; Curry, *Confusion of Prophets*, 13–15, 66. One notable prosecution, of J. Bradshaw in 1843, was discussed in *Zadkiel's Almanac for* 1844 and 1845.

be traced in writings on the weather. In 1851, William Joseph Simmonite, who published an almanac in Manchester that featured weather predictions, began to refuse personal interviews. He declined to answer inquiries about theft or missing objects (a common category of astrological advice) and claimed to be no more than a conduit for correspondence with a London astrologer (the "Mercurius Herschel").[25] Such measures tried to distinguish his work from astrological practice and to avoid entrapment by the law. Although there is no evidence of any prosecution of astro-meteorologists taking place, even a potential threat marked the philosophy.[26]

The criminalization of astrology made it essential for contemporaries to define the relationship of astrology to the science of the weather. Astro-meteorologists generally outlined a long heritage for their philosophy, while insisting on its connection to phenomena that were central to modern scientific thought, like electricity or magnetism. Astro- and lunar meteorologists constantly compared their theories with other contemporary approaches to meteorology. Debates about the nature of the influence of the sun, moon, and planets therefore reinforced debates about reliable methods of natural inquiry in the science.

Astrology was an ancient practice closely intertwined with the history of natural philosophy in general and with meteorology in particular. In Britain, astrology had peaked with the career of William Lilly during the Civil War and Interregnum. Following the Restoration, astrology fell into a gradual decline as religious and philosophical authorities, particularly the new Royal Society, chose to denounce it. Subsequent reforms in astrology took the science in two directions: some cosmological and medicinal elements were assimilated into respectable circles of thought, while judicial astrology and horoscopes found a sustaining niche in popular almanacs. One of the principal paths by which reforming astrologers of the late seventeenth century sought to restore their science was meteorology. John Goad's *Astro-Meteorologica* of 1686, for example, drew on classical authorities like Ptolemy and on Francis Bacon's plea for a purified, scientific astrology.[27] Following Goad's lead, Victorian astrologers saw in meteorology an attractive framework within which to present their philosophical claims. Astro-meteorologists argued that the position of the planets, as well as that of the sun and moon, affected the atmosphere by a form

25. Simmonite, *The Meteorologist for* 1851, 37.

26. Simmonite, *The Meteorologist for* 1851.

27. Capp, *English Almanacs 1500–1800*; Curry, *Prophecy and Power*, 61, and *Confusion of Prophets*; Howe, *Urania's Children*; Taub, *Ancient Meteorology*.

of influence related to gravitation: the planets influenced the atmosphere by their attraction, so major weather fluctuations could be connected to planetary motion. A planet's effect on the atmosphere was stronger when it was aligned at angles of thirty, sixty, or ninety degrees with respect to the earth. The intensity and nature of the influence depended on significant combinations of planets and positions. Jupiter in combination with Mars, for example, raised the average temperature, whereas the solar conjunction of Venus brought cloudy or rainy weather. Finally, each planet exerted different effects on the atmosphere because of the different speed at which each moved past the earth. Mercury, which traveled quickly and remained only briefly in one of the significant positions, produced sudden and violent changes, whereas Venus and Mars, moving at a speed closer to that of the earth, imparted slower and sustained influence, and were useful for predicting a whole season's character.[28]

Victorian astro-meteorologists sought to connect these claims to the natural philosophy of their day. Although there can be no doubt that many astro-meteorologists were astrologers,[29] some maintained that astrology and planetary weather science were distinct, and they called the latter "astrono-meteorology" in order to stress the connection to astronomy rather than astrology. The 1862 *Journal of Astronomic Meteorology and Record of the Science and Phenomena of the Weather*, for instance, claimed to promote "the study of atmospheric phenomena on practical astronomic principles."[30] Typically, such astro-meteorologists insisted first of all on the force of reasoning by analogy. If the moon exerted a gravitational effect on the oceans, it could do the same by creating atmospheric tides. The actual mechanism varied: some thought the moon disturbed electrical tension of the atmosphere; others looked to gravitational forces or vortices put in motion by the relative proximity of the moon. The strategy of analogy then sought to integrate lunar or astro-meteorological accounts of the weather with other approaches to meteorology. Frederic Pratt, for instance, who published one of the most detailed accounts of astro-meteorology in the *English Mechanic and World of Science* in 1871, argued that general expectations provided by the astro-meteorological theories

28. Pratt, "Science of the Weather," *English Mechanic*, February 3, 1871, 457–59, and February 10, 1871, 484–85.

29. As two examples, William Joseph Simmonite of Manchester published his weather predictions alongside nativities, political prognostication, and advertisement for personal astrological services, while Alfred J. Pearce authored both *The Weather Guide Book* (1864) and an influential *Textbook of Astrology* (1879).

30. [Editorial], 1 (April 1862): 1. The *Journal* and associated "Copernican Society" appear to be an offshoot of the Astro-Meteorological Society, discussed later in this chapter.

should be coupled with "leading weather and instrumental readings" for each locality. The weather observations of official meteorology complemented an astro-meteorological approach, and Pratt's article enacted the argument by linking a description of the theory of planetary influences with tables giving barometric curves.[31]

But astro-meteorologists also conventionally attacked the meteorology of standard authorities as a low, empirical form of knowledge and futile pursuit of data. By contrast, they insisted, astro-meteorology had succeeded in linking theory and observation. Popular weather prophets participated, then, in the debates about scientific methodology that characterized meteorology as a whole in the nineteenth century; indeed, the fact of their participation was part of what made those debates so critical.

This brief account of ideas about lunar and planetary influences on the weather has suggested their persistent role in the development of the science of meteorology. Leading philosophers and provincial almanac writers both speculated about the nature of such influences, and their connections to electrical and magnetic phenomena. The legal prohibition of astrology, moreover, intended to set some activities and persons outside the bounds of respectability, had the contradictory effect of underlining any possible relationship between the ideas of a John Herschel or a William Simmonite. Astro-meteorologists skillfully exploited debates over how to advance meteorological science, representing their work as guided by theory in distinction to Baconian fact-collecting. Indeed, the circulation of astral theories about the weather clearly contributed to a suspicion of all theorizing in meteorology. In 1841, John Hind, then a junior employee at Greenwich Observatory and later a distinguished astronomer, warned some colleagues interested in developing a meteorological society to keep their distance from "wild ideas" like the "supposed influence of the planets." Give yourselves entirely to "observations & research," urged Hind, "& touch (at present) upon no theory whatever."[32]

Yet beyond these factors, the crucial significance of astro-meteorology for the development of a science of weather prediction centered on the medium in which its theories characteristically circulated: the almanac. Considered as a genre with a particular and revealing history in nineteenth-century print culture, the almanac helped define the interactions between elite and popular scientific knowledge. Hind's own story points to its influence, for he had himself experimented with theories of the weather and an almanac. For three

31. Pratt, "Science of the Weather," *English Mechanic*, February 3, 1871, 456.

32. J. R. Hind to W. H. White, May 18, 1841, Royal Meteorological Society Papers.

years, 1839 to 1841, *The Atmospheric Almanac* outlined his ideas and laid out a year's worth of weekly predictions. Though his advice of 1841 suggested that the experiment had failed, the interesting point is that he made the effort at all. Why had a young, aspiring meteorologist and astronomer turned to almanacs and weather predictions in the first place? To answer this question we need to explore the place of the almanac in the early Victorian world of print, and its special promise as a voice of science in popular culture.

Profits and Prophecy: Tales of the Almanacs

The Victorian almanac reached backward to a long tradition of popular literature. Its hallmark was age: a stability of form connected the almanacs of the present to those of centuries past, with their calendars, astronomical positions, and prophecy. But the key to understanding the almanac in the nineteenth century is to see how far this air of antiquity was spurious. Many of the best-selling almanacs played to this tradition, retaining established titles and "old" authors, and celebrating their longevity as did *Old Poor Robin* in 1828, with "One Hundred and Sixty-third Edition" prominent on the title page. Almanacs were thoroughly modern publications. Almanacs flourished for the same reasons that other periodicals did: the emergence of concentrated and literate populations, technologies that facilitated paper making, printing, and illustration, and the political dismantling of taxes on the press. Almanacs were a distinctive class of publications, however, for several reasons that make them both important and neglected resources. One reason, as earlier discussions have shown, was the absolute scale of their audience. Lifting the taxes on almanacs in the 1830s unleashed a flood that had already proved difficult if not futile to control. Yet as important as sheer numbers was the variety. To associate almanacs with rural readers of chapbooks and to emphasize unduly the largest and "lowest" titles can be misleading. Almanacs reached through all social classes and were marketed to every conceivable niche of the population. The *Mirror of Literature* noted in a review of almanacs in 1824 that there was "variety enough to suit all tastes."[33] It became increasingly common for other periodicals to put out their own almanac monthly or as a separate annual publication, so that almanacs wove themselves into the general diversity of the print marketplace.

33. "Spirit of the Annual Periodicals," 402. For some useful discussions of readership, see Perkins, *Visions of the Future*, 23–44; Capp, *English Almanacs* 1500–1800; Harris, "Astrology, Almanacs and Booksellers"; Harris and Myers, eds., *The Stationers' Company and the Book Trade, 1550–1990*; Vincent, *Literacy and Popular Culture*.

Almanacs, furthermore, approached these transformations of audience, disciplines, and technologies as established resources for natural knowledge. Here the stability of the genre becomes an important consideration. Almanacs are both simple to define and difficult to categorize. They were yearly publications containing a calendar and diverse other contents that varied enormously. They could be issued in sheet form—a single closely printed page, perhaps for posting in some public place—or "book" form, which ranged from a few to a few hundred pages. But the arrangement of the typical contents of an almanac expressed its individual character. To accompany its calendars, tables, chronologies, and lists, editors of book almanacs often included articles on subjects of interest. Prophetic political commentary was the most notorious feature. Yet, almanacs also included notes or lengthier articles on astronomy and statistics, history and folklore, domestic economy and gardening. Within this miscellany, the most essential components of the almanacs were calendars. Calendars linked the natural and human worlds, connecting the sequence of seasons and planetary motions to worldly cycles of academic terms, legal sessions, and fairs. In short, the dominant purpose of the genre was instruction and reference, and it was centered in accounts of time and the activity of the cosmos. The regular features of an almanac thus often provided opportunities to record and respond to changing accounts of the natural world.

The almanac as a type was thus Janus-faced, equally evoking its reputation for "useful information" and "imposture" (the *Mirror of Literature* referred to *Old Poor Robin* as a series of "gross libels on public taste").[34] For contemporaries, judgments about the almanacs hinged above all on the question of profits and circulation. Louis James in his *Print and the People 1819–1851* called almanacs "the most widely diffused and least known type of printed ephemera in our period," and commented that "even cottages without a broadsheet or chapbook would be likely to have a sheet almanac pinned to the wall."[35] A more recent study of the almanac, Maureen Perkins's *Visions of the Future: Time, Almanacs and Cultural Change: 1775–1870*, concurs. Almanacs were mass literature, and promised striking profits. The legitimate market for almanacs was dominated by the Stationers' Company, a London-based group of shareholders, which held a monopoly on this type of publication (as well as on the Book of Common Prayer) until 1775, when a court decision put sole rights over the publications into jeopardy. After that date and until 1834, its control was sustained instead by high duties and take-over tactics, or buying up the rights to successful

34. "Spirit of the Annual Periodicals," 403.

35. James, *Print and the People*, 53.

rival products. The Stationers' list, known as the English Stock, included 25 titles in 1801, ranging from titles such as the *Gentleman's Diary* and *Ladies Diary*—respectable, even intellectual, but of limited circulation—to *Moore's* (or, *Vox Stellarum*)—astrological, crude, and immensely profitable. According to the 1801 "Statement of Almanacks" in the records of the Stationers' Company, *Gentleman's Diary* sold 2,648 copies and made one shilling profit in 1801; *Ladies Diary* sold 8,671 copies and brought in more than £54; whereas *Moore's* sold 362,449 copies and made a profit of nearly £2,600.[36]

The growth of popular literature, with new audiences, more printers, and especially a proliferation of provincial print culture, changed the monopoly but not the profits. By 1833 a survey produced by the publisher Charles Knight for the Society for the Diffusion of Useful Knowledge (SDUK) estimated that legitimate almanacs made up only a small percentage of sales. Knight, referring principally to *Moore's*, decried "the two shilling's [*sic*] worth of imposture" in almost every home in "Southern England."[37] Considering that the typical legitimate almanac price of two shillings and threepence was being challenged by prices as low as sixpence or twopence for the unstamped publications, this estimate seems entirely plausible. In 1834, the Stamp Acts, which had imposed duties of one shilling and threepence on almanacs since 1797, were repealed. Prices dropped sharply, and the variety of almanacs rose, as more publishers could afford to float an almanac without the heavy overhead of the duties. The happy position of the Stationers' Company following these changes underscores the value of the market. *Moore's* alone sold around 517,000 copies in 1838. In 1839, the total almanac output of the Stationers' Company (including the giant *Moore's Vox Stellarum*) was nearly 700,000; but it is estimated that at least that many again were sold outside its control, mainly in provincial markets. Although it had fought the repeal, the Company continued to improve its profits afterward. In 1833 it recorded profits of about £4,000 spread over eleven titles; in 1835, it cut prices, doubled sales, and made profits of £5,000.[38]

These figures are noteworthy because they demonstrate the size of the audience for almanacs. They also show how the scale of the market was central to contemporaries' understanding of the publications. Almanacs were

36. Perkins, *Visions of the Future*, 238; Myers, *The Stationers' Company Archives*.

37. Perkins, *Visions of the Future*, 93, 107.

38. "Spirit of the Annual Periodicals"; Knight, *Passages of a Working Life*, 2:59–65. Knight describes the survey, 2:64. The records of the Stationers' Company give figures of copies printed, copies sold, and profits made (Perkins, *Visions of the Future*, 238; Myers, *The Stationers' Company Archives*).

"reading for the million," and evidence suggested to contemporaries that the lower the tone, the higher the circulation. The proliferation of almanacs created the circumstances for a second, distinctive aspect of the genre. This was their place in debates over cultural reform in the 1830s and 1840s.

As quintessential mass reading, on the one hand, and as inheritors of astrological and radical traditions on the other, almanacs presented a vital target for reformers seeking to discipline modern culture like Charles Knight. Knight, a London publisher, who began a campaign to reform the almanac as soon as he joined the SDUK as its publisher in 1827. In targeting the genre, Knight and his supporters seized on the hypocrisy and venality in the Stationers' Company. That institution, as Knight saw it, put profit before principle. Prophetic almanacs like *Moore's Vox Stellarum* sold much more widely than the mathematical *Ladies Diary*, and the company saw no incongruity between sponsoring both almanacs for gentleman and ladies and flagrantly astrological publications. As late as 1848, the *Athenaeum* identified the Stationers' Company as "the Astrologer's College of our day."[39] An old ritual of the company ironically underlined the mixture: each year, the company sent its new almanacs ceremoniously in a barge down to Lambeth Palace, seat of the Archbishop of Canterbury, for presentation to ecclesiastical authority.[40] The company's almanac trade thus appeared to represent a corrupt failure of cultural leadership. In 1828, Knight published the first *British Almanac* and its *Companion* of feature articles. The *British Almanac*, its publication costs heavily subsidized, was designed to lift the almanac from its atmosphere of superstition and ignorance. Weather predictions were for Knight the defining symptom of vulgar irrationality. The first pages of explanatory remarks in Knight's new *British Almanac* tackled the "injurious" and "absurd" practice of weather prediction and denounced the "cunning cheats" who made their living from the foolish human desire for certainty. Assisted by high-ranking men of science, Knight replaced weather prophecy with careful statistical discussion of average weather observations.[41]

These complex associations of the almanacs with commercial success, popular culture, and the modern world of print can be traced in the work of

39. *Athenaeum*, December 16, 1848, 1262–63.

40. Weale, ed., *London and Its Vicinity Exhibited in 1851*, 328–39.

41. "Observations on the Weather," *Companion to the British Almanac* 1 (1828): 4; cf. a later discussion, "Natural History of the Weather," *Companion to the British Almanac* 3 (1830): 68–92 and 4 (1831): 28–36. Knight described the list of scientific men on which he, with the assistance of Henry Brougham, was able to draw (e.g., John Frederic Daniell, John Lubbock, and John Wrottesley) (Knight, *Passages of a Working Life*, 2:62, 123, 126, 129, 179).

one of the master critics of the age, Thomas Carlyle. His *Past and Present* (1843) was linked in a characteristically allusive way to the contemporary reputation of the almanac. The structure of Carlyle's text (which despite the binary title commented in fact on past, present, *and* "prognostications" for the future) and the denunciations of quack remedies to treat the national crisis both suggestively linked his cultural critique to the almanac.[42] Morison's Pills, the patent medicine that served as Carlyle's symbol of a misguided, misdoctored nation, were sold in the same newsagents and print shops that had sprung up to distribute popular literature, including almanacs. In the same year that Carlyle wrote *Past and Present*, Herbert Ingram, a publisher with strictly commercial (in distinction to political) ambitions for his newspaper, had bankrolled the new *Illustrated London News* with the profits from his promotion of a direct competitor to Morison's Pills, Old Parr's Life pills, boxes of which lined his first London premises. Newly arrived in London from Nottingham, Ingram published *Old Moore's Almanack*—one of many knockoffs of the successful *Vox Stellarum*—primarily as an advertising medium for Parr's pills. The patent medicine connection was a notorious ingredient in Ingram's success.[43] This situation underlined the fundamentally commercial relations of modern print enterprises with the new reading classes—relations that Thomas Carlyle, Charles Knight, and others watched with much unease. When Carlyle wrote in *Past and Present* of patent nostrums, the flimsiness of contemporary opinion, and the emptiness of modern authorities, then, the almanac was part of his subtext. Almanacs presented the disturbing picture of popular influence out of all proportion to their financial and intellectual weight. Perhaps it was no coincidence that Carlyle repeatedly misspelled the name of the pill maker. Rather than Morison, he called him Morrison—the name of one of the most notorious almanac authors and astrologers of the same period.

It makes sense, then, that the young astronomer John Hind should have turned to the almanac in 1839 to 1841, not only for the very real prospect of making some money from it, but also because the almanac itself was a particularly appropriate vehicle for the promotion of scientific ideas. Weather prediction, and the questions it raised about methods and utilitarian goals as well as the nature of atmospheric phenomena themselves, became a key to defining the character and values of scientific culture.

The careful management of meteorological science in one publication indicates how the almanac developed into a critical medium for such debates

42. Altick, "Past and Present."

43. Grogg, "The Illustrated London News 1842–1852," 30–31, 105, 124, 139.

about natural knowledge and its place in society. *Illustrated London Almanack* was produced by the *Illustrated London News* starting in 1845, that is, at the end of the first season of the parent publication. Larger than most almanacs, the *ILA* was a folio of eight by eleven inches, but it was a typical length of sixty-four pages and sold for a shilling, a fairly moderate price. After 1858, when George Cargill Leighton took over as printer and publisher for the *Illustrated London News*, the almanacs were increasingly elaborately illustrated, with stunning color engravings on the covers.[44] Its contents (including tables of cab fares in London, a description of how to write a will, or lists of popular excursions) suggested an urban, domestic audience.[45] The most striking feature of the *ILA*, however, was its scientific content. From its inception, it had extensive sections on natural history and astronomy. Scientific notes became increasingly central to the identity of the almanac. In the issues of the first two decades, a visible decision to concentrate on scientific content of a particular sort can be traced. Changes to the almanac's content and format show a publisher honing the direction of his work, with the intent of establishing through science a product that was simultaneously useful, moral, popular, and visually innovative.[46]

The first issue of the *ILA*, in 1845, was scattershot, addressing history, folklore, scientific subjects, sports and leisure, domestic matters, and statistics. It consisted of a two-page spread per calendar month (the first twenty-four pages) followed by about forty pages of articles, interspersed with densely packed pages of information and tables. Scientific material was prominent. Each of the monthly pages had a calendar on the verso and natural history notes on the facing recto. Two long articles on astronomical subjects followed on the heels of the calendars: first, a detailed account of the time ball at the Royal Observatory, Greenwich; and second, an article on "New Comets," with special focus on the discovery of Neptune the previous year. In the first issue there was a weather table and a poem on weather signs in folklore: familiar

44. Many of the illustrations were republished in Leighton, *Pictorial Beauties of Nature*.

45. Perkins, *Visions of the Future*, 39–45, speculates that the almanacs were seen as appealing primarily to women. Although I am not convinced by this as a general claim, I note that the *ILA* seemed to be addressed to a female readership. It included a regular domestic hints column, occasional articles on recent domestic inventions, descriptions of popular excursions in and around London, and lists of fruits, vegetables, and meats in season. According to a study of the *ILN*, the *ILA* sold excellently (circulation 50,000 in 1849) at a time when its parent was the best selling weekly newspaper (circulation 140,000 in 1852). Grogg, "The Illustrated London News 1842–1852," 124, 139–40.

46. The *ILN* showed similar shrewd experimentation with its features under Henry Ingram. Grogg, "The Illustrated London News 1842–1852," iv.

fare for almanacs. The meteorological content of the *ILA*, however, trod a middle path, endorsing neither the prognostication of the popular almanacs, nor the rational tone of the *British Almanac*, with its meteorological averages and columns of observations. The poem on weather signs detailed all the traditional rules for predictions for the following year. It described how to anticipate the season's weather from certain saint's days even while it ended by exhorting the reader to "Let no such vulgar tales debase thy mind, / Nor [St.] Paul, nor [St.] Swithin rule the clouds and wind!" (Predicting the season's weather by the weather on St. Swithin's day was one of the widely known traditional prognostication rules.) The 1845 weather table gave a similarly equivocal message. Its conclusions, as the accompanying text noted, derived from "many years' actual observation," and the text acknowledged with due philosophic caution a (rather ample) room for error in that "the weather . . . is more uncertain in the latter part of the Autumn, the whole of Winter and the beginning of Spring." At the same time the table listed predictions based on ideas about lunar influence that linked it to popular folklore and astrology rather than to a modern rational meteorology. Like the broad array of topics in the rest of the first issue, the meteorological coverage aimed to please all tastes.[47]

In the following year, however, a much more focused strategy emerged in the *ILA*. Meteorological discussions disappeared, and the format was changed to put its sound sister science, astronomy, front and center. The monthly pages now numbered four, with a calendar page, a page on how to observe the stars and planets, a page of seasonal notes, and finally the natural history notes. Over the next two or three years the observing notes became more detailed and the calendar more elaborate, with historical notes, anniversaries, gardening, or cookery squeezed out in favor of more chronological and scientific data. There was regular discussion of astronomical discoveries, especially of new planets and comets.[48] For two years there was no mention of the weather in the almanac at all. In 1848, an article on weather observations reappeared, but with a significantly different emphasis. The *ILA* now stressed the rigorous scientific character of its weather observations, based on "averages as calculated from the observations taken at the Royal Observatory at Greenwich every two hours, night and day, for four years." The meteorological notes would avoid predictions and speak only of "the general character" in a month-by-month description.[49] Despite this disciplined (re)introduction, meteorology

47. *ILA for* 1845, 5, 40.

48. Clerke, *A Popular History of Astronomy*, 89–110.

49. *ILA for* 1848, 2.

remained in distant second place to astronomical subjects. The next extensive reference occurred a further three years later in 1851, when the almanac called attention to the formation of the British Meteorological Society, a society itself designed to purge meteorological investigations of any popular or astrological tendencies.

Fortunately, we can trace something about what or, rather, who was behind these significant editorial changes. From the second year of its existence, the *ILA*'s calendar and the astronomical and the (occasional) meteorological notes were produced by James Glaisher, in charge of the magnetical and meteorological department at the Royal Observatory. Making use of Glaisher, the *ILA* increasingly emphasized its scientific content as its leading feature. Throughout the 1850s the gardening and natural history notes, written by the well-known naturalist Jane Loudon, metaphorically as well as literally took second billing to the astronomy notes. Glaisher, as the preface in 1854 noted, had care of the almanac's "vital parts."[50] The almanac increasingly identified itself as an astronomical reference tool for the amateur. It described how to construct an inexpensive telescope and gave monthly viewing charts. It emphasized the contributions of the "private observatories" that sprang up in the 1830s and 1840s, and the developments in optical science that were making a new age in astronomy. By the end of the 1850s the *ILA* presented color engravings of astronomical photographs in ways that rivaled the much more obviously "picturesque" color engravings of flowers, birds, or fish.

The picturesque qualities were at the core of the identity of this almanac, and naturally so, considering its parent publication was the principal illustrated weekly of the era. The illustrations offer examples of experimentation with visual records in science, in the particularly valuable context of the illustrated newspaper—that is, a context in which artistic contributions were acknowledged to be essential to the finished literary product.[51] As color printing techniques developed, the front covers became visual tours de force, while inside there were more sumptuous natural history prints, engravings of astronomical photographs, and unusual graphs and ways of presenting the calendar. Most significantly the visual aspects of the *ILA* suggest an explicit parallel with the hieroglyphics of the prophetic almanacs. Here were offerings

50. *ILA for 1854*, 2. Glaisher continued to provide materials for the *ILA* in the latest issue I have seen (1889), although the astronomical emphasis of the journal diminished in the 1870s.

51. Starting in 1856, the *ILN* included regular color plates printed using techniques developed by George Cargill Leighton. The *ILA* had color prints and covers starting in 1858, the date that Leighton became the printer and publisher for the *ILN*. See McLean, *Victorian Book Design and Colour Printing*, 191–94; Anderson, *The Printed Image*; Muir, *Victorian Illustrated Books*, 148–78.

designed for "that part of the public which is more open to receive information from pictorial representation than from tabulated numbers."[52] The engravings then represented a conscious effort at "responsible popularity," in much the same way that the *Illustrated London News* converted a disreputable format—the illustrated weekly—to respectable middle-class reading.[53] The striking cover of 1860, for instance (fig. 2.2), transformed the zodiac into a window into the natural world. The zodiacal signs, cushioned in chrysalides, merely frame the light flooding in from an exterior sky and the insect life swarming on the page. The potentially alarming associations of the zodiac have been displaced by the beauty and variety of the natural world, and nature has taken over the almanac—literally settling down on the title. Such examples indicate the careful design of the almanac, its idea of audience, and its integration of scientific work with that design and audience.

Of particular interest here in the history of the *ILA*, however, was the treatment of astronomy and meteorology. Because of the attack on the almanac by Knight and others as a symptom of the worst possibilities of popular literature, and the use of weather prediction as a gauge of its errors, meteorology, by its presence or absence, clearly delineated an almanac's ideological position. The solution of the *ILA*, to demote meteorology in favor of astronomy, revealed something interesting about the scientific predicament of the almanac. How could a publication that was regularly rendered both obsolete (like a calendar) and false (when predictions were fraudulent or simply inexact) become a suitable forum for scientific knowledge? The *ILA* argued that, since all fields of knowledge were now developing so quickly, almanacs were no more ephemeral than any other "repository of fact."[54] By implication this placed the almanac on an even footing with the most learned productions of the culture. Astronomy was crucial to the second problem, that of the almanac as a catalogue of errors. The *ILA* became a different sort of object: the reader has certainties, and "the predictions of one year are now founded on so secure a foundation that they become the facts of another."[55] Instead of ephemerality, it provided perpetual reference, instruction, and entertainment in a rapidly changing age. Through astronomy, an almanac, one of the most insubstantial of publications, was transformed into a solid, enduring work. Weather prediction

52. *ILA for* 1852, 2.

53. On the *ILN* and its pursuit of respectability, see Grogg, "The Illustrated London News 1842–1852," 167–95; the quotation is from 119.

54. *ILA for* 1853, 2.

55. Ibid.

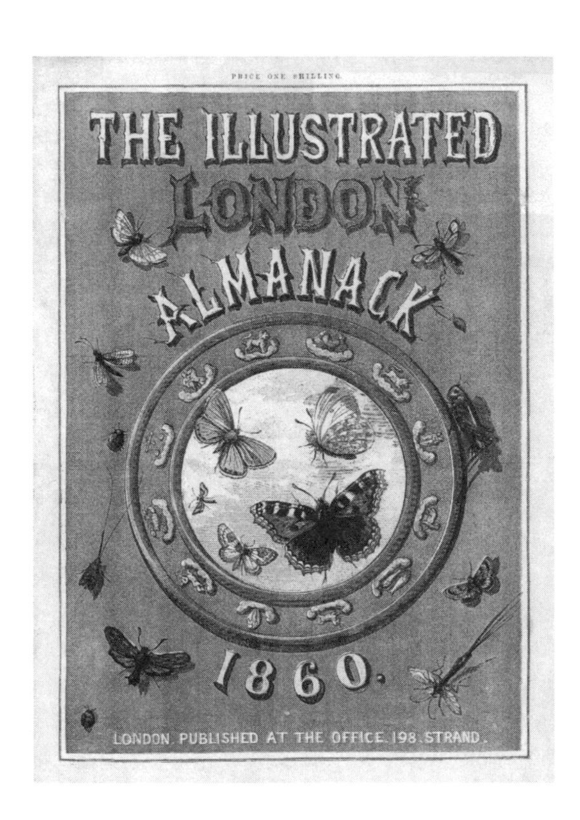

FIGURE 2.2 The *Illustrated London Almanack* appeared in 1845 and was as visually innovative as its parent, the *Illustrated London News*. It shaped its identity through its scientific content. After 1858, when George Cargill Leighton took over as printer and publisher for the *Illustrated London News*, the almanacs were increasingly elaborately illustrated, with stunning color engravings on the covers. (Reproduced by permission of the Ruari McLean Collection, Robertson Davies Library, Massey College, University of Toronto.)

was left to one side, then, because it represented insubstantial knowledge of dubious scientific value.

Yet of course the *ILA* was not typical of the genre. It was a middle-class response to the great appeal of popular almanacs, in which weather prediction continued to thrive. The best avenue for exploring almanacs and weather science, then, is not an elite publication such as the *ILA*, but the kind of almanac that it sought to replace. The brief but sensational *Murphy's Weather Almanac* in 1838, and the enduring appeal of *Zadkiel's Almanac*, the work of Richard James Morrison, both demonstrate the importance of the almanac in the history of Victorian weather prediction.

In 1838, *Murphy's Weather Almanac*, one of dozens of similar almanacs specializing in weather prediction,[56] predicted that the coldest day of the year would fall January 20th. When this prediction was confirmed by extreme cold, his printers were swamped with the demand for copies of the publication. Satires and imitators popped up everywhere. *Cruikshank's Almanack*, a popular satirical publication, published a barometer of gullibility marked in pounds rather than inches of pressure (fig. 2.3), and showed a host of readers pushing eagerly into a bookshop. In a parallel spoof, Thomas Hood's *Comic Annual* told the tale of a visit from the Man in the Moon by balloon. News of "a profiting Prophet below" had compelled him to visit with copies of his own lunar theories of the weather, dedicated to Sir John Herschel "now at the full in celestial fame."[57] By all accounts Murphy profited enormously. Murphy had demonstrated the strong financial appeal of weather prediction. J. H. Maverly, a member of the Meteorological Society, noted in disgust some months later that the almanac was "a complete failure, an imposition on the public and a mere catchpenny publication; but [one] which I believe was so cunningly managed by the printers as to gain 8 or 10,000 £."[58]

The almanac was not Murphy's first venture into print. For several years previously, Murphy had published a number of works on meteorology and astronomy, beginning with *An Inquiry into the Nature and Cause of Miasmata* (1825) and *Rudiments of the Primary Forces of Gravity, Magnetism, and Electricity, in Their Agency on the Heavenly Bodies* (1830). In 1834 and 1836, treatises focused on meteorology appeared: *The Anatomy of the Seasons* and *Meteorology Considered in Its Connexion with Astronomy, Climate and Geographical Distribution of Animals and Plants Equally with the Seasons and Changes of the Weather.*[59] Both of the latter volumes presented a theory of planetary influence on the atmosphere via a combination of electrical, magnetic, and gravitational effects. Yet, significantly,

56. E.g., Simmonite's *The Meteorologist for* 1851; Peter Legh's *Ombrological Almanac*; Orlando Whistlecraft's *Whistlecraft's Almanac*; *Orion's Prophetic Guide and Weather Almanac*; *The British Weather Almanac and Rural Diary . . . On the Principle of Solar, Lunar and Planetary Reflection*; George Shepherd's *Meteorological Almanac and Monthly Weather Ephemeris*; Henry Doxat's *Lunar Almanac*; and *Bushell's Weather Almanac*.

57. Hood, "A Flying Visit," 145. For the estimates of profit by contemporaries, see Baker to W. H. White, May 14, 1839, Royal Meteorological Society Papers. Cf. Perkins, *Visions of the Future*, 205.

58. J. H. Maverly to W. H. White, July 4, 1838, Royal Meteorological Society Papers. Murphy charged one shilling and sixpence for his almanac. Weather almanacs in this period typically cost one shilling: Zadkiel began charging two shillings for his *Herald of Astrology* in 1830 but dropped to one shilling by 1835 and to sixpence in 1847.

59. Murphy, *The Anatomy of Seasons* and *Meteorology, Considered in Its Connexion with Astronomy*.

162 JANUARY [1839.

AL-MANIAC DAY.—A RUSH FOR THE MURPHIES.

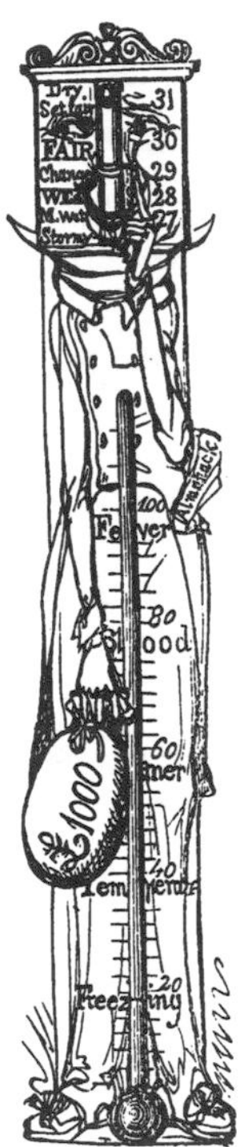

MYSTERIOUS Murphy, whose transcendent skill
Makes hail, rain, vapour,
Come forth obsequious to your will,—
At least on paper,—
Tell us what famous college
Bestow'd your wondrous knowledge!
Perchance your learned sconce found it *at once;*
Perhaps *by degree* of T.C.D.
Some say the Prince of Evil has been too civil,
And that, in change for all your knowledge boasted
You're doomed—like other murphies—to be roasted.
Some think, like me for one,
You've kissed the Blarney Stone;
But though your blunders make a pretty rout,
Sure, if you're right, by *second* sight,
You well may be, *at first*, a little out.

But cock your weather eye athwart the sky,
Of wind and storm disclose your store,
For one year more,
And tell us true.—
Led by your lies the ships *lie to*,
Or snugly *arbour'd*, with *bower anchor* ride,
And lose the tide—
Their funnies near, the watermen look sad,
Short cut or shag alone their sorrow lulls,
In sunshine read your page of weather bad,
And shake their heads, for no one wants their sculls.
But, sad to think, the washerwoman's pain,
Praying for rain,
And vainly hoping, as for showers she sniffs,
To fill her *butts* with your delusive *ifs*.
Ah, me! I sought the throngs in Beulah's bowers,
Seduced from home by your *fair* fiction,
But found none *out*, amid the drizzling showers,
Save my sad self and your prediction.
Now if again the weather's care you take on,
Don't try your flam on,
But if you wish to save your bacon,
Give us less gammon.

FIGURE 2.3 George Cruikshank's "Al-maniac Day—A Rush for the Murphies" satirized the spectacular success of Patrick Murphy's 1838 almanac after a "lucky hit" when its prediction of extreme cold on January 20th came true. Murphy reputedly made a fortune from the sales: the thermometer of gullibility to the left of the verses shows Murphy, a sly finger to the side of his nose, clutching a money bag. (George Cruikshank, "Al-maniac Day—A Rush for the Murphies," in *Comic Almanack, an Ephemeris in Jest and Earnest: 1835–1843* [London: J. C. Hatton, 1860], 162.)

the discussion about weather prediction that was sparked by Murphy's success endured more successfully in magazines than in philosophical treatises. *The Gardener's Gazette*, edited by the horticulturist and journalist George Glennys, published a weekly column from January 27, 1838 (one week after Murphy's success), until the journal changed hands in 1844. Until January 1840 the column compared the forecasts of several prophets in table form and debated their respective merits in the accompanying text.

Among the prophets evaluated in the *Gazette* was the notorious astrologer Richard James Morrison. The career of Morrison, even more than that of Murphy, displays the significant place of popular prophecy in the history of weather prediction. His work demonstrates the range of ways that discussion of meteorology circulated in Victorian society and shows how that circulation made claims about the place of natural knowledge in society. Morrison's activities indicate how meteorology exemplified the potential of scientific knowledge as the foundation of a disciplined community—but also summed up the challenges of controlling that community.

Morrison was a man full of projects, developing inventions, floating coal mining stock and life insurance schemes, pleading for annuities from the Admiralty, and organizing a giant exhibition telescope company.[60] But his almanac was the most important achievement of his career. Morrison was a naval lieutenant, retired from active service, who had studied under a London astrologer called Dixon. In the 1820s, Morrison became a member of an occult association known as the Mercurii, headed by Raphael (Robert Cross Smith), who himself was the author of a widely read almanac called *The Prophetic Messenger*. By the late 1820s Morrison had founded his own almanac, the *Herald of Astrology*. Renamed *Zadkiel's Almanac* in 1836, it flourished until the end of the century (fig. 2.4).[61] His success presented a history in microcosm of the effects of the changes to market and publication laws for almanacs. The *Herald of Astrology* was published at a typical price of two shillings from 1829 to 1833, but then dropped to one shilling after the repeal of the Stamp Acts.

60. His inventions, according to Cooke, included a bell buoy and improvements to electrical telegraph transmission and to the screw propeller for steam vessels. Cooke, *Curiosities*.

61. Accurate figures are hard to obtain, since critics cited alarmist figures while astrologers had reasons to exaggerate their circulation as well. Zadkiel's own estimates in his *Almanac* were 22,000 in 1849; 32,000 in 1851; 44,000 in 1861; 58,000 in 1864; and 70,000 in 1870. After A. J. Pearce took over in 1870, sales increased further (according to the new editor), to 150,000 copies. For comparison, Curry suggests that *Raphael's Almanac*, one of the most widely circulating of the Victorian almanacs, was selling 100,000 per issue in the early Victorian period, and approached 200,000 by the end of the century. Curry, *Confusion of Prophets*, 60, 107, 111.

In 1847, *Zadkiel's Almanac* lowered its price again to sixpence and claimed a circulation ranging from 22,000 to 32,000 for a forty-eight- to sixty-four-page issue. Morrison continued to publish the almanac until his death at Kingston, Surrey, in 1874.[62] The title then—in a process typical of successful almanacs—was sold to another leading astrologer.[63]

Writing to the secretary of the Meteorological Society in 1839, Morrison claimed that he had been "obliged to tack astrology to my almanac even to insure a sale to pay my expenses; but I now hope to devote myself entirely to the service of agriculturists &c by letting all my attention be given to foretelling the weather."[64] This claim that the astrological elements of his meteorology were included merely to gather in a popular audience deserves to be received with skepticism. On the contrary, there is plenty of evidence that the promotion of astrology was Morrison's fundamental motive. In other words, Morrison turned to meteorology in pursuit of a professional, scientific respectability for astrology. Weather prediction fell into the category of Mundane Astrology, he told his readers in 1832, which foretells the weather and all national events and is the most difficult branch of astrology.[65] Effective weather prediction would provide a demonstration of astrological theories. Nearly thirty years later, he again emphasized this account of the relationship of astrology and meteorology, commenting to a friend that "if Astrology ever make its way with the public it must be through the means of Astro-meteorology as an introduction."[66] Yet in an important sense, both Morrison's interpretations of the content of the almanac were disingenuous: the publication was neither astrological, tinged with meteorology, nor meteorological, tinged with astrology. Rather, the combination of astrology and meteorology—dramatic, useful, and all-encompassing natural knowledge—represented the progress of the sciences and their importance in modern society, just as the medium of the almanac itself exemplified the new, expanding means of circulating such knowledge.

62. *Oxford Dictionary of National Biography*; Cooke, *Curiosities*.

63. Cooke, *Curiosities*. Much information about Morrison's life comes from an annotated copy of *Curiosities* held at the Meteorological Office Library, Bracknell. The date of the annotations is unknown, but there are references to Morrison's death in 1879. (There is a second, similarly annotated copy at the British Library.)

64. Richard James Morrison to W. H. White, February 25, 1839, Royal Meteorological Society Papers.

65. *Herald of Astrology for* 1832, 40. The others were Genethliacal Astrology, or nativities, and Horary Astrology, or answers to particular questions.

66. R. J. Morrison to C. Cooke, January 8, 1861, in Cooke, *Curiosities*, 271.

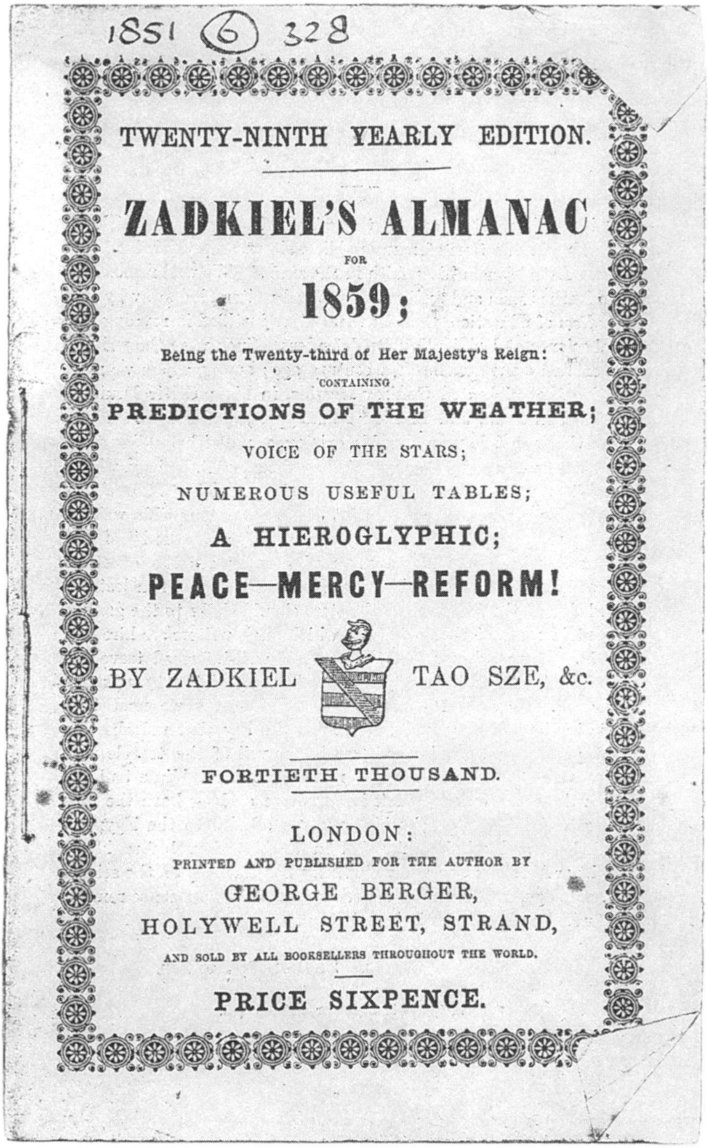

TWENTY-NINTH YEARLY EDITION.

ZADKIEL'S ALMANAC

FOR

1859;

Being the Twenty-third of Her Majesty's Reign:

CONTAINING

PREDICTIONS OF THE WEATHER;

VOICE OF THE STARS;

NUMEROUS USEFUL TABLES;

A HIEROGLYPHIC;

PEACE—MERCY—REFORM!

BY ZADKIEL  TAO SZE, &c.

FORTIETH THOUSAND.

LONDON:

PRINTED AND PUBLISHED FOR THE AUTHOR BY

GEORGE BERGER,

HOLYWELL STREET, STRAND,

AND SOLD BY ALL BOOKSELLERS THROUGHOUT THE WORLD.

PRICE SIXPENCE.

FIGURE 2.4 The cover of *Zadkiel's Almanac for 1859* listed the contents, prominently displaying in capital letters its predictions of the weather. *Zadkiel's Almanac* was an astrological publication that sought respectability, emphasizing its usefulness and loyalty, as in the subtitle here: "PEACE—MERCY—REFORM!" (Reproduced by permission of the Syndics of Cambridge University Library.)

From the beginning *Zadkiel's* positioned itself as part of the expansion of natural knowledge. Indeed, one of the more striking efforts Morrison made to identify himself with Victorian scientific culture was the formation in 1844 of the "British Association for the Advancement of Astral Science &c and Protection of Astrologers" (BAAAS), a society founded to campaign against the legal penalties attached to astrology.[67] In his almanac, Zadkiel/Morrison pointedly rejected the impious reputation of astrology and presented readers with a version that was fully reconcilable with the natural theology and with the free will of the Christian soul. He saw himself as the guardian of an ancient science, reformulated for a new age. For Zadkiel/Morrison, this involved not only reasserting the old principles of astrology (his pseudonym of Zadkiel Tao-Sze was also that of the seventeenth-century astrologer Lilly) but also demonstrating the connections between astral influence and the new sciences of the Victorian period, such as mesmerism, electricity, astrophysics, and spectroscopy. Coming out in November each year, *Zadkiel's Almanac* could respond to whatever currents of scientific debate flowed at the more well known British Association (BAAS) meetings in late summer or early autumn. For instance, *Zadkiel's* wrote about meteorological observation procedures and advertised special rain gauges and thermometers in wake of discussions of telegraphic weather forecasting at the BAAS in 1859.[68] The exchange between his almanac and the BAAS presented a model for open debate about the merits of astrological science.

From the late 1830s, then, the notoriety of figures like Murphy and Zadkiel meant that the pursuit of meteorology in general and weather prediction in particular was associated with popular prophecy and with astrology. The impact of these ideas and personalities can be charted in the history of the English meteorological societies.[69] The Meteorological Society of London, founded in 1823, collapsed after one season, when its most prominent member, Luke Howard, left London for Yorkshire. The society revived in 1836, a few months before Murphy's famous "hit"; its members printed an expensive first volume of transactions in 1839 and commissioned new instruments.[70] Murphy wrote

67. Curry, *Confusion of Prophets*, 64.

68. *Zadkiel's Almanac for* 1859, 77–78.

69. For details of this history, see Symons, "History of the English Meteorological Societies."

70. The instrument maker Robert Wood sold the society an extravagant and flawed set of instruments and produced a complicated and equally expensive frontispiece to the first volume of transactions.

to the secretary, donated his weather register, and asked for a hearing.[71] By this point Morrison had joined the society, and he sat on council until 1843, when the society collapsed again. In 1848 to 1850, some version of the society briefly revived, with Morrison as president. Some of the society's problems were financial, but the divided opinion of the members about astro-meteorology and weather prophecy—synonymous terms at that point—was obvious from its revival from the 1830s on. One member warned the secretary in 1837 that a mere hearing of Patrick Murphy could give the society a bad name, making it seem that "we are enthusiasts, instead of careful and laborious recorders and investigators . . . we must carefully avoid the appearance even of promulgating individual deductions."[72] In July 1839 an offshoot, the Uranian Society, was founded to openly discuss the question of planetary influence: its members included the Meteorological Society secretary, W. H. White, Morrison, and Simmonite.[73] Finally, in 1850 a new society emerged, which has survived as the Royal Meteorological Society. This latter group, made up of gentleman observers and established meteorologists, produced a revealing manifesto that stressed instruments and precision, and defined as their goal "the advancement of the aerostatical branch of physics."[74] In the small world of London science, the history of its predecessors was certainly well known,[75] and the Royal Meteorological Society worked hard in its early history to avoid controversy. It subsequently kept its distance from the disputes surrounding the introduction of official weather forecasting in the 1860s. Prophecy of any sort seemed to counter serious "aerostatical" investigations.

As he navigated amid these disputes, Zadkiel's career indicates the obvious

71. Murphy, *The Anatomy of Seasons*, 86–88; *Meteorology, Considered in Its Connexion with Astronomy*, iii, 19–23. See R. J. Morrison to W. H. White, December 17, 1836; and J. H. Maverly to W. H. White, July 4, 1838, Royal Meteorological Society Papers.

72. Whityer to W. H. White, July 10, 1837, Royal Meteorological Society Papers.

73. Details about the society are from its transactions. White was chairman of the society, J. M. Cavalier the honorary secretary. It apparently ceased meeting in 1841.

74. The founding members of the society were Dr. John Lee, Rev. Samuel King, Rev. Joseph Bancroft Reade, Rev. Charles Lowndes, James Glaisher, Edward Joseph Lowe, Vincent Fasel, John Drew, and William Rutter. Glaisher was appointed Honorary Secretary and Dr. Lee, a keen astronomer at whose estate, Hartwell House, the first meeting was held, became Honorary Treasurer. Symons, "History of the English Meteorological Societies," 88–89; Walker, "The Meteorological Societies of London."

75. Leonard Jenyns Blomefield, a well-connected clergyman-naturalist and a keen meteorologist, declined to join the 1850 incarnation of the society, and in his memoirs quipped that his "forecast" had been wrong. Blomefield, *Chapters in My Life*, 94.

significance of meteorology and the almanacs to both supporters and critics of astrology. Reformers like Knight had presented a purified, statistical meteorology to replace what they considered the rank superstition of weather predictions. In a parallel fashion, Zadkiel turned to meteorology as a way to define the standards of science as he saw them. When the Meteorological Department of the Board of Trade in London began issuing storm warnings and weather forecasts, Zadkiel emerged as a skeptical judge of elite science. He saw himself as an outsider, able to represent the people's voice and limit the arbitrary and exclusive concerns of the scientific establishment. And some of his audience agreed. "I wish Zadkiel would publish his lectures delivered lately in London, & in a cheap form," one member of the Meteorological Society wrote to Secretary White. "How many flowers are born to blush unseen in the intellectual world!"[76] Weather prediction became a critical part of this populist stance. He argued in *Zadkiel's Almanac for 1852* that meteorological work had a responsibility to "the hard earned labour of British industry" to apply its knowledge to the advantage of the people—which meant pursuing forecasts rather than accumulating statistics.[77] When, in 1867, on the advice of the Royal Society, the government meteorological office suspended its forecasting efforts, Zadkiel expressed outrage. He denounced the "helpless old ladies, who figure away" and called for a renewed commitment to weather prediction.[78] Both astrology and weather prediction, as he saw it, were victims of an exclusive scientific culture.

Murphy's coup in 1838 as well as Morrison's sustained criticisms of the Meteorological Department in the 1850s and 1860s revealed what was at stake in the almanac trade: a disciplined communication of natural knowledge. The contradictions and possibilities of the almanac exposed concerns with reform, audiences, and the marketplace in Victorian culture. Was the almanac a symptom of and encouragement to vulgar superstition, or was it a modern vehicle of culture with which reformers could reach vast audiences? Without disciplined communication, even the most exalted scientific leaders such as Herschel could be ridiculed in front of a vast, impressionable audience.

The problem of weather prophets and the almanacs, then, exposed the intertwined arguments about social and intellectual respectability that shaped scientific culture. Examining the almanacs as part of print culture leads to

76. Baker to W. H. White, May 14, 1839, Royal Meteorological Society Papers.

77. *Zadkiel's Almanac for* 1852, 61–64.

78. *Zadkiel's Almanac for* 1868, 74; *Zadkiel's Almanac for* 1869, 72–74.

broader questions about identity and expertise in scientific life. The range of publishing options for an author and the conventions surrounding the use of almanacs intensified the difficulty of establishing the authority of a scientific theory or a scientific individual. The weather prophets were known by their names, as individual authorities, or would-be authorities, and their approach stood in stark opposition to the quiet, relatively anonymous labor of observation.[79] In the world of Victorian science, this contrast came to shape the way men of science responded to the public world in which they worked. The distinction centered on questions of personality in science. Personal reputation was intimately bound up with an intellectual or scientific reputation, and the public world of print was the sphere where the two intersected most visibly. As one anonymous participant in astro-meteorological debates argued, weather prophets were not only jumped-up fortune tellers, but, rather than rendering it respectable, their scientific pretensions and pedantry had taken all the amusement out of the subject. "Who are they, the Prophets? men of any education, of any information, of any authority? Murphy? Hind? Zadkiel? Seed? François? Simmonite? Matchless list!" he grumbled. "A gypsy fortune-teller, who would be punished by law, deserves more praise. In her there would be some shrewdness, and wit, and fun, of all which the prophets are destitute, and over whom dullness and falsehood jointly reign." White, in his column on meteorology in *Gardener's Gazette*, responded by outlining the credentials of several contemporary astro-meteorologists. Murphy was an "athletic lad from the land of Erin, of some talent and more enthusiasm." Another was "a Parisian of some considerable rank, and has the honour of calling an Arago and a Biot among his intimate friends." Simmonite was "a gentleman; though to me personally unknown, I know his locality, and many of his scientific associates, who are men of talent and character." Zadkiel, finally, was "not a 'mercenary' scribbler, but a gentleman of high mathematical attainments and astronomical requirements, as both his personal friends (of whom I am not ashamed to be styled one), the readers of his scientific papers, and his hearers in lecture rooms, can bear ample testimony."[80] The careful rankings established by such comments were a commonplace of contemporary scientific activity, putting forward the attributes of a gentleman—independence, favorable private connections, assurance and eloquence in public settings—in defense of the scientific claims of astro-meteorology.

79. *Gardener's Gazette*, December 21, 1839, 834.

80. *Gardener's Gazette*, January 4, 1840, 11.

An analysis of astro-meteorology and the almanacs leads, therefore, to the management of private and public character, one of the most interesting problems for science in Victorian culture. Obviously, as White's careful descriptions showed, gentlemanly credentials helped confirm the standing of scientific work. Yet on the other hand, many claimed that, as a system of thought, science must rise above mere personality. It must refuse to recognize the lowest common denominators of charisma and individual opinion. The dilemma was that to downplay personal authority contradicted the public roles that men of science increasingly claimed for themselves as leaders of modern cultural life, and made it more, not less, difficult to defend the boundaries between a weather prophet and a man of science. After mid-century changes in the public world of science shifted the pressures on reputation, personal character, and intellectual authority.[81]

For an instance of these changes, we can turn to Captain Stephen Martin Saxby, an instructor of steam engineers who taught at the Steam Reserve College in Greenwich. Saxby began publishing on weather prediction in the *Nautical Magazine*, a monthly journal for the merchant marine, at the same time the government Meteorological Department was founded. Saxby was a lunarist, holding that the moon's position relative to the equator could be used to foretell "dangerous periods" for storms years into the future.[82] Saxby repeatedly sought the attention of the authorities, corresponding for instance with the Astronomer Royal, who counseled him to drop his investigations.[83] Undeterred, Saxby approached Lloyd's, the marine insurers, sending storm warnings to their board in 1861.[84] Shortly afterward, he visited the Meteorological Department, hoping to persuade them to examine the lunar theory. He was lent instruments, but otherwise rebuffed. These setbacks only served to convince Saxby that there was "a binding thong strapped across the very mouth of science, which the greatest philosophers of this and previous ages . . . have by degrees so tightened, that we have in the question of meteorology a

81. Hays, "The Rise and Fall of Dionysius Lardner." A related discussion of character and public life appears in Collini, *Public Moralists*, esp. 91–121.

82. See Saxby, "The Coming Winter and the Weather," "Lunar Equinoctials," *Saxby's Weather System*, and "Weather Warnings."

83. S. M. Saxby to George Airy, December 28, 1860, Airy Papers, RGO 6/702/25–31; Saxby, *Saxby's Weather System*, 14.

84. Lloyds Committee Minutes, January 23, 1861.

perfect gag to be gotten rid of, before a weather system of any public use can be fairly investigated and encouraged."[85]

In some respects Saxby's experiences indicate a scientific community well able to repel enthusiasts. But Saxby's metaphor of speech and silence suggests another less secure feature of scientific life. Like Zadkiel, Saxby felt publicity was the key to fair investigation. By the middle of the nineteenth century, the means for pursuing public conversations about science were more varied and easier than ever to use—societies, lectures, books, articles in journals, letters to the newspaper, and even, as we will see, the courts. The development of a mass print culture made possible a flood of information and argument on all manner of topics, but scientific discussions in particular seemed critical as a force for shaping a rational society. In these concerns about useful knowledge and social reform, popular weather prophecy represented either the quicksand of ancient superstition and ignorance, or the dramatic new opportunities for participation in scientific culture.

The tactics of the lunarist Captain Saxby as he tried to gain a hearing for his lunar theories gives a revealing example of how this public world shaped the definition of a scientist. Saxby actively manipulated the parallels between private character and intellectual authority. In his writings, Saxby put forward both his private and public reputation as credentials simultaneously. Since he was a public figure, he declared, he was risking himself in promulgating theories some ridiculed. Moreover, he had always been open about his views: "I have dealt in no mysterious charlatanerie; reasons for all my opinions have been freely given to the public and in . . . plain untechnical language." Yet, as a private individual, Saxby escaped significant constraints: "the combination of facts, the leisured unprejudiced contemplation of such facts . . . is equally open to the professor or the private individual . . . nor need we concede that all the experience and ability of the country is concentrated in one mind." By contrast, the Meteorological Department had been fettered by its "fear of responsibility" and the "tramlines" of routine.[86] Saxby delineated this best when presenting his credentials in his journal articles on lunar meteorology. As a private individual, he claimed, he was independent—free of numbing bureaucracy or political and scientific censorship. Saxby also made claims about his public role: attaching his name to his views placed him at risk, and thus confirmed his honorable intentions. To have one's personal character so intimately aligned with the status of the work could be a liability, however,

85. Saxby, "The Coming Winter and the Weather," 664.

86. Saxby, *Foretelling the Weather*, 8.

rather than a resource. In thus juggling his private and public respectability, sometimes in contradictory fashion, Saxby was not simply clutching at straws. Such attention to reputation and measures of self-publicity often are pictured as the unconventional resource of those outside standard channels for developing professional credibility. However, the controversy over weather forecasting indicates that Saxby's assumptions about the relationship of the private individual and public office, and the kinds of knowledge each produced, were quite conventional.

The most important example of the tensions between popular and official interest in forecasting, remained Zadkiel, or Morrison. In 1860 Morrison founded yet another society in a further effort to bring "astral science" within the pale of scientific discussion.[87] By the end of 1861, one member thought this Astro-Meteorological Society was "pretty well known," through advertisements and circulars as well as the journal.[88] It is safe to say that the society fell from whatever small notoriety it achieved pretty rapidly, but before long another much more public event brought Zadkiel into the news. Zadkiel was sued for libel in 1861, in a heavily publicized trial. This incident demonstrated the distressing notoriety of Zadkiel and suggests how unpleasant any comparison to Zadkiel would have been to those anxious to uphold the position of scientific research in the government. Above all, the trial exposed the intimate connection between individual and intellectual credibility informing the Victorian understanding of scientific work. It inevitably filtered into the controversy over forecasting; although weather science was not directly on trial, the question of genuine versus fraudulent prediction certainly was.

The story of how an astrologer mounted a libel case is itself a marvelous vignette of Victorian mores, giving us glimpses of outraged patriotism, the salon entertainments of upper-class London, and, not least, the newspapers' complicated record of that world. In *Zadkiel's Almanac for 1861* (which, we should recall, routinely contained astrological political predictions as well as those about the weather), Zadkiel predicted ill health for the Prince Consort. When Albert obliged by dying of typhoid on December 14, 1861, Zadkiel's prestige and his sales shot up. The *Daily Telegraph* in loyal indignation mounted an attack against the pernicious influence of astrology and Zadkiel in particular. But the newspaper seemed unaware whose name the pseudonym protected, until it was informed by a letter from an anonymous reader ("Anti-Humbug").

87. White, ed., *The Journal of Astronomic Meteorology and Record of the Science and Phenomena of the Weather*. The society lasted some sixteen months.

88. Cooke, *Curiosities*, 233.

The writer urged that Morrison be "ferreted out" and prosecuted as a rogue and a vagabond (the statute under which astrologers had been punished since 1824).[89] Morrison eventually discovered and sued the author of the letter, Rear Admiral Sir Edward Belcher, for libel.

Belcher was known as an explorer and surveyor of the Pacific Ocean and the Bering Strait and, most recently, as chief of the expedition sent to discover the whereabouts of Sir John Franklin, who disappeared while trying to discover the Northwest passage.[90] Interestingly, given the trial's object to preserve reputation, Belcher had been involved decades earlier in a sordid divorce trial of 1834, which had held his own character open to question. (He was accused of deliberately transmitting a venereal disease to his wife.)[91] It was never mentioned in 1863, but suggests an intriguing subtext for at least one of the participants in the 1863 affair. The trial of *Morrison v. Belcher* took place at the end of June 1863 and highlighted the currency of astrology and "scrying"—viewing a crystal ball—in society circles. Many eminent witnesses testified on Morrison's behalf. Lady Harry Vane, the Bishop of Lichfield, Lady Egerton, and the Earl of Wilton took the stand (all under subpoena) to agree that Zadkiel/Morrison had received no money for his ventures into London's salons. Morrison was cleared of intent to defraud, but was awarded insignificant damages.[92] The outcome, however, was less important than the publicity for astrology and fraudulent prediction. The *Saturday Review* estimated that "without a doubt, there are a million of English people with some sort of confidence in Zadkiel; certainly there is ample encouragement to them in the countenance afforded to Zadkiel by the many great, and wise, and learned of the land."[93] The *Morrison v. Belcher* suit gave astrology a high profile in London society and newspapers at the same time as weather forecasting got under way. Even though Zadkiel's interest in weather was not under direct discussion, the trial could not have benefited efforts to establish a storm warning service by the government's Meteorological Department. While the Astro-Meteorology Society met in Pall Mall and Zadkiel publicized astrological theories of the

89. Reprinted in *Zadkiel's Almanac for* 1864, 63.

90. It seems very likely Belcher knew Fitzroy, the director of the Meteorological Department, and he spoke warmly in later years of Fitzroy's efforts. Edward Belcher to Secretary of the Royal Meteorological Society, August 1, 1874, Royal Meteorological Society Papers.

91. "Belcher against Belcher," *Times*, May 2, 1834.

92. Curry has compiled a detailed account of this trial based on newspaper reports; see Curry, *Confusion of Prophets*, 85–108. Morrison also published a full description in *Zadkiel's Almanac for* 1864, 62–79.

93. *Saturday Review*, July 4, 1863, quoted in *Zadkiel's Almanac for* 1864, 77–78.

weather in his *Almanac*, the Victorian press at all levels was discussing the experimental storm warnings of the government and Herschel was defending himself against the imputation that he, too, was a lunarist after the publication of his *Good Words* articles.

The libel trial of 1863 displayed a mixture of claims about identity, respectability, and reliable knowledge. In the first place, the suit itself suggests that the Zadkiel pseudonym was no token gesture, since the *Telegraph* was unable initially to identify Zadkiel as Morrison despite the nearly thirty years Morrison and Zadkiel had both been publishing. It remains unclear how successfully the pseudonym had been preserved during Morrison's earlier involvement with the London Meteorological Society. Yet the effort was clearly there—for instance, Zadkiel and Morrison were cited separately in White's meteorology column in the *Gardener's Gazette*.[94] The suit moreover suggests the importance of the pseudonym as a means of preserving respectability; ironically, Morrison was prepared to further publicize his dubious identity in order to defend his reputation. The society witnesses who testified to Morrison's/Zadkiel's independence were also important. For Morrison to have refused money for his appearances represented claims both about his personal status and about the authenticity of his science. Any evaluation of his predictions fundamentally involved judgments about personal respectability. Echoing this familiar lesson, a pair of contrasting portraits demonstrates that the image of Morrison's/Zadkiel's person was a significant resource in the public controversy over forecasting (fig. 2.5). *Punch*'s 1863 cartoon showed a devilish figure with simian features, a beard, and pointed donkey ears, dressed in black robes and wearing a peaked sorcerer's hat around which a snake is curled. Zadkiel as represented in a portrait enclosed with the 1855 *Zadkiel's Almanac*, however, was a respectable and prosperous-looking gentleman with broad brow and far-seeing eyes, who held an almanac while leaning on a globe. The latter portrait interestingly mingled Morrison's/Zadkiel's identity. Although titled merely "Zadkiel, Tao Sze" with no mention of Richard James Morrison, the caption lists some of both Zadkiel's and Morrison's publications, and the figure is wearing medals, presumably those Morrison received as an officer in the Coast Guard from the National Institution for Preservation of Life from Shipwreck (later the Royal National Lifeboat Institution) in 1829. The

94. White typically listed his correspondents of the week to acknowledge receipt of their communication at the end of his column; he sometimes listed both Zadkiel and Morrison. He did not identify Zadkiel as Morrison in his discussion or charts (*Gardener's Gazette*, 1839–42, passim). On conventions of anonymity and pseudonymity and the case of *Vestiges*, see Secord, *Victorian Sensation*, 17–24, 178–80, 364–400.

FIGURE 2.5 This pair of contrasting portraits shows how authority and identity were constructed in print. *Punch* mocked Zadkiel as a magician, with simian visage and a snake twined around his wizard's cap, gesturing at the moon. Looking on is a small ignorant worker in rural smock. In his own almanac, Zadkiel represented himself as a learned gentleman, prosperously dressed, holding his almanac in his hand. Interestingly, although the pseudonym is not openly dropped, the portrait cites the works that "Zadkiel" published under his own name of Richard James Morrison, and it showed the medal that Morrison won for his work in the Coast Guard in 1829. (*Left*: *Punch*, May 5, 1863. *Right*: *Companion to Zadkiel's Almanac for 1855*, 3 [British Library].)

management of personal character that Morrison and Zadkiel both undertook demonstrates the kind of malleable and important resource it represented in the establishment of intellectual authority.

An 1838 comic play about weather prophets made the same argument more light-heartedly. *Murphy's Weather Almanac*, a one-act farce performed at Sadlers Wells Theatre, was a stock comedy about mistaken identity. An old gentleman and enthusiast for scientific learning mistakes Patrick Murphy, hot potato vendor, for Patrick Murphy, the meteorologist of the day, after overhearing an exchange of casual remarks about the weather. Patrick Murphy MNS (Member of No Society) and BTM (Baked Tatey Merchant) is taken home to view the gentleman's collection of philosophical instruments, where a manservant takes the opportunity to steal some silver, intending to blame its disappearance on the peculiar visitor.[95] (See fig. 2.6 for a contemporary image inspired by the play showing Murphy as a potato-head, smiling at the sour-faced moon, whom he has replaced as chief weather prophet.) The spoof has a number of obvious targets—public gullibility, scientific eccentricity, and the Irish being some of them. But it also clearly comments on the real Murphy's science through the presentation of identity confusion, gullibility, and opportunistic theft. The pretensions and social climbing of both Murphy and astro-meteorological theories are equated and mocked.

Conclusion

This history of popular weather prophecy shows how by mid-century the presentation of personal and scientific authority had acquired a sharper edge. The play with names, pseudonyms, and credentials that can be seen in the almanacs was only partially a function of astrology's dubious status: it also expressed more broadly based negotiations that surrounded the conventions of personal identity. In seeking a popular audience, for instance, authors had to sidestep charges of the motive of personal gain, a charge facing Murphy in 1838 as well as Morrison in 1863. Countering this charge, almanac authors evoked gentlemanly standards of conduct, such as openness, or risk of personal honor, and characterized their opponents as hiding behind the false authority of their official or institutional positions. For Morrison/Zadkiel and Saxby, publishing "in cheap form" was perhaps the only way to make visible the flower of their intellect. But for others in more established circles, this tactic was not the only one, and it was fraught with risk. Yet neither a low opinion of the masses' ability to judge or the allied danger of contagion by disreputable ideas materially changed this feature of scientific life.

In this sense, perhaps the most notable feature of the exchanges between popular and elite knowledge was their persistent uncertainty. The formation of the various meteorological societies, or the foundation of a government office to pursue meteorology did not close off conversation by identifying expertise. On the contrary, such developments stimulated debate, bringing new interests and new participants to weather science. The case of John Herschel provides a clear example. It would be hard to imagine a more recognized scientific authority than Herschel. He was one of Britain's most respected scientific

95. Rogers, *Murphy's Weather Almanac.*

FIGURE 2.6 A print, "Murphy the Dick-Tater, Alias the Weather Cock of the Walk" showed Patrick Murphy as the new weather prophet. The print circulated at the same time as a play about Murphy called attention to the social pretensions of prophets and the gullibility of public opinion. (Satirical Prints [1838]. Reproduced by permission of Guildhall Library, Corporation of London.)

gentlemen, and was especially known for his physical observations in astronomy, magnetism, and meteorology. Yet this renown made his family name circulate from private conversation to letters and newspapers in ways that he could not control. Indeed, even as he complained about the misinterpretation of his views, and tried to set boundaries between speculation and knowledge, private views and public expression, he contributed to the exchange with his letters and his popular articles on meteorology.

Although the exchange between popular knowledge and elite science was remarkably persistent from the early Victorian period through the 1860s, however, it was not static. Other particular developments shifted meteorology into increasing public prominence. Popular meteorology, although it reached mass audiences in the almanacs, was associated with individual prophets like Zadkiel (or Herschel). But in other circles, meteorology was developing as a science of collective observation, with networks of observers supplying data to a central institution. In Britain, with the formation of the government department for meteorology, collective observation defined the official face of the science. As Fitzroy insisted, "the measures practised daily [in predicting the weather] . . . do not depend on any *one individual*, but are the result of facts exactly recorded, and deductions from their consideration, for which rules have been given."[96] Meteorology thus became a science in which the nature of collaboration, institutional authority, and public responsibility could be debated. Evaluating meteorological science increasingly involved evaluating the social organization and values of collective science. It became a science not of weather prophets but of weather networks, managing the connections among leaders, observers, and instruments.

96. Fitzroy, *Weather Book*, 171, original emphasis.

CHAPTER THREE

Weather in a Public Office

DISCUSSING the requirements of the new government department of meteorology with George Airy, the Astronomer Royal, in 1853, Robert Fitzroy sketched a composite picture of the science. Ideally, he commented, meteorology required a mastery of "Tides—Currents—Winds—Temperature—Magnetism, Electricity and the Atmosphere" or, in other words, "a Herschel—Whewell—a Rennell a Reid a Sabine and a Faraday combined under Humboldt." He, himself, he assured Airy deferentially, could not hope to fill all these shoes, but with proper supervision and encouragement from Airy and others, he might "grapple with minute portions—and like an ant struggle along with a load."[1] In this description, Fitzroy moved awkwardly between claims resting on the scientific authority of an individual and the authority that derived from collective endeavor, like the industry of social insects, ants or bees, which so fascinated Victorian society. Adolphe Quetelet, director of the national observatory in Brussels, drew attention to the same tension when he called for greater system and uniformity in European meteorological observations. "Above all, both the savant and the nation itself should abandon the notion of individuality," Quetelet urged.[2] This tension defined Fitzroy's work in the Meteorological Department, as he distributed instruments to ships' officers and gathered in

1. Robert Fitzroy, memorandum on American oceanographer Matthew Fontaine Maury's proposals, November 5, 1863, Met. Office Papers, PRO BJ7/113. The archival holdings relevant to the Meteorological Department are mainly found in class BJ7 of the Board of Trade papers at the Public Record Office, hereafter referred to as PRO BJ7.

2. "Il faudrait, avant tout, oublier l'individualité et du savant et de la nation." Quetelet, *Science mathematiques et physiques*, 10.

their logs, established a telegraphic observation network, and published storm warnings and general forecasts. An account of the science of meteorology, then, needs to organize itself around leaders like Fitzroy but also around those institutions, collaborations, and networks they organized.

The nature and practices of collective science charted the development of a professional scientific culture. Whether this meant a society for sharing research, an observing network, a school for geologists, or a government office, collective practices were crucial. The prestige of science depended on making clear who were the leaders and who the followers in science. The formation of the British Association for the Advancement of Science in 1831 is the classic instance, organized to "promote the intercourse of the cultivators of science" and to give "systematic direction" to science.[3] Similarly, the formation of the Meteorological Department two decades later promised to extend the sphere and values of professional science into the management of public life. The creation of scientific posts within government addressed several professional goals simultaneously. They provided income for worthy individuals, they formalized the informal practices of support and advice from scientific leaders, and they gave a visible indication of the national importance of science.[4] Like the formation of scientific societies, the work of science in government office defended the values and talents of an increasingly energetic and self-conscious scientific elite, and offered a model for the kind of social relationships that would sustain those values.

Yet science as a collective practice also sharpened the challenges of building authority of science. From its early years, the BAAS was ambivalent about its popular audience. It organized public lectures and social events and annually used the widely reported presidential addresses to trumpet the accomplishments of science and influence opinion. But many members were uneasy when fashion and interest, rather than accomplishments, began to determine attendance—an anxiety about public participation that was signaled, for instance, by the presence of women at lectures.[5] The same ambivalence shaped

3. Orange, "Idols of the Theatre," 43. Morrell and Thackray, *Gentlemen of Science: Early Years*, 517–23.

4. There is a solid literature on government and science in the Victorian period: see Alter, *Reluctant Patron*; Cardwell, *Organisation of Science*; Chapman, "Science and the Public Good"; Hall, *All Scientists Now*; Heyck, *Transformation of Intellectual Life*; MacLeod, ed., *Government and Expertise*; MacLeod, *Public Science and Public Policy*; MacLeod and Collins, eds., *The Parliament of Science*; Morrell and Thackray, *Gentlemen of Science: Early Correspondence*; Morrell and Thackray, *Gentlemen of Science: Early Years*; Turner, "Public Science in Britain"; and the rejoinder, Hall, "Public Science in Britain."

5. Morrell and Thackray, *Gentlemen of Science: Early Years*.

the experience of scientific men in government positions. Governments sought scientific advice on the basis of its independence; men of science valued independence as the badge of knowledge and a necessary condition of its production. Nevertheless, public funding for science could place its independence in doubt and seemed to open scientific affairs to the play of opinion rather than expertise.

A government office for meteorology then became a catalyst for questions about scientific organization and public life. Fitzroy's experiments with forecasting polarized the differences between a science of model observations and a science that addressed practical and popular concerns. Forecasting multiplied audiences for the science, turning it into "meteorology for the millions."[6] With its telegraphic networks and public warnings, it highlighted the hierarchies of scientific observation. Finally, weather prediction sparked debates over utilitarian values in science. Material benefits made possible the cultural claims of science, and they provided an obvious standard by which to judge the national value of science. Yet at the same time they linked science with commercial values in ways that many men of science found humiliating and destructive. To John Herschel, writing in 1830, there was no more deplorable question than "cui bono: to what practical end and advantage do your researches tend?"[7] Practical advantages might stimulate interest in science, but they dragged it away from a philosopher's ideals.[8]

In the struggle to define the sources of scientific authority and resolve the place of practical science, the evaluation of government meteorology turned to the question of individual character and leadership with new intensity. As the examination of popular weather prophets has demonstrated, personal and scientific integrity were interwoven concerns in Victorian Britain. Professional science—which in the vision of Francis Galton was, by the 1870s, "a sort of scientific priesthood"—depended on moral assertions about judgment and the courageous pursuit of truth.[9] How to translate such standards to the machinery of institutions or the ensemble of scientific research was a critical question of modern society.[10] Collective forms of work embodied the systematic direction

6. *Punch*, April 19, 1879, 177.

7. Herschel, *Preliminary Discourse*, 10. Cf. [Whewell], "Preliminary Discourse on the Study of Natural Philosophy," 405.

8. MacLeod, "The 'Practical Man'"; Ashworth, "Calculating Eye"; Appleton, *Endowment of Research*.

9. Galton, *English Men of Science*, 260; cf. Desmond, "Redefining the X Axis," 13; Secord, "'Be What You Would Seem to Be.'"

10. Collini, *Public Moralists*.

that was the model of scientific organization, but how could moral and intellectual leadership truly express itself within a collective body? The Royal Society tried to defend science by dismissing weather prediction. But to critics, that decision seemed a failure of leadership. It evaded moral responsibility and retreated to narrow special interests.

Observation and Scientific Discipline

For the Victorian natural philosopher, the weather inspired a particular call for scientific discipline. As William Whewell put it in 1831, "The people have been collecting facts for a very long time (ever since Noah) and are now just beginning to get a notion of the general laws and properties into which the mass is to be resolved. I do not know any subject which is at present in so instructive a condition."[11] From the first, then, meteorology was a special object of attention for the BAAS leadership. Writing to James David Forbes, who soon after took the chair of natural philosophy at the University of Edinburgh, Whewell was once more enthusiastic about the opportunities for meteorology within the BAAS. Forbes had been commissioned to survey the field of meteorology for the BAAS meeting in 1833. "I do not think there is any subject on which we are so likely to do something as yours," Whewell began.

> I should think it likely that we may get such cloud maps, as you speak of, drawn for a given time. Another point which appears to me to require consideration is the theory of clouds and rain . . . Again, what is the immediate cause of the conversion of clouds into rain? . . . Is there any chance of obtaining information of the condition of the higher regions of the atmosphere? How high does the direct current of the common winds extend? Can any facts be collected to verify or test Herschel's theory of hurricanes? Again, will any analyst solve for us the problem of the pressure of an elastic fluid *in motion*, so as to apply to the barometer?[12]

Meteorology, with its "multiplied and extensive fagging," seemed to offer an ideal set of "good things" for an organization like the BAAS to do, Whewell thought.[13] As Whewell indicated, bringing together observations was hardly a new idea. If not quite "since Noah," plans for cooperation and exchange had offered a way forward in the science ever since instruments for quantitative

11. William Whewell to Rev R. Jones, July 23, 1831, in Todhunter, *William Whewell*, 2:124.

12. William Whewell to Edward Forbes, June 10, 1833, in Todhunter, *William Whewell*, 2:165–67.

13. William Whewell to Roderick Murchison, October 10, 1851, in Morrell and Thackray, eds., *Gentlemen of Science: Early Correspondence*, 83.

measurement, like the barometer and thermometer, had been developed in the mid-seventeenth century. In England, the precedent for coordinating weather observations through a scientific society dated to that era: Robert Hooke had suggested it as an activity of the Royal Society members in 1663, and in 1723 the secretary of the Royal Society, James Jurin, made another, grander effort to collect and publish observations from an international network of correspondents. Later, there was the impressive work of the Societas Meteorologica Palatina, centered in Mannheim, Bavaria, which had collected observations from fifty-seven locations across Europe (and from a few remoter sites, such as Greenland and Massachusetts) from 1780 to 1795.[14] Whewell's ambitions thus built on those of his predecessors, translated to the global scale and to the new instruments of the nineteenth century. In its first two decades, the BAAS funded the purchase of instruments, studies of subterranean temperatures, hourly observations at remote locations like Inverness and Kingussie, and an expensive series of observations at Plymouth. As interest in magnetic influences on the atmosphere and electrical phenomena grew in the 1830s, the BAAS also supervised a magnetic survey of the British Isles in 1834–36 and sponsored magnetic and meteorological observatories in Canada, New Zealand, St. Helena, and the Cape of Good Hope, which were managed by the Board of Ordnance.[15]

Driving all this activity was the motivation of scientific reform. As both Whewell and Forbes certainly knew, the dismal state of meteorology had been one of the strongest complaints made by scientific reformers in the 1820s and 1830s. *Elements of Meteorology* by John Frederic Daniell, the professor of chemistry at King's College (first published in 1823 and out in a third edition in 1845), began with a call for reform of English practices. For evidence of the weakness of the nation's weak, dilettante science, according to Daniell, one need only look at the series of meteorological observations undertaken by the Royal Society. The society's instruments were old and poor, the "observer's nightcap" dictated the irregular hours when observations were taken, and the wind vane was situated in a sheltered alley where nothing but a gale could disturb it.[16] In 1833, Forbes echoed these complaints about observations in his BAAS survey. "Meteorological instruments," he complained, "have been for the

14. Feldman, "Late Enlightenment Meteorology"; Frisinger, *The History of Meteorology*; Kington, *The Weather of the 1780s over Europe*, 1–26; Rusnock, "Correspondence Networks and the Royal Society."

15. Cawood, "The Magnetic Crusade."

16. Daniell, *Meteorological Essays and Observations*, xii–xv.

most part treated like toys, and much time and labour has been lost in making and recording observations utterly useless for any scientific purpose."[17]

Daniell, Forbes, and Whewell all expressed the definition of expertise centered on precision observations that was developing within the British scientific community. But how did the rhetoric translate into practice? A closer look at the meteorological work presented at one BAAS meeting, at Swansea in Wales in 1848, can give a picture of the standards guiding meteorology at mid-century from the perspective of the scientific elite, the organizers of the BAAS.

Following the pattern established early in the previous decade, the Swansea meeting gathered together scientific men from all over Britain for several days in mid-August for a junket of talks, dinners, and excursions. The meeting's president was Spencer J. A. Compton, the Marquess of Northampton, at the time president of the Royal Society. The meeting attracted fewer participants than either the preceding meeting at Oxford or the following in Birmingham, probably because of its rather remote location: most of the participants had to travel by steamer from Bristol. But it was nonetheless well attended by scientific notables. The public excursions at the meeting reflected the popularity of geology, which went far beyond any other science, including meteorology. There were visits to limestone districts, smelting works, and an anthracite mine, and attendance peaked on the day that Gideon Mantell, the geologist, lectured on *animalcula* fossils found in the local chalkstone. Meteorological talks were given in section A, the physical sciences, which as usual drew sparser crowds than the geological section C. The few meteorological papers were pedestrian, by figures of little eminence. However, individual papers did not represent the significant meteorological work in the BAAS. Meteorology assumed a much larger place in the researches connected with the business of the institution as a whole, through its officers or granting system. The small number of talks on meteorology at this meeting indicated not indifference; rather, it pointed to the other ways meteorology was pursued within the institution: through extended and collective observation.

At Swansea, the premier example of meteorological research was the work of the German meteorologist, Heinrich Dove, which was presented by Edward Sabine, the general secretary. Sabine, a Royal Artillery officer whose career to this point included scientific voyages to the Arctic and tropics, longitude calculations with John Herschel, and the magnetic survey of the British Isles,

17. Forbes, "Report on Meteorology," *Report of the BAAS*, 1833, 197. Cf. "Supplementary Report on Meteorology," *Report of the BAAS*, 1840, 37–155.

was the chief coordinating figure behind Britain's global terrestrial magnetism project, known in scientific circles as "the magnetic crusade." He gave an enthusiastic account of Dove's global temperature maps, which offered a model and a justification for the collective work of associations and observatories. The dominant figure in meteorology in the mid-nineteenth century, Dove had been a student of H. W. Brandes, author of the first synoptic charts of the weather in 1820, using data of the late eighteenth-century Palatine society.[18] Stressing mechanical as well as thermal forces, Dove argued that European weather was the product of two air currents, equatorial (warm, moist) and polar (cold, dry) in origin. Their global circulation was due to both the earth's rotation and the thermal effects of the sun—a scenario familiar in outline since the work of Edmund Halley on the trade winds in the seventeenth century. Dove added a model of characteristic pressure, temperature, and humidity to the account, making these then the functions of wind direction. Storms, he argued, resulted from the friction between the two air masses when they met, and the rotating effect, which he called gyration, or wind veering, was the result of the air currents alternatively displacing each other, rather than a true rotary action.[19] Dove's *Law of Storms* was eventually translated into English in 1860, and his theories had obvious and direct implications for weather prediction. But, typically, Dove won accolades from British men of science not for these theories but for his extensive studies of global temperature. In 1853, a few years after Sabine's presentation of his work at Swansea, the Royal Society of London awarded Dove its highest honor, the Copley Medal, for his maps, an impressively "laborious" enterprise of data collection. Dove was commended for his "true spirit of inductive inquiry," and the "enlarged and comprehensive views by which we are enabled to recognized real order in the midst of apparent confusion."[20] Showing the same appreciation in 1848, the BAAS ordered 500 copies of Dove's maps.

The support of the BAAS for the English researches of William Birt was another signal of the values guiding their direction of meteorology. Birt, a former assistant of John Herschel, pursued a study of barometric fluctuations in the western hemisphere from 1843 to 1849, receiving annual grants from the BAAS. He was looking for monthly or fortnightly regularities in the data that would supply evidence of great currents (atmospheric waves) in the

18. Kington, *The Weather of the 1780s over Europe*, 21–46.

19. Dove, *The Law of Storms*; Kutzbach, *The Thermal Theory of Cyclones*.

20. Parsons (Earl of Rosse), "Address Delivered before the Royal Society [1850–54]," 6:353–54, and "Address Delivered before the Royal Society [1843–50]," 5:865.

atmosphere. These waves, according to Birt and Herschel, might even suggest an explanation for cyclonic storms, since a revolving local disturbance might develop when two of the immense waves of barometric pressure crossed each other. Birt's researches, it should be noted, were in some sense an answer to the American controversies over storms. In the preceding years, several American meteorologists, notably Robert Hare, William Redfield, and James Espy, had traded arguments about the production of storms.[21] By 1848, the controversy over the origin of storms—electrical, thermal, or gravitational effects—was well over a decade old and still unresolved. To many British men of science, exchanges on storms thus seemed to represent a sterile meteorological topic unless fertilized, as the atmospheric waves analysis offered to do, by very extensive data. Both Birt and Herschel entertained hopes that the analysis of barometric patterns would bring a breakthrough, based on sufficiently widespread and expert observations. Even though Birt by 1848 was succumbing to pessimism about the project, then in its fifth year, the BAAS's reaction was not critical. No strong connections between the pattern of barometric waves and other atmospheric conditions had emerged, and Birt was increasingly, and uncomfortably, aware that without such links, the term "wave" in his promising empirical analysis merely described the curve traced on a graph, rather than any motions in the atmosphere that might lend themselves to a hydrodynamical analysis. Rather than ending the program outright, the BAAS leadership instead assigned Birt the task of analyzing observations of atmospheric electricity, the next promising angle of computation. Birt's experiences were more evidence that the calculator's approach to meteorology was the accepted path.

A third critical aspect of meteorology at the Swansea meeting was the regular summary of the Kew instruments. Kew Observatory was the most significant of the BAAS's endeavors in meteorology. This institution was a notable example both of the resolved pursuit of precision instruments and of the awkward balance between private and public funding. Kew had been built in 1769 as a private observatory for George III, but it had fallen into disuse and was to be abandoned in 1840. Offered the use of the observatory by the government as a gift, the Royal Society appointed a committee to consider the use of the windfall. This committee—made up of Charles Wheatstone, the well-known physicist and inventor of the telegraph, Herschel, and Sabine—quickly proposed a series of potential uses, from a storage facility for instruments and other bulky property, to a site for magnetic observations, to a location

21. Fleming, *Meteorology in America*, 23–54.

for a standards department for instruments. Yet by March 1840, the Royal Society had decided against adopting the facility, deterred by the prospects of a costly establishment. Presumably it also saw acquisition of the observatory as a departure from its historical role as a body of scientific men dedicated to intellectual exchange, rather than an office of scientific research. Sabine, however, had just commenced his twenty-year tenure as general secretary of the BAAS (1839–59), and Herschel and Wheatstone were also members. If the Royal Society would not undertake the responsibility, they immediately saw an alternative.

In the president's chair at the 1840 BAAS meeting, Whewell called for the association to become a new, voluntary "Solomon's House" in the manner of Francis Bacon's *New Atlantis*. A few months later, the BAAS took over the management of the Kew Observatory. Yet the decision hardly displayed Bacon's utopia of state-supported science, since the BAAS thereby promptly incurred bills for "wages, repairs, furniture and sundries" in addition to the expenses of new instruments and researches. The maintenance of Kew was shortly the largest regular item among the BAAS expenses, and by 1858 it absorbed £500 annually. Nevertheless, in 1842 the acquisition of the observatory represented to the association an example of ideal relations between state and science: "the collective science of a country should be on the most amicable footing with the depositories of its power . . . free from undue control and influence . . . but enjoying its good will and favour."[22] In the following years, grumblings surfaced frequently about costs, but the observatory remained in the hands of the BAAS for nearly thirty years. (It was returned in 1871 to the Royal Society, its maintenance guaranteed by a generous endowment from a fellow.)[23]

As Herschel, Sabine, and Wheatstone had proposed, Kew Observatory became a center for standard instruments and played a leading role in developing precision instruments especially for meteorological and magnetic observations. The first superintendent, Francis Ronalds, at Kew from 1842 to 1859, experimented in 1847 with a kite and instrument platform for upper

22. Scott, "The History of the Kew Observatory," 52.

23. See the cumulative "General Statement of . . . Grants for Scientific Purposes" published in every annual *Report of the BAAS* for the amounts paid to support Kew Observatory work. Peter Gassiot left £10,000 to the Royal Society for the observatory in 1872. Also see Scott, "The History of the Kew Observatory"; Jacobs, "The 200-Years' Story of Kew Observatory"; Barrell, "Kew Observatory and the National Physical Laboratory." A detailed description of the instruments for measuring atmospheric electricity is found in Kämtz, *A Complete Course of Meteorology*, 551–54. Illustrations of the instruments in use at Kew toward the end of the century can be found in "Weather Watchers and Their Work."

air observations; his successor, John Welsh, made four balloon ascents from Vauxhall Gardens in London to carry out observations in 1852. Equally novel, and almost as newsworthy, were Kew's photographic instruments, that is, instruments whose registrations were recorded by photographs. In 1845 and 1846, Francis Ronalds experimented with a photo-barometer, -thermometer, and -electrometer, and by 1848, his photo-barometer prints were being developed daily by a photographer in Regent Street, from where they were distributed to interested meteorologists. Ronalds's contributions were part of a concerted effort followed in several institutions in the 1840s to develop self-registering instruments. Wheatstone himself supplied Kew with several instruments of his own design, including the continuous electromagnetic meteorological register he developed in 1844 (fig. 3.1). This was a six-foot high contrivance, regulated by a clock, which could record observations from up to five instruments. Every several minutes, to a total of 1,008 observations a week, the clock triggered an interruption of the electrical current between the mercury in the thermometers or barometer and a delicately weighted wire balanced on the mercury's surface. The time before the wire resumed contact with the surface of the mercury, reestablishing the current, corresponded to the height of the mercury columns, allowing variations as small as a tenth of an inch to be recorded.[24] By 1851, Kew had opened a verification department, again following the original suggestions of Herschel, Sabine, and Wheatstone. Individuals could send their thermometers to Kew to be checked against a standard instrument provided and specially installed by Henri-Victor Regnault, the Paris authority on accurate instruments in the physical sciences. Verification of barometers and hygrometers was added in 1853. In little over a decade after the BAAS assumed its management, then, Kew had become a world authority for magnetism instruments and the acknowledged source of thermometers and barometers for expert observation.

The meteorological work of the BAAS in evidence at Swansea in 1848 showed clearly the kind of scientific endeavor it supported. Detailed observations, sophisticated instruments, and standardized measurement were the keys to the development of meteorology. Meteorology called for a collective approach because it required observations coordinated on as large a scale as possible, like the global isothermal maps of Dove, or Birt's atmospheric wave data. In social terms as well, then, the science seemed to need just the sort of pruning and systematic direction that the collective authority of the BAAS could supply. The progress of meteorology, according to leading men

24. Kämtz, *A Complete Course of Meteorology*, 590–93.

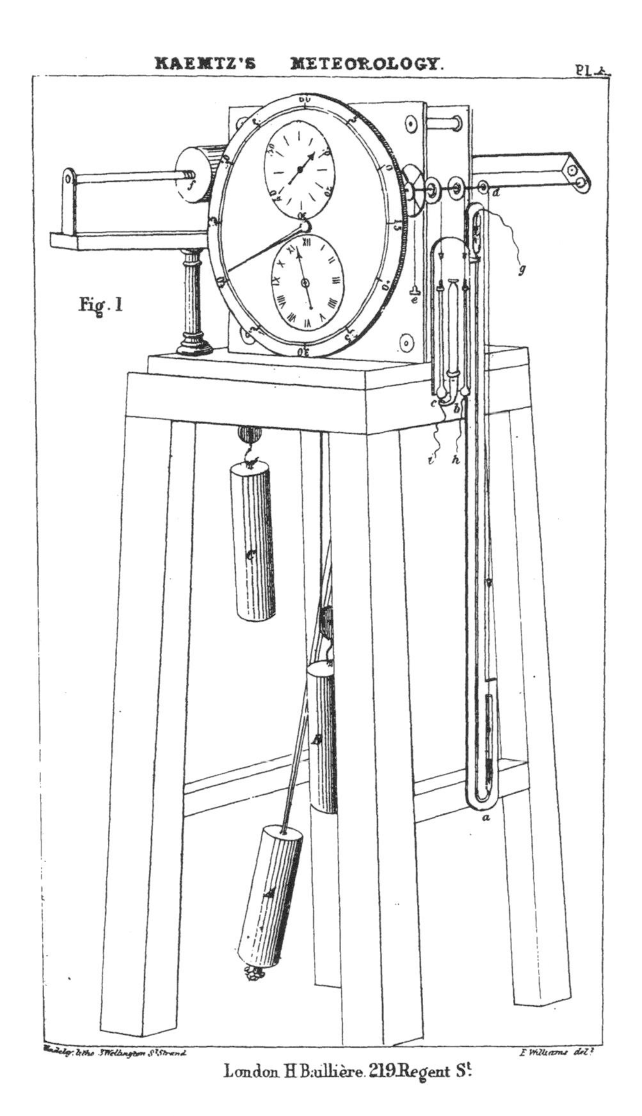

FIGURE 3.1 Kew Observatory, managed by the British Association for the Advancement of Science from 1842 to 1871, became a central authority that could knit together the collective enterprise of meteorology both through its production of standardized instruments for other observers, and through its own sophisticated self-registering instruments. Charles Wheatstone constructed the six-foot high electromagnetic meteorological register for Kew Observatory in 1843 with a barometer (a), thermometer (b), and hygrometer (c) connected to a clock. Automatically recording 1,008 observations a week, the instrument offered a tangible expression of Kew's intended place as the pinnacle of a hierarchy of observers and instruments: "None seem more likely to advance the knowledge of meteorological phenomena, with giant strides." (Ludwig Friedrich Kämtz, *A Complete Course of Meteorology*, translated by C. V. Walker [London: H. Nailliere, 1845], 590–91.)

of science, therefore required a kind of detailed drudgery, both mechanical and human, exemplified by the instruments of Kew or the calculations of Birt. Daniell, for instance, was refreshingly frank about the tedium of observations, commenting in his textbook that he "had no wish to continue in all my life a mere registrar of the weather."[25] Such values built an intellectual hierarchy that confirmed the leadership of the BAAS. As Whewell noted, "There can be no want of employment for any zealous labourers . . . who are willing to be directed by others."[26] Careful management of observations would allow the natural philosophers to concentrate on questions of a higher order.

Weather Clerks

Although we can read about the instruments, the calculator's tasks, or the maps, it is much harder to detect in the records of the BAAS actual encounters between leaders and followers. In that respect, we know much less about networks in the physical sciences than we do in natural history. Throughout the century, local natural historians supplied scientific men with their detailed knowledge and specimens of such varied objects as fossils, sea slugs, and mosses. The relationships between such local experts and elites were complex ones in which class, patronage, and gender mingled with ideologies of intellectual exchange and independence. However important their resources, the contributions of local authorities often received little formal acknowledgment and were assimilated into the body of work produced by an elite scientist.[27] In contrast, the collective nature of the science was much more visible in meteorology, perhaps because networks were a necessity dictated by the scale of the phenomena. The worth of the information gathered from the network depended both on the size of the operation—the more observers the better—and on control of the individual parts—by inspection of the observing stations, for instance. For the same reason, because the elements of a meteorological network were more acknowledged as parts of a whole, the job of managing the network was also more obvious. Meteorological networks coordinated observation times; they exchanged forms and instruments. But they also negotiated

25. "Leading Principles of Meteorology," 49; Daniell, *Meteorological Essays and Observations*, xi.

26. William Whewell to William Harcourt, September 22, 1831, in Morrell and Thackray, eds., *Gentlemen of Science: Early Correspondence*, 74.

27. Alberti, "Amateurs and Professionals in One County"; Allen, *The Naturalist in Britain*; Gates and Shteir, *Natural Eloquence*; Gooday, "'Nature' in the Laboratory"; Secord, "Science in the Pub" and "Corresponding Interests."

the balance of individual contributions and central authority and became a critical site for the encounter of differing visions of the value and purpose of science. It is not surprising, then, that meteorological networks were not all the same. There was more than one model of collective science.

This variety is evident if we examine two of the leading figures of meteorology, James Glaisher and George James Symons. Glaisher, as earlier discussions have demonstrated, was in charge of meteorological observations at the Royal Observatory, Greenwich, but he also engaged in a fascinating and diverse range of meteorological activities beyond Greenwich. He sometimes emphasized the "concert," as he called it, of his endeavors, but just as often, he emerged as an individual figure, even a flamboyant one, ascending in balloons or writing newspaper columns. Symons, unlike Glaisher, worked on a wholly private basis, creating an enormous group of observers across the nation. The success of his enterprise left Symons in a somewhat paradoxical situation. Although for him and his observers, the independent, voluntary nature of the organization epitomized the best of the scientific spirit, its vast size and the need for strict coordination meant that constant direction from a single authority was required. The experience of both these men with their collective projects raised questions about the public and private grounds of scientific work.

James Glaisher had begun his career as a surveyor in Ireland, then in 1833 joined the Cambridge Observatory as assistant to George Airy. When Airy moved to Greenwich as Astronomer Royal in 1835, Glaisher accompanied him, and in 1840, when the department of magnetic and meteorological observation was founded, Glaisher became its first superintendent. The Greenwich department rapidly developed at the same time and along the same lines as the Kew Observatory. His staff initiated photographic registration of instruments from 1845, jockeying with Ronalds for priority. But through Glaisher, the Greenwich Observatory was linked to a series of other meteorological enterprises. Like Kew, Glaisher's department became a center for standard thermometers and barometers; in 1854, for instance, he ran comparisons of three hundred thermometers with the Greenwich standard, mainly for British Meteorological Society observers. As this verification work indicated, Glaisher was deeply involved in the scientific world beyond Greenwich. He was secretary of the British Meteorological Society for most of the period from 1850 to 1873, he was active in the Astronomical, the Microscopical, the Meteorological, the Aeronautical, and the Photographic Societies, and he became a Fellow of the Royal Society in 1849. In the Great Exhibition of 1851, Glaisher was juror and reporter for Section X on philosophical instruments, a tribute to his role in promoting rigorous observation and exact instruments. More dramatically,

FIGURE 3.2 James Glaisher took his identity as meteorologist of the Royal Observatory into other more public enterprises like ballooning, lecturing, and publishing for newspapers. Glaisher's career suggests one definition of a Victorian scientific network, that of a diverse range of activities and audiences connected by an energetic and enterprising individual. (Glaisher Papers, Royal Astronomical Society.)

Glaisher became an aeronautical pioneer (fig. 3.2), making twenty-nine ascents in four years (1862–66) to record temperature and humidity in the atmosphere. On one highly publicized occasion in 1862, he and a fellow balloonist ascended to record heights, losing consciousness at the apex but surviving to bring the records back to the earth.[28]

This catalogue does not exhaust the list of his meteorological activities. Glaisher became associated early in his career at Greenwich with the General Register Office. Beginning in 1844 the reports of births and deaths published by that body (founded six years earlier) included a summary of the Greenwich data. For the next fifty-six years, for a small stipend, Glaisher produced detailed quarterly summaries of barometer, thermometer, and hygrometer readings, and the difference from the mean, the force and direction of the wind, the amount of rain, the electrical state of the atmosphere, temperature readings from the Thames, and the amount of cloud. In the wake of the cholera epidemic of 1853–54, the General Board of Health (founded in 1854) commissioned a study from Glaisher of the connection between atmospheric conditions and cholera. His report concluded that weather conditions (above-average temperatures, prevalent mist and still air) during the outbreaks of 1832 and 1849, as well the recent epidemic, had certainly encouraged the spread of cholera, even if the weather conditions could not conclusively be said to have caused disease.[29] Glaisher's network of observers, about fifty "zealous meteorologists or servants and gardeners of landed gentlemen and noblemen,"[30] sprang from his work for the General Register Office. His network overlapped with the network of the British Meteorological Society—thus underlining the fact that many weather observers in that body began their work as contributors to projects on national health, sanitation, and agricultural statistics. In 1860, Glaisher also placed his expertise at the service of the Royal National Lifeboat Institution, a national body founded in 1854 to promote a voluntary coastal rescue service, and with the president of the British Meteorological Society, Thomas Sopwith, he superintended the installation of public meteorological instruments at fourteen stations on the northeast coast.

Through his General Register Office work, Glaisher developed close connections with the press, writing articles on the Greenwich Observatory, astronomy, and meteorology for various periodicals. One of the earliest of these was an

28. On Glaisher's ballooning, see Tucker, "Voyages of Discovery."

29. Glaisher, *Report on the Meteorology of London*.

30. As described in 1877 by William Farr of the General Register Office, *Report of the Treasury Committee*, 794 (Parliamentary Papers).

1844 account of his department at Greenwich and its novel instruments for the *Illustrated London News*. The editor of the *ILN*, Dr. F. K. Hunt, shortly moved to the new *Daily News*, and Hunt enlisted Glaisher to produce weather reports for the *Daily News* during the harvest weeks in 1846. The newspaper continued to publish signed daily reports from Glaisher for several years. Glaisher enlisted the assistance of leading railway companies and made inspection visits in 1849 and 1850 to chosen observation sites. Station masters entered morning readings into a form printed by the *Daily News*, then forwarded them by the first available train to London, where they were collected by messenger at midnight and published in the next day's newspaper. (The same network posted data for a daily weather map at the Great Exhibition in 1851.) The *Daily News* work marked only the beginnings of Glaisher's regular ventures into print and publishing. He contributed scientific data and articles to almanacs throughout the 1850s and 1860s, and in 1861, James Glaisher, Sopwith, and some others formed a short-lived "Daily Weather Map Company," which combined the offerings of science and journalism, proposing to furnish daily maps alongside literary and scientific news.[31]

Many of Glaisher's activities then were carried out in full view of an interested national audience, rather than buried in the rooms of the Greenwich Observatory. The range of Glaisher's activities creates a daunting picture of Victorian energy and enterprise, but its breadth led to some potential for conflict. Most importantly, Glaisher was the Greenwich meteorologist, but that authority cascaded to his other positions as government advisor, secretary of the British Meteorological Society, recipient of BAAS grants, and scientific writer. On occasion, the different roles collided. As Greenwich meteorologist, Glaisher had an interest in supporting the preeminence of that institution; as participant in the BAAS he had some stake in the development of a competing Kew Observatory. He was at Greenwich, technically, a government official and, sometimes, from the independent vantage point of a scientific society, a critic of government scientific policy. The most important source of tension, however, was George Airy's dislike of his journalism. In 1844, the Astronomer Royal was appalled that Glaisher had signed his own name, and that of the observatory, to his early piece for the *Illustrated London News*—the article, he considered, should have been left anonymous.[32] Either Airy withdrew his objections, or

31. Marriott, "An Account of the Bequest of George James Symons," 258–59; the others involved were Dr. Tripe (a medical man), Mr. Beardmore (an engineer), and Mr. Perigal. Richardson, *Thomas Sopwith*, 285–86, describes Sopwith's design for the map.

32. Hunt, "James Glaisher," 245.

Glaisher ignored them, because he continued to list himself in many of his writings as "James Glaisher, Esq. F.R.S., Greenwich Observatory," a typical and revealing catalogue of the sources of his personal authority. Airy was not alone in finding Glaisher too quick to promote himself. The botanist Joseph Hooker later referred to Glaisher as one of "those cattle, who live by self-glorification."[33] These criticisms were as much social as they were scientific, reflecting a distaste for the entrepreneurial associations of some of Glaisher's work. But they also underscored the difficulties involved in managing a science that was based on collecting others' observations and that repeatedly addressed vital practical questions of broad public interest.

Glaisher clearly strove to make himself a national point of reference for meteorological work. The variety of Glaisher's projects, from public institutions like the national observatory and the General Register Office, to charitable enterprises, scientific societies, journalism, and ballooning exploits, were characteristic of the scope of Victorian science. Glaisher was typical, then, of one vision of scientific work: heterogeneous enterprises and overlapping networks connected to each other by the dynamic leadership of individuals. But in meteorology, more than many other sciences, it seemed obvious how much depended on those networks. Glaisher once described his work as that of "a vast engine which sets in motion a thousand wheels."[34] And, although one suspects he located that engine principally in his own person, he certainly recognized the effort required by other meteorologists. It is possible to trace the work of creating such concerted activity in a network organized by George James Symons. Like Glaisher, Symons was very aware of working within a large machine. The British Rainfall Organisation, which he created, was one of the most remarkable scientific associations of the century. Observers from literally thousands of locations across the British Isles made daily records, which they turned in monthly or annually. Symons and the BRO coordinated scientific work across a range of talents, interests, and social classes. Such a large group required careful management, and Symons was an unusually skilled leader. Examining his stewardship of the system shows some of the difficulties of creating and retaining a set of observers.

Symons (fig. 3.3) was the only child of a London businessman, who died when his son was a young man. Briefly engaged in trade, Symons saved enough to pay for a course of lectures in physical sciences offered by John Tyndall at the Normal School, South Kensington in the late 1850s. Symons ranked first

33. Joseph Hooker to Charles Darwin, May 2, 1865, Darwin Papers, Cambridge University Library.

34. Glaisher, *Philosophical Instruments and Processes Represented in the Great Exhibition*.

in Tyndall's class, and Tyndall was impressed enough to help him find a post in the Meteorological Department at the Board of Trade.[35] Working as one of Fitzroy's assistants, Symons's interest in rainfall statistics was stimulated by the droughts of the 1850s and contemporary speculation about a climate change. On his own initiative, he began to organize a network of rainfall observers.[36] Symons's network probably first depended on contacts made within the British Meteorological Society and the General Register Office observers. But his organization developed quickly. At the end of 1863, Symons sent a letter to the *Times*, listing desired locations for observers, offering to subsidize costs of instruments, and calling for observers "of both sexes, all ages, and all classes."[37] The rapid expansion of the network convinced Fitzroy that Symons's official duties would be neglected, and Symons left the Meteorological Department in late 1863. He soon enlisted hundreds of volunteers: some 1,300 observers in 1867, and 2,100 by 1887. The numbers rose steadily, prompted by further newspaper campaigns, calls for old meteorological diaries, and recruitment by members, who each received an extra set of forms each year, ready to hand over to like-minded acquaintances.[38] By Symons's death, in 1900, he received data from 3,408 stations. Symons described the network in 1879 as an "amateur organisation, without either State aid, magniloquent title, managing council or pecuniary resources except the voluntary contributions of the staff." Its members, "nearly every social grade from peer to peasant," were scattered across the nation, and remained largely unknown to each other, yet worked "with a regularity and heartiness which is beyond praise."[39] The tribute described an industrious, self-governed model of national scientific intercourse.

The management of such a large volunteer movement required a judicious balance of cajolery, reprimands, and opinion polling. Indeed, careful cooperation was indispensable for more than one reason, because the BRO ran on observers' financial contributions as well as their data. In return for these efforts, they received the *British Rainfall* publications (not free, but at cost),

35. "George James Symons [obituary]"; Bilham, "George James Symons"; Mill, "George James Symons"; *Oxford Dictionary of National Biography*.

36. See, for example, Steinmetz, *A Manual of Weathercasts*, 130–35; and Symons, *Rain: How When Where and Why It Is Measured*.

37. See Symons's letters written to the *Times* of November 24 and December 12, 1863, reprinted in *British Rainfall 1863*, 6–7.

38. Symons, *Rain: How When Where and Why It Is Measured*. For numbers of observers, see also the lists in each annual volume of *British Rainfall*.

39. *British Rainfall* 1879, 5–6.

FIGURE 3.3 George James Symons, here in a photograph showing him as the benign patriarch of the Royal Meteorological Society, exemplified a different sort of leadership and network, building the British Rainfall Organisation into a model of national scientific intercourse. (*Quarterly Journal of the Royal Meteorological Society* 26 [1900]: opp. 154.)

purchased the charts and forms on which to record the data which they turned back in, and provided a ready-made market for *Symons's Monthly Meteorological Magazine*. Despite a small annual grant from the BAAS from 1865 to 1875, the occasional bequest or Royal Society grant, or some income from Symons's consultant work for sanitary engineers, the organization was funded largely by voluntary contributions. In 1887, Symons estimated that about 300 or 14 percent of his observers could not afford a small contribution; of the remaining

1,800, about half subscribed small amounts, mostly from £1 to £5 pounds, to the organization.[40]

Yet it would be misleading to imply that Symons's consideration for the observers was self-serving. Although Symons may have gathered some income from the BRO, as seems likely in later years, when the network did turn a small profit,[41] it seems impossible that it represented Symons's sole or even principal means of support. Symons worked, as did his observers, out of dedication to the project. If he constantly urged them for more funds, he also showed enormous tact. In beginning his network in 1862, he called his observers "authorities,"[42] and his first list of "rules" for observers, published in 1868, included a disclaimer of the "dictatorial" tendencies of such rules. Any objections or suggestions, he promised, would be published in the next issue of his *Meteorological Magazine*, whereupon "the rule shall be altered or not, as decided by the observers themselves."[43] In 1870, perhaps taking one such suggestion, he further refined the tone by renaming the rules "Suggestions for Securing Uniformity of Practice among Rainfall Observers."[44] Symons carried out these egalitarian assurances faithfully in addressing a disputed question about the time of observation and the recording of data. After taking a poll, Symons decided that the contents of the rain gauge were to be measured (at least) once daily, at 9 a.m. The observers were almost split, however, on whether to apply that record to the current date or the previous day. Symons initially supported the current date solution but acknowledged the slight majority who favored the logic that the 9 a.m. reading belonged to the previous 24 hours and thus mostly to the previous date. Both the British meteorological societies (based in London and Edinburgh) endorsed this view, and Symons thereafter printed their resolutions in his "Suggestions" as well, to emphasize the collective spirit of the decision.

In Symons's view, such care from its "chief" was indispensable to a voluntary system, and the system only functioned because of that voluntary character. His network, he felt, embodied the spirit of the "English people" (he was,

40. *British Rainfall* 1887, 10.

41. See the "Financial Statements" in the front of each issue of *British Rainfall*. In 1865, for example, expenditures were £132 and the receipts from sales of publications and two grants of £50 each from the Royal Society and the BAAS came to £266. *British Rainfall* 1865, 1. Cf. *Report of the Treasury Committee*, 823 (Parliamentary Papers).

42. *British Rainfall* 1862, 5.

43. *British Rainfall* 1868, 16.

44. *British Rainfall* 1870, 128.

characteristically, careful to clarify that he referred to language and not geography), who valued independence and self-help.[45] In 1875, the BAAS annual grant ended, in part because Symons rejected a proposal to transfer the work to the Meteorological Office. He "flatly refuse[d]," Symons told observers, to push the work into "an obscure corner in some Government office."[46] Not only would the move undermine his own "enormous expenditure of money, time and physical and mental energy" but it would mean that the BRO "esprit du corps would be extinguished."[47] The weight of the Meteorological Office would distort the network. Instead, that organization continued independently, supported by participants, until some years after Symons's death. The tributes at that time to his "power of making friends," his geniality, and his "wide knowledge of human nature," were also tributes to the collective effort involved in the BRO.[48]

Symons and Glaisher each adopted strikingly different approaches to the work of managing a network of observers. Although he acknowledged the work of observers, Glaisher saw himself as the driving force, and he was not willing to place his personal authority behind an institutional screen, whether of the Greenwich Observatory or a meteorological society. Airy himself has been compared by historians of astronomy to a factory boss, running the observatory with an industrialist's concern for uniform practices and disciplined labor.[49] Glaisher was an entrepreneur of a different sort, one whose expertise was personal and could be carried from one activity to another. His networks, of course, involved much smaller numbers than Symons and often consisted of members who already had established credentials as gentlemanly scientific observers, either from an association with the sanitary movement and the General Register Office, or the Meteorological Society. Symons's net was cast much wider, and he deliberately emphasized the action of the group over his own leadership, overtly pursuing an ideal of civilized scientific intercourse.

How do these differences throw light on the controversies that developed in the Meteorological Department? If some of Glaisher's work was government science, he was nevertheless positioned within a venerable institution, founded in 1675. Under Airy, the Greenwich Observatory had worked out a balance

45. *British Rainfall* 1879, 5–6.

46. *British Rainfall* 1875, 7.

47. *British Rainfall* 1875, 8–9.

48. Mill, "George James Symons," 166, 168.

49. Ashworth, "Calculating Eye"; Chapman, "Private Research and Public Duty"; Schaffer, "Astronomers Mark Time"; Smith, "A National Observatory Transformed."

between public service and a scientific research program, and it preserved considerable independence in its dealings with the government. Symons, as earlier examinations have shown, felt strongly about such independence. He had left public service as his work with the British Rainfall Organisation steadily consumed more of his time, and by the end of his life he saw the principles of his group in stark contrast with those of a government office. In refusing the government offer, Symons may well have been thinking of a recent controversy between the Scottish Meteorological Society, based in Edinburgh, and the London Meteorological Office. In the SMS, as in Symons's network, competent observers in remote and underpopulated locations were highly prized.[50] Indeed as his network grew, and gaps in its coverage seemed more and more important to Symons, the exceptions to his own emphasis on voluntary contributions were a few paid rainfall observers in particularly remote and wet regions in which he could find no volunteers. The Scottish network similarly possessed several such valuable observers, and knew their importance. But in 1872 as the central Meteorological Office extended its observation network, it enlisted a Scottish observer at Stornoway. In the process they transformed him from volunteer to a paid assistant who then, having learned his worth to London, would not supply the same data *gratis* to Edinburgh. From the viewpoint of the embittered SMS, the national office was deliberately using public funds to undermine their independent efforts.[51]

In these networks, claims about social organization and knowledge were mutually reinforcing. Observation networks and the "concert" of their data mirrored the complexity and scale of the phenomena. Glaisher's reference to the machinery with a thousand wheels, Symons's tributes to the principle of voluntary cooperation for a national purpose, or Dove's maps of 800 global stations all stressed size, distance, and the interrelation of parts. Taken individually, a meteorological fact was isolated and trivial; worked into a collective scattered across a larger territory, the individual became valuable. Observers overwhelmed by the grand and expansive phenomena could take heart. "If

50. According to Alexander Buchan, secretary of the Scottish Meteorological Society, the 104 observers in 1877 included 54 lighthouse keepers, 34 gardeners, 18 doctors, 11 landed gentlemen, 10 schoolmasters, 7 clerks, and 5 farmers as well as an asylum keeper, a policeman, and a few merchants. *Report of the Treasury Committee*, 810–11 (Parliamentary Papers).

51. Burton, "History of the British Meteorological Office to 1905," 121–33, 196–207; *Royal Commission on Scientific Instruction and the Advancement of Science: Fourth Report*, 547, and *Royal Commission on Scientific Instruction and the Advancement of Science: Eighth Report*, 482–83 (Parliamentary Papers).

each does what he can, and contributes his mite, much may be accomplished," announced Rev. Charles Clouston, pursuing his own meteorological observations in a remote area of Scotland.[52] Scientific networks were not unique to meteorology, but the representation of those networks as a model of natural knowledge was particular and emphatic. Study of the weather was an enterprise of coordination, bringing different observers and scattered facts together. But it also divided them, establishing distinct hierarchies of value based on notions of expertise and opinions about the purpose and direction of meteorology.

Fitzroy and the Forecasting Controversy, 1860 to 1867

The new government office for meteorological statistics called renewed attention to the collection and distribution of facts, and therefore to the relationship between scattered observers and central authority. But the nature of government science created new expectations in other ways as well. Beginning in 1854, Fitzroy had the task of fostering collective science in a department that was relentlessly and often uncomfortably in the glare of politics and public life. The introduction of forecasting at the end of the decade became the catalyst for several different issues. It sharpened the distinction between the scientific value of accurate observations and the immediate usefulness of such observations. And, because weather predictions were fallible, forecasting raised questions about the authority of Fitzroy and his department. Critics simultaneously condemned Fitzroy's failures of judgment and the impropriety of a public office issuing such judgments.

Best known to posterity for offering a berth to the young naturalist Charles Darwin on the *Beagle* voyage of 1831 to 1836, Robert Fitzroy was an autocrat by birth and temperament (fig. 3.4). A descendant of Charles II, he was the youngest son of the Lord Charles Fitzroy and a nephew of the Duke of Grafton. He went to sea in 1819 at the age of fourteen after training at the Royal Naval College. Taking his lieutenant's examinations five years later, he received the highest possible marks, early confirmation of his dedication and intelligence. His first command was the *Beagle*, gained when he was 23, halfway through a South American surveying voyage. On the *Beagle*'s second voyage, Fitzroy circumnavigated the globe, surveying the coasts of South America and lands of the Pacific. The trip confirmed his own high standing as a hydrographer

52. Clouston, *Popular Weather Prognostics of Scotland*.

and announced him as one of the navy's most eminent scientific sailors. He won the Royal Geographical Society's gold medal in 1837, held membership in the Astronomical, Geographical, and Ethnological Societies, and was elected a fellow of the Royal Society in 1851. Yet, after the *Beagle* voyages, Fitzroy's career was checkered. He first entered political life, serving on a committee investigating shipwrecks in 1838 and winning a Durham seat for the Tories in 1841. A short-lived career as governor of New Zealand followed. Arriving in 1843, Fitzroy promptly antagonized colonial interests by supporting the aboriginals, and the Colonial Office called him home a humiliating twenty-six months after he sailed from England. Then, in 1849, Fitzroy was appointed captain of an experimental screw-driven steamship, the *Arrogant*, a novelty for the technologically conservative navy. After Fitzroy gave up the command of the *Arrogant* in 1850, citing ill health, he damaged his chances of receiving another command and effectively retired from active service. The *Arrogant* command, the New Zealand fiasco, and the fractured party politics of the Crimean War era all made it difficult for Fitzroy to find another position. In 1852, his first wife, Mary, died, leaving him with four young children. When plans for the Meteorological Department emerged, he was more than eager to involve himself.[53] He entered into the work with characteristic energy, dedicating himself to meteorology until his death in 1865.[54]

Fitzroy was deeply religious. A member of the Church of England, he held evangelical beliefs, coupling faith in the regularity and predictability of the material world with active work for the moral reform of individuals.[55] (He took a personal interest in the education of four indigenous people that he had captured in Tierra del Fuego, hoping to return them as missionaries, and was much disappointed by the rapid disappearance of their "civilized" attire and equipment shortly after their return in 1835.)[56] He supported too a literal interpretation of the Bible, writing letters to the *Times* in late 1859 that attacked recent archaeological discoveries placing human beings in the

53. Fitzroy, *Memorandum*.

54. Mellersh, *FitzRoy of the Beagle*; Sulivan, *Sir Bartholomew James Sulivan*. Darwin, skeptical and liberal, described Fitzroy's first wife unflatteringly as "so beautiful and so religious a lady" (Desmond and Moore, *Darwin*, 284–85). Fitzroy married again in 1854, to Maria Smyth, a second cousin, but suffered another loss when his eldest daughter died in 1856. Following his retirement from active service he continued to be promoted, under the usual practice, to rear admiral in 1857, and vice admiral in 1863.

55. Hilton, *The Age of Atonement*; Best, "Evangelicism and the Victorians."

56. The Fuegian encounter and its results are discussed in Mellersh, *FitzRoy of the Beagle*, 48–49, 88–103; and Beer, "Travelling the Other Way."

FIGURE 3.4 Robert Fitzroy, in his vice admiral's uniform, in a portrait by Francis Lane (1882) from an earlier photograph, c. 1863. (Royal Naval College, Greenwich, London, UK/Bridgeman Art Library.)

geological record.[57] But his own faith was subject to crises. The *Narrative* of the Beagle voyage ended with two chapters on the origins and geographical distribution of humankind and a discussion of the book of Genesis, in which he earnestly counseled skeptics against infecting youth with their doubts, describing himself as "having suffered much anxiety in former years from a disposition to doubt, if not to disbelieve, the inspired History written by Moses."[58] Science offered Fitzroy spiritual consolation. He assured a colleague in 1863 that meteorology led him to "glimpses of the unapproachable and utterly

57. Fitzroy wrote these protests under the pseudonym Senex. For an account of the episode, see Mellersh, *FitzRoy of the Beagle*, 270–74.

58. Fitzroy, *Narrative of the Surveying Voyages*, 2:658.

incomprehensible Power originating and maintaining all that can be seen—felt—or known by the mind."[59] With these beliefs guiding him, Fitzroy was a forceful advocate for his chosen science.

Fitzroy's department was aligned with several institutions and government departments, and his work had to navigate between distinctive, sometimes opposed visions of science. The Meteorological Department of the Board of Trade was founded in 1854 as an office for coordinating marine meteorological statistics, following an international meeting in Brussels. Most of the delegates at that meeting were naval officials of a scientific bent, rather than philosophers. Marine observations offered a fresh sphere for meteorological work: there it might be possible to achieve uniform observations and bring "every part of the Ocean . . . within the domain of philosophic research."[60] The function of the office was outlined in detailed consultations between the Board of Trade and the Royal Society, whose council members laid out a comprehensive program for data collection. Funds for the new office came from both the Admiralty and the Board of Trade, and observations were collected from both men-of-war and merchant navy. Another factor increased the independence of the department: in 1857, Captain Sulivan, a former officer under Fitzroy on the *Beagle*, won the position of chief naval officer in the Marine Department of the Board of Trade, and Sulivan tactfully asked that the Meteorological Department not come under his authority.[61] From the first, then, the intricate relationship of the department to the government, the navy, and merchant marine, and to the scientific community complicated Fitzroy's directorship. By inclination and experience, Fitzroy was a leader, used to exercising solitary command. By the terms of his new official position, he was an anonymous civil servant, transcribing and collating the doings of Nature. But when he introduced storm warnings and weather forecasts, Fitzroy added another dimension, producing science for a national audience.

Fitzroy laid out the most obvious boundary between different visions of science in his first official report (1857), distinguishing between two objects of the department: collecting "accurate and digested observations for the future use of men of science" and aiding navigation in the present by applying meteorological knowledge.[62] This distinction shaped the work of the office and marked out the controversy over weather prediction that developed in

59. Robert Fitzroy to John F. W. Herschel, January 15, 1863, J. F. W. Herschel Papers, HS 7/262.

60. *Abstract of the Report of a Conference Held at Brussels*, 450 (Parliamentary Papers).

61. See Burton, "History of the British Meteorological Office to 1905," 36.

62. *Report of the Meteorological Department*, 1857, 290 (Parliamentary Papers).

subsequent years. The decision to go beyond data collection to navigational assistance was Fitzroy's. It shifted meteorological work toward a much larger audience. The department became the busy center of a telegraphic network, and as Fitzroy oversaw the ebb and flow of information within that network, more and more questions arose about his role in the process, and the balance between his private authority and that of his office.

Fitzroy distributed standard instruments to the navy and to the merchant marine, and collected and analyzed the records that observers returned to London. Captains received a supply of high-quality instruments (a good compass, sextant, and chronometer as well as the Kew marine barometer and three air and water thermometers) and valuable wind charts. In return, they recorded a long list of data: barometric temperature, air and sea temperature, humidity, wind, cloud, soundings, currents, magnetic variation, and unusual conditions like auroras, wind spouts, ice, or shooting stars. The observations were reinscribed in London in separate books for each type, with the book's pages representing successive "squares" of the ocean. Fitzroy also learned quickly to manage his observers with tact and respect. Fitzroy faced constant negotiations with local marine boards to solicit good sea-going observers and relied on the goodwill and dedication of those who agreed to take instruments and logs. In 1860, for instance, the customs officer at Hull reprimanded Fitzroy for not writing personally to acknowledge the logs turned in to the Meteorological Department. He told Fitzroy that one captain, having received "no thanks," after "considerable inconvenience" in making "his observations as perfect as he was able," had refused to undertake another log for subsequent voyages.[63]

Despite such challenges, collecting the marine observations quickly became routine. Once the statistical work was on track—forms designed, instruments and procedures in place—Fitzroy turned to other projects. Early in 1857, Fitzroy applied for the post in the Marine Department of the Board of Trade that was instead given to Sulivan, a former lieutenant, as mentioned earlier. When he was rejected, Fitzroy put his heart into the department's second object: practical meteorology. In quick succession, Fitzroy designed a barometer and manual to promote weather observation (fig. 3.5), undertook to represent American oceanographer Matthew Fontaine Maury's ocean data in what he felt was a form more suited to mariners, began a program of loaning barometers to fishing villages, and arranged for translation of Dove's influential monograph

63. Zebedee Scaping to Robert Fitzroy, February 9, 1860, Met. Office Papers, PRO BJ7/406.

on the theory of storms, the *Law of Storms*.[64] Most importantly, however, he began to campaign for the use of the telegraph for storm warnings.

His efforts were prompted by a devastating storm in late 1859. On the 25th of October, winds off the Cornish coast developed to gale force, and heavy rain lashed at the boats in harbor. During the following day, the center of the storm moved right over the mainland of Britain. Reaching Yorkshire and the North Sea, it then headed northeast again, up the Scottish coast, arriving at the Shetlands and Norway on the 27th. High winds and seas battered fishing fleets and other coastal craft from Tyneside to Banffshire. The most famous incident of the storm, however, occurred on the west coast, about seventy miles north of Liverpool. The captain of the *Royal Charter*, an iron-clad ocean steamer nearly home from an Australian voyage, decided to anchor, with all sails lowered, to wait out the storm. During the night, the winds veered sharply. By early the next morning, this ship was driven onshore, pounded into three fragments and sunk, with the loss of about five hundred lives.[65]

The *Royal Charter* was only one of 343 ships wrecked on British coasts during the storm, a count that made the losses of that single catastrophic gale almost reach the total number of wrecks for the previous year.[66] News of the wreck and the recovery of the cargo appeared in the *Times* almost daily until the end of the year. Charles Dickens, who had recently dealt with sensational shipwrecks in fiction, encountered this real drama directly. On a tour of public readings, he traveled between Oxford, Cheltenham, and Birmingham in the rain and mud as news of the wreck unfolded. His wife's cousins—two brothers, and the wife and child of one—were passengers on the *Royal Charter* and all four drowned. Several weeks later, at the New Year, Dickens himself visited the shore where the *Royal Charter* had foundered, and a solemn account appeared the next month in *All the Year Round*, his weekly family journal.[67]

Dickens's involvement was representative of contemporary interests, and not simply personal. In several ways, the wreck of this particular ship took on a larger significance for Victorians, dramatically underlining the national and imperial significance of scientific navigation. Within ten miles of the *Royal*

64. See Robert Fitzroy to John Herschel, May 4, 1858, J. F. W. Herschel Papers, HS 7/252.

65. Fitzroy, "On British Storms," *Report of the BAAS*, 1860, 41; Lamb, *Historic Storms*, 135–36; McKee, *The Golden Wreck*.

66. *Returns of Wrecks, Casualties, and Collisions*, 1861 *to* 1870 (Parliamentary Papers).

67. Dickens, "The Shipwreck"; the piece was reprinted in Dickens's *The Uncommercial Traveller*. With Wilkie Collins, Dickens produced the successful melodrama of Arctic exploration, shipwreck, and betrayal, *The Frozen Deep* (1856), and the same year published a tale of a ship wrecked as it headed out to the California goldfields (*The Wreck of the Golden Mary*).

FIGURE 3.5 One of the tangible ways that weather forecasting was associated with the first director of the Meteorological Department: a Fitzroy barometer, probably from the late nineteenth century, with observation instructions on the face. First made in the 1860s, the barometers showcased the authority of Fitzroy's name until the early twentieth century. (Reproduced by permission of the Science Museum, London, Science and Society Picture Library.)

Charter, a second ship, under American ownership, was tacking in the same squalls. Interpreting the storm as a cyclone, and hence expecting winds to veer, her captain decided to drive offshore as the gale intensified and was saved. Ironically, the *Royal Charter* had seemed an outstanding example of modern science and marine technology. In 1855, the ship was much in the news as George Airy and other authorities developed techniques to adjust magnetic variations caused by the iron in the ships, using the *Royal Charter* and its compass as an experimental subject.[68] The ship was of interest as well because its passenger list included many who were bringing back fortunes from the Australian gold rush. In the aftermath of the storm, tales circulated of desperate men drowned as they tried to swim to shore weighed down with nuggets. "Thus from a deficient attention to Meteorology," as one writer put it, "those who had laboured for years under a scorching sun . . . were doomed to an awful death, after a voyage of 14 000 miles, within sight of, and almost touching, their native shore."[69]

The *Royal Charter* storm swept the science of meteorology into public view. Over the winter months, Fitzroy's proposal moved forward more rapidly. French support was also important: in April 1860, LeVerrier wrote to Airy outlining his own proposal to collect telegraphic information and asking for assistance. (LeVerrier downplayed the prospect of foretelling the weather—a cautious attitude that, as Fitzroy noted, was belied by the fact that his network, like Fitzroy's proposed network, used ports as observation points.) By the following December, Fitzroy was receiving observations from thirteen stations around Britain and a daily report of five sites from the Continent. His British observers were telegraph clerks from offices near the coasts. The Meteorological Department did not pay for observer's services, but the clerks were issued instructions and standard instruments. Six days a week, all the observers took observations of rainfall, barometric pressure, wind, and temperature at 8 a.m. and telegraphed them to London at a discounted rate. Meteorological Department staff analyzed the records in combination with continental observations, also telegraphed daily, drew conclusions, and prepared a table for insertion in the second editions of the London newspapers.[70] As a related project, Fitzroy developed a set of coastal signals to indicate that a gale was imminent. Often, the same telegraph clerks were placed in charge of these ball and cone signals, hoisting them near the station, in view of seamen (fig. 3.6). Fitzroy first warned

68. Winter, "'Compasses All Awry.'"

69. Plant, *Meteorology*, 5. Plant was associated with the Birmingham and Midland Institute.

70. "Notes on British Storms."

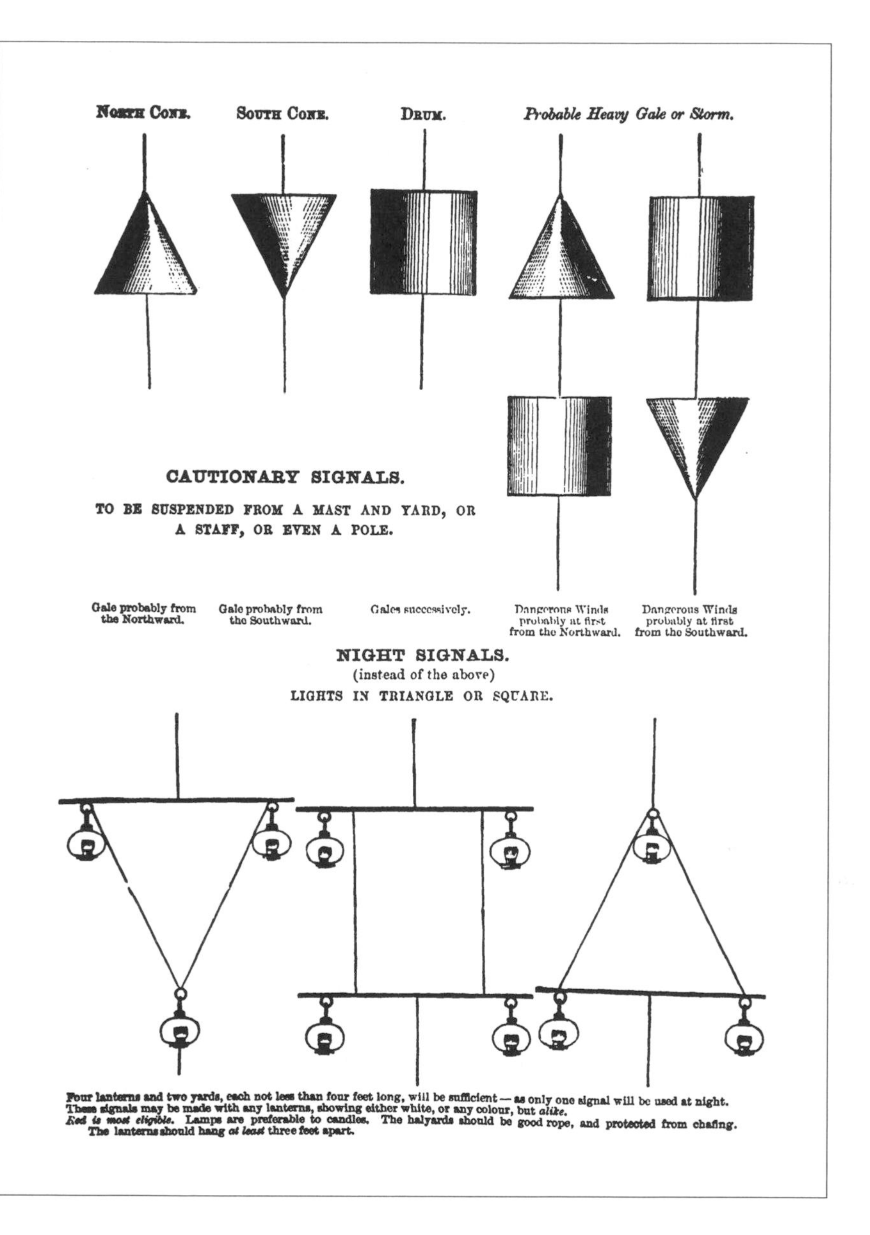

FIGURE 3.6 Fitzroy developed this set of canvas signals to be hoisted on a headland or in a port to warn the community of a coming storm. A cone alone indicated that a gale from a northerly or southerly direction was probable; a drum alone indicated cyclonic or veering winds; whereas a drum and cone together indicated "dangerous" winds, with the cone indicating whether to expect a polar or northerly wind (coming initially from WSW to ESE), or a tropical or southerly wind (coming initially from ESE to WNW). (Robert Fitzroy, *Weather Book*, 2nd ed. [London: Longman, 1863], 350.)

British seamen of an oncoming gale in February 1861, on the northeast coast. Over the next several weeks, he issued eight further warnings, completing his first season's work.[71] In August he added daily forty-eight-hour forecasts (wind force and direction, plus a brief statement of coming weather) for five weather regions in the British Isles.[72] In the space of a few years, then, Fitzroy had decisively turned from collection and discussion of observations, months old by the time they reached his hands, to the collection of data from coastal and inland observation points by telegraph, and the synoptic analysis of such data for immediate weather forecasts.

The storm warnings and forecasts transformed the Meteorological Department. In the first place, weather prediction dramatically changed the department's expenses. Telegraphy had been the most expensive item in the meteorology budget since 1861–62 (Fitzroy always envied the French their state-owned system, which meant that LeVerrier could transfer his observations to Fitzroy free of charge). But extending weather prediction meant increasing telegraphy and costs grew from £3,240 in 1855 to £5,460 in 1865. Perhaps more significantly, telegraphy made the department a center of activity in a way that it had not been before. As a statistical office, distributing instruments and charts and collecting logbooks, a single center was an administrative convenience rather than a necessity. Instruments were prepared and shipped from Kew, miles away from the rooms at No. 2 Parliament Street.[73] By 1863 the bulk of the oceanographic charts work had been shuffled off to the Hydrographic Office in the Admiralty. What the department needed as a priority was physical space for storing all the logbooks and laying out the various books in which to reinscribe their data. Already by 1856, the department had fifteen rooms and about four thousand square feet. These had made the department a material fact, but nevertheless an inconspicuous one. But the introduction of forecasting presented other needs: a telegraph service to hand, space for a chart to post synoptic data, and a messenger service to relay the information quickly to newspapers. With a telegraphic observation network and public communication of the data added to the coordination of logbooks, the Meteorological Department became a much more intensive observation network.

From the beginning the department had distributed its authority along the

71. Burton, "Robert FitzRoy," 161.

72. The districts were North Britain (Moray Firth to mid-Northumberland), Ireland, Central (Wales to Solway), East Coast (Northumberland to Thames), and South England (Thames round to Wales).

73. The first offices were in South Kensington (adjacent to the present-day Science Museum); the office moved to 118 Victoria Street in 1870. Jacobs, "A Short History of Former Homes."

coasts. It had to select agents who would find suitable captains and ships, organize the instrument loans, inspect their placement on board, and collect the books. Agents, who received a fee per vessel, included harbormasters, local opticians, retired naval officers, and, in one case, the director of a local astronomical observatory (Liverpool). With the coming of official weather reports and forecasting, that network was dramatically extended. In addition forecasts gave meteorology a place in the daily news. This public profile can be easily traced in commentary and satire. "To many persons," the *Athenaeum* remarked in 1865, "a daily bulletin of the weather has become as much a necessity of their existence as a daily newspaper."[74] From an obscure study, pursued by reclusive and pedantic figures who paid "mysterious reverence" to wet and dry bulb thermometers "for the benefit of remote posterity," meteorology had now emerged as a subject of "unbounded curiosity," and storm signals had become "a household word."[75] *Punch* reveled in the novelty. A political cartoon in 1863 featured Britannica hoisting a storm signal, and another satirized the fashion for the crinoline by comparing it to Fitzroy seaside cone warnings (fig. 3.7).[76]

Yet, as the *Athenaeum*'s remarks indicated, forecasting most of all exposed critical differences in how to evaluate the value of science. By undertaking weather prediction, Fitzroy had overtly aligned himself with the practical interests of navigation over the abstract and remote interests of men of science. In his letters to colleagues, he repeatedly denied that he was "a truly scientific man" like Herschel or Airy.[77] On occasion, nevertheless, he entered vigorously into theoretical debates. He repudiated the theory of atmospheric waves, work that was closely connected with Herschel and the BAAS. He also successfully tackled William Whewell on tides, and he championed the outsider, William Scoresby, against Airy in an important controversy over compass deviation in iron ships (1847–55).[78] Invariably, in these disputes he appealed to practical experience and the public value of his work to justify his positions in scientific controversies. "Like seamanship," he argued, "the ability to foretell is acquired

74. "Progress of Meteorology," *Athenaeum*, April 15, 1865, 523.

75. *Athenaeum*, June 9, 1866, 770.

76. "The Storm-Signal," *Punch*, October 17, 1863, 159; and "Crinoline Storm Signal: A Warning to Young Ladies at the Sea-side," *Punch*, October 31, 1863, 218.

77. Fitzroy, *Weather Book*, 432. Cf. *Athenaeum*, November 24, 1860, 710.

78. See Fitzroy, *Weather Book*, 95–97, and *Meteorological Papers*, ii–iv; also Met. Office Papers, PRO BJ7/831–49. On atmospheric waves, see Birt, "On Atmospheric Waves and Barometric Curves"; and Jankovic, "Ideological Crests versus Empirical Troughs." On tides, see Deacon, *Scientists and the Sea*, 266–67. On Scoresby and compasses, see Winter, "'Compasses All Awry.'"

FIGURE 3.7 Popular accounts indicate the high profile of Fitzroy's storm signals. In one political cartoon, Britannica hoists a signal that indicated dangerous winds, coming successively from all directions, with anticipated conflicts between Russia and Poland, France and Russia, Prussia and Austria, Germany and Denmark, Mexico and the United States. Another, a social satire, pointed to the similarities between Fitzroy's cone signals and the fashionable crinoline. (*Above*: "The Storm Signal," *Punch*, October 17, 1863, 159. *Facing*: "The Crinoline Storm Signal: A Warning to Young Ladies at the Seaside," *Punch*, November 28, 1863, 218.)

THE CRINOLINE STORM SIGNAL; A WARNING TO YOUNG LADIES AT THE SEA-SIDE.

by degrees, practically."[79] Not surprisingly, he resisted the derogatory way that some scientific authorities treated a "mere" observer. In an early exchange of documents with his superior at the Hydrographic Office, Fitzroy repeatedly replaced the adjective "simple" as in "simple mariner" or "simple instructions," with the word "practical"—a pointed refusal to denigrate the expertise of sailors in favor of that of scientific men.[80] In basing his expertise on a notion of practical experience, Fitzroy took an inclusive rather than an exclusive stance on scientific work.

The association of science with practical benefits, however, risked connecting the philosopher to the tradesman.[81] Here the choice to place the Meteorological Department within the Board of Trade took on new significance. The Hydrographic Office in the Admiralty supervised many of Britain's efforts in maritime science, and the Greenwich Observatory already had a meteorological department and an established scientific reputation. Both of these homes were considered, but in the end the Meteorological Department instead went to a young government office (formed only in 1830) that was responsible for a rapidly growing list of economic and technological controls. Besides railway regulations and standard weights and measures, the Board of Trade regulated the merchant marine. The Merchant Shipping Act of 1854, a massive reorganization of laws connected with ownership, qualifications, and safety in shipping, passed into legislation a few months after the Meteorological Department was founded.[82] Fitzroy's science, then, developed within a distinctly utilitarian framework. The weekly *Intellectual Observer* compared weather forecasts to the kind of information that would circulate within shipping insurance offices: "A Fitzroy announcement partakes of the character of a mercantile letter of advice, but instead of 'shipped from so and so, so many chests of tea' it tells that air currents, dry or moist, have passed over certain places, and may be expected to visit other localities as soon as time permits."[83] Thus, from one perspective the Meteorological Department developed in a setting where the practical applications of science, especially concern with navigation, seemed obvious; from another, this setting only increased the importance of defining scientific activity scrupulously and according to the standards of elite philosophers.

79. Fitzroy, *Weather Book*, 432.

80. Ms. draft of *Report of the Meteorological Department of the Board of Trade for 1855*, Met. Office Papers, PRO BJ7/344.

81. See Yeo, "An Idol of the Market Place," on Baconian science in the Victorian period.

82. Prouty, *The Transformation of the Board of Trade*.

83. "Attempts to Foretell the Weather," 104.

And these standards were clear: utilitarian concerns destroyed the character of science. By the time of the first daily forecasts, opposition had begun to emerge. Several months after Fitzroy's first storm warnings, the Wrecks Department of the Board of Trade began to collect observations after every warning, in order to provide an independent check of results. In India, a gentleman comparing projected and actual weather from the *Times* at a leisurely distance, announced that Fitzroy's system was of "so inaccurate and haphazard a character, as not to be of any true scientific *value*."[84] In response to his critics, Fitzroy spent the summer of 1862 writing a treatise on meteorology for the general reader. Published in the autumn, the *Weather Book* received critical reviews, and Fitzroy anxiously defended the speculations aired in the work to Airy and Herschel.[85] By 1864, Fitzroy felt beleaguered. In an undated note to his closest assistant, Thomas Henry Babington, he complained about "Philistines and land sharks" (perhaps the Treasury officials?) visiting the office.[86] Other difficulties crowded professional ones. Fitzroy was going deaf and overworked himself; accounts of the 1860 BAAS meeting describe him as an elderly man, though he was only fifty-five at the time.[87] More importantly, perhaps, he was hard-pressed financially. According to the *Times* obituary, Fitzroy died £2,000 in debt. He had earned a salary of £800 a year at the end of his period at the Meteorological Department, but the Admiralty had never reimbursed him for expenses he incurred on his own authority for the hydrographic work in the 1830s, which had dented his private fortune. His unfortunate excursion to New Zealand was also costly, not least because he gave up lucrative sinecures to take the post. During a breakdown in February 1850, he had sent anguished letters to his superior officer in the navy, citing his "intense anxiety—and the impaired state of my mind—as well as health of body," and asked to resign his command.[88] By the end of April 1865, he was deep in another of his black depressions. On his last evening, he met with Maury, about to leave England after a long visit conducting experiments with electrical mines and torpedoes

84. Burgess, *On Admiral Fitzroy's Storm Signals*, published originally in *The Times of India*, reprinted as a pamphlet, original emphasis.

85. Robert Fitzroy to George Airy, December 24, 1862, Airy Papers, RGO 6/703/27–28; id., January 15, 1863, Airy Papers, RGO 6/703/25–26; id., May 7, 1863, Airy Papers, RGO 6/703/31–32; id., June 11, 1863, Airy Papers, RGO 6/703/33–34; Robert Fitzroy to John Herschel, April 21 and 25, 1862, J. F. W. Herschel Papers, HS 7/259–60.

86. Robert Fitzroy to Thomas Babington, Met. Office Papers, PRO BJ7/835.

87. Desmond and Moore, *Darwin*, 495.

88. Robert Fitzroy to Commodore Martin, commanding the Lisbon squadron, February 3, 1850, Add. MSS 41417 f. 3–4, British Library.

and soliciting political support on behalf of the Confederate navy. Whether the evening consisted of political or scientific conversation is unknown, but Fitzroy returned from the encounter much agitated. On the morning of the 30th, he cut his throat with his razor and bled to death in his dressing room.[89]

Meteorological Judgments

Fitzroy's suicide plunged the Meteorological Department into a welter of inquiries, committees, and public dissension. Or perhaps it would be more accurate to say that Fitzroy's death lifted the lid on doubts and investigations already proceeding. Above all, the disputes demonstrated the ease with which claims about natural knowledge translated into concerns with social and political responsibility, government and leadership. In a remarkable manner, this personal tragedy exposed concerns about the scientific status of weather forecasting and showed how personal character entered Victorian evaluation of the role of government science.

Fitzroy's suicide has been interpreted by some historians as a despairing response to Darwinian ideas. Fitzroy had denounced Darwin's theory at the BAAS in 1860, and published further criticisms in 1863.[90] This interpretation distorts events by suggesting an intense relationship between two men, Fitzroy and Darwin, whose ways had long since parted. It also implies that Fitzroy's response was unusual and desperate, instead of a part of broad range of responses to Darwinism. Making sense of suicide was a compulsion for Fitzroy's contemporaries too, but, unlike historians, Victorians in 1865 knew Fitzroy best as a meteorologist, not simply as Darwin's captain. The manner of Fitzroy's death became part of the debate over weather prediction rather than evolution. (Indeed, it should be noted that Fitzroy was principally at the BAAS to present his synoptic map of the *Royal Charter* storm and drum up support for his warning system, not to confront Darwinian theory.) Sympathetic contemporary accounts did not hesitate to attribute the death to his scientific work. Fitzroy's anxiety for those under his care, such reports argued, had unhinged his mind, and the admiral died as "a victim to the incalculable mental stress and exertion" of his office.[91] Undoubtedly, the portrait of Fitzroy

89. "Suicide of Admiral Fitzroy," *Liverpool Journal,* May 6, 1865, 12; Lewis, *Matthew Fontaine Maury,* 174–86.

90. Mellersh, *FitzRoy of the Beagle,* 270–74.

91. *Lifeboat,* July 1, 1865, 712.

as a martyr to meteorology was in part a generous attempt to overcome the stigma of suicide.[92] Yet this reaction was also genuinely a comment on the nature of Fitzroy's scientific work. The *Athenaeum* obituary, written by James Glaisher, described it in harrowing terms:

> that almost constant uneasiness experienced in the receipt of the daily weather telegram, from which he had to deduce the most probable coming weather, . . . his anxiety when the data were insufficient to speak with certainty of the future . . . the feeling of responsibility after this, for fear that some points had not been sufficiently weighted—then the successive weather telegrams repeatedly received during the day in stormy periods and eagerly dwelt upon.[93]

For meteorologists of his generation, then, Fitzroy's death became a sign of the hazards of the science. Several years later, Buys Ballot responded in an international survey on the value of weather telegraphy that "we must remember that any one who has to forecast weather, if he does it earnestly and conscientiously, is in great danger of going off his head through nervous excitement."[94]

The verdict of the inquest was death while in "an unsound state of mind," but suicide in Victorian England, whether interpreted as sin or madness, compromised the moral and mental integrity of the individual.[95] As a combination, then, practical science, popular prophecy, and suicide were triply damning—all marking a collapse of the gentlemanly values that were indispensable to science. Fitzroy's forecasting and his death thus undermined the extension of scientific claims that his office had supposedly embodied. Only such a strong reversal of expectations can explain the strikingly hostile account of Fitzroy's work that appeared in the *Edinburgh Review* in 1867. The *Review* called all forecasting "vulgar and fallacious" (an allusion to astrological weather prophecy) and portrayed Fitzroy as a bankrupt—a gentleman fraud, who had assumed a higher station in life than he could really sustain. This revealing critique is worth quoting at length:

> A gentleman of some ability and tact . . . as well as of much self-confidence, finds himself in a high social position and is the possessor of a fine estate and of

92. See for instance, "A National Debt of Honour," *Punch*, March 31, 1866, 129.

93. *Athenaeum*, May 6, 1865, 622 (attribution from marked files of *Athenaeum*, University of North London).

94. *Report on Weather Telegraphy and Storm Warnings*, 59.

95. "Suicide of Admiral Fitzroy," *Liverpool Journal*, May 6, 1865, 12. On suicide, see Anderson, *Suicide in Victorian and Edwardian England*; Oppenheim, *"Shattered Nerves."*

> considerable reputed wealth. He assumes all that belongs to such a position, manages his own estates, asks little advice, believes he can show his neighbours how to manage their estates, keeps up all the social habits befitting his reputation and maintains his name among the foremost of his class. He becomes an oracle and a monitor and is listened to with deference, for no one doubts his sagacity.[96]

The well-known circumstance that Fitzroy had died significantly in debt gave this analogy a truly sharp edge. Without mentioning suicide directly, yet alluding to the lesser but linked failing of bankruptcy, the *Review* indicted both Fitzroy's character and his science. Fitzroy's fraud, for this writer at least, had betrayed the scientific community and the responsibilities of its "high social position."

The responses to Fitzroy's death showed how Victorians interpreted science within a framework of personal morality and public responsibility. *Good Words*, a family weekly journal, noted the "imperative duty" that Fitzroy had felt, characterizing his scientific work in markedly evangelical terms.[97] As Fitzroy had remarked himself in his *Weather Book*, "plain common sense may allow for perception of material relations as fairly at least as moral conditions."[98] Just as the opportunity and responsibility to act morally extends to each individual in society, the study of nature, Fitzroy thought, must be considered accessible to all. To avoid forecasting seemed to evade this duty. In this way Fitzroy's death further polarized the values of practical science and abstract science.

The tension between practical and abstract science, with its implications for defining the place of economic questions, popular audiences, and responsible leadership in science, governed the history of the Meteorological Department in the immediate aftermath of Fitzroy's death. It can be tracked in the comments of the Royal Society leadership about weather forecasting from the early 1860s, to a formal investigation conducted by a government committee in 1865–66, and the reorganization of the meteorological work in 1867. Officially, the reaction of the Royal Society to weather prediction before 1865 had been circumspect. In 1863, responding to government inquiries, the council of the Royal Society returned a noncommittal reply, simply citing Fitzroy's

96. [Liefchild], "Weather Forecasts and Storm Warnings," 83. Liefchild was a regular contributor on scientific subjects, an evangelical, and the author of several works on popular science. In light of this paper's subject, however, the anonymity of the article is an important feature to recall: anonymous authors spoke as "the *Edinburgh Review*."

97. "Admiral Fitzroy," *Good Words*, June 1, 1866, 406.

98. Fitzroy, *Weather Book*, 201.

own comments about the general success of warnings.[99] Their response neatly evaded the requested opinion of the daily forecasts by focusing on finances. According to Fitzroy, the Royal Society noted, forecasts involved no extra expense because the observations needed to be collected and analyzed anyway for storm warning decisions. Fitzroy's death changed the situation. Sabine, now serving as president of the Royal Society, returned a long memorandum in reply to a formal request for advice from the Board of Trade. He conceded that Fitzroy's second-in-command, Thomas Babington, could continue the storm warnings for the meantime, as they showed steady improvement in accuracy. Sabine, "decline[d] expressing any opinion" about the forecasts, a silence that spoke volumes, but he also argued that the Meteorological Department would now be better employed, from a scientific point of view, in land meteorology, particularly as there were some indications that the Admiralty wished to take over the ocean data analysis.[100] With a memorandum of this detail and length, Sabine hoped to provide a blueprint for reforming the department. In response, the Board of Trade opened a formal investigation into the Meteorological Department, with representatives from the Admiralty, the Board of Trade, and the Royal Society, chaired by Francis Galton.

Francis Galton, Charles Darwin's cousin, is well known to historians for his interest in statistics, inheritance, and eugenics. But in 1866 most of that work and renown lay ahead of him. His scientific reputation at the time was based on his travels in southwest Africa. His explorations from 1850 to 1852 earned him a gold medal from the Royal Geographical Society. Following his explorations, Galton married, established a home near the South Kensington complex of scientific buildings, and, financially independent, began to immerse himself in London scientific life. A friend of Edward Sabine, he joined the managing committee of the Kew Observatory and had undertaken a study of European weather conditions, the results of which he presented to the Royal Society and published separately in a series of maps, *Meteorographica* (1863). These social and intellectual qualifications secured him the position on the committee of inquiry as the nominee of the Royal Society. The Admiralty representative was Captain Frederick Evans, chief naval assistant to the Hydrographer of the Navy, and the final member was Thomas Farrer, a senior government official

99. W. Sharpey, secretary of the Royal Society, to T. H. Farrer, March 27, 1863, *Report of the Meteorological Department*, 1863, 97 (Parliamentary Papers).

100. Edward Sabine to T. H. Farrer, June 15, 1865, "Correspondence between the Board of Trade and the Royal Society," 316. Cf. "Further Correspondence between the Board of Trade and the Royal Society"; and Sabine, "Note on a Correspondence."

from the Board of Trade. He was a keen botanist himself, connected to the Darwin-Wedgwood family, and his own scientific publications begin in the late 1860s, around the time of the forecasting controversy.[101] But it is hard to say from these alliances where his own scientific opinion of forecasting lay. Farrer deferred to the Royal Society and seemed concerned only to act as a government official, with his skepticism about meteorology firmly fixed on financial and organizational questions.

Who actually penned the report on Fitzroy's work in 1866 is unclear, but Galton, the only one of the three without other professional responsibilities, seems a likely choice. In any case, the report bore the stamp of Royal Society opinions. The 1866 report insisted on Fitzroy's personal responsibility. It underscored the shift from meteorological statistics to forecasting. "There is no indication," declared the report, "that it was a part of the functions of the Department, as originally instituted, to publish undiscussed observations on the one hand, or to speculate on the theory of meteorology on the other. Still less can it be considered a part of these functions to attempt to prognosticate the weather."[102] In other words, Fitzroy had hopelessly distorted the aims of his office. This conclusion was one Fitzroy's critics had repeated for years. The forecasts, as described to the committee, were dismayingly empirical and unscientific. "No notes or calculations are made. The operation takes about half an hour and is conducted mentally." Rules for interpreting the data were not "reduced to any definite and intelligible form of expression . . ." nor were they "capable of being communicated in the form of instructions."[103] This was a far too individual proceeding. The criticisms contained an implicit description of proper science. It required unvarying rules, preferably left evidence of its logical processes in the tangible form of notes or calculations, and lent itself to command of observers and observations. The empiricism and imprecision of forecasts undermined the processes of legitimate scientific work.

Since forecasting was not based on sound scientific principles, the report concluded, there could be no grounds for its continuation. The decision also of course reflected anxiety about comparisons with "ordinary weather prophets who attempt to connect the weather with the stars." Making this recommen-

101. Farrer's first wife, whom he married in 1854, was a Wedgwood; his second, married in 1873, was another Wedgwood, the niece of Charles Darwin; and, in 1880, Farrer's only daughter, Ida, married Horace Darwin, the son of Charles and Emma (*Dictionary of National Biography*).

102. *Report of a Committee Appointed to Consider Certain Questions Relating to the Meteorological Department of the Board of Trade* [Galton Report], 6 (Parliamentary Papers).

103. Ibid., 20.

dation in the face of public support for forecasting, the committee decisively pronounced on the proper relations between government, scientific work, and the public. "The practice of issuing daily official notices of the weather, the truth of which is warranted neither by science nor by experience," the report announced, "is inconsistent with the position and functions of a Government department, and must be prejudicial to the advancement of true science." It would cause the public "to confuse real knowledge with ill founded pretences, and, in the end, to despise the former because the latter proved to be unfounded."[104]

The report referred here only to the daily weather forecasts issued by Fitzroy beginning in August 1861. It agreed that the collection of weather data by telegraph should continue and that while occasional weather reports—that is, remarks or observations concerning these data—were not objectionable, regional forecasts of probable weather were and should stop. Storm warnings, they stated, were a separate matter. This conclusion contradicted Fitzroy's repeated and logical assertion that the principles underlying the two were identical, an assertion that the Royal Society had accepted in 1863. Far from wholeheartedly endorsing storm warning, however, the committee wished to both restrict the duration of the warning (from seventy-two to forty-eight hours) and confine the indication to force alone, rather than force and direction.[105] But Fitzroy's chief assistant, Babington, resigned in November. This resignation, coupled with the severely critical report, resulted in the cancellation of the warnings. Weather prediction then halted altogether, as Farrer noted: "We cannot appoint a new prognosticator when we are told on the highest scientific authority that there is no grounds for prognostication."[106] Farrer, who had been objecting to the Meteorological Department's escalating budgets for the preceding five years, was relieved but irritated by the report and its aftermath. Spared the prospect of placing "our concerns and the public purse blindfold in the hands of another Admiral Fitzroy," he was nevertheless frustrated that the Royal Society had not seen fit to express its views so clearly at an earlier stage.[107]

Yet the concerns of the 1866 report with the proper scope of individual judgment and responsibility echoed those expressed years earlier, when Fitzroy's experiments began. After two years of forecasting, at the end of 1862, officials

104. Ibid., 34, 24.

105. Ibid., 38.

106. Thomas Farrer memo, November 18, 1866, Met. Office Papers, PRO MT9/29/W4962.

107. Farrer's memo commenting on Royal Society decision to stop warnings, October 27, 1866, Met. Office Papers, PRO MT9/29/W4653.

at the Board of Trade had considered the idea of taxing ports or ships for the meteorological information, a contribution principle similar to that guiding the establishment of lighthouses and lifeboats. Such a tax would rein in costs *and* the role of government; it was dropped as a scheme more complicated to implement than the revenue would justify.[108] Some of Fitzroy's earliest correspondence about storm warnings with Farrer had explored the question of individual versus official responsibility. Fitzroy was sensitive to the question of authorship and proposed that forecasts be signed with the initials of those who made them, "to avoid compromising the Higher Authorities who might thus sanction such experiments."[109] As the prediction service emerged, Farrer suggested a further distinction, between "facts" and "interpretation." The government, that is, Fitzroy's department, could enter only into the "correct relation and transmission of facts." Fitzroy, acting in a private capacity, or as "a member of the British Association or any other scientific body," could interpret these and publish the warnings—but preferably not on notepaper headed "Board of Trade." Farrer noted with good humor that "when the thing has become acknowledged and absolutely certain like the eclipses or the tides, the President of the Board of Trade will be glad to steal from Admiral Fitzroy the name and credit of the thing." Yet the distinction between the individual and the office could seem overly subtle. As Fitzroy commented in stiff tones to Farrer, "Responsibility I have never shrunk from. My humble name may be attached to what I believe true and right." But, he continued more warmly, he could not understand "why more credit should be given to me individually than is due—seeing that I am nothing except by my office here."[110] From the first, this point had been obvious to those interested in the development of meteorological science. John Locke, a Dublin gentleman endorsing storm telegraphy, wrote to Fitzroy in 1859, entrusting his own "isolated and humble aid" to "*your* hands with the advantage of your scientific ability, official position, and 'Office force.'"[111]

In an especially striking move, the question of individual judgment also

108. Met. Office Papers, PRO MT9/32/M595 is a file of exchanges between Thomas Farrer, Milner Gibson (president of the Board of Trade), and Fitzroy from December 4, 1862, to January 21, 1863. Fitzroy wanted a graduated tax on vessels, collected through the customs officials; Farrer and Gibson wanted an annual port tax of fifteen pounds. Both plans were dropped.

109. Robert Fitzroy memo to Thomas Farrer, May 4, 1860, Met. Office Papers, PRO MT9/12/5380.

110. Thomas Farrer to Robert Fitzroy, February 16, 1861; Robert Fitzroy to Thomas Farrer, February 18, 1861, Met. Office Papers, PRO MT9/13/M2883.

111. John Locke to Robert Fitzroy, December 13, 1859, Met. Office Papers, PRO BJ7/726, original emphasis.

merged intellectual authority with economic ideology. Masters and owners of ships, for example, occasionally complained that Fitzroy's predictions represented an effort to tamper with individual market decisions. Similar concerns had surfaced with regard to private weather predictions in the 1840s, when John Hind, the young almanac writer discussed in chapter 2, complained,

> It is really ridiculous to hear the length to which the sceptical, on point of predictory science, carry their prejudice. Our scientific Nottinghamites seem to imagine that the discovery of a true theory of the weather will produce bad effects on trade, causing almost a "stagnation." Not thinking it worthwhile to reply to these philosophical reasonings I have continued my predictions in face of all & they probably feel not a little chagrinned, that the Nottingham trade should thus be placed in imminent danger of depression or stagnation! [112]

Twenty years later, Fitzroy took similar accusations remarkably seriously and defended himself against the free trade lobby explicitly. For those without other sources of knowledge, he asserted, "an idea of the kind of weather thought *probable* cannot be otherwise than acceptable provided that he is in no way *bound* to act in accordance with such views, against his own judgement."[113]

The efforts to find a solution to the sensitive question of responsibility for weather prophecy continued after 1866. One of the first and most revealing decisions was to avoid another director like Fitzroy. Following the recommendations of the 1866 report, the Board of Trade turned meteorology over to the management of a committee of the Royal Society. This committee designed the role of the Meteorological Office (as it was renamed) to increasingly resemble that of a subordinate calculator or machine for creating observations. Interestingly, Glaisher's application for the post of director, which he made a few days after Fitzroy's death, was dismissed. He, like Fitzroy, had a leader's and not a follower's character.

The significance of the shift in management of the office was obvious to contemporaries. The public response was to view the cancellation of warnings and forecasts as competition between Fitzroy and the Royal Society. The popular scientific press was also almost universally hostile to the decision. The *English Mechanic* denounced those who attacked Fitzroy before his body was scarcely cold ("the gnats have stung a noble being to death"), and emphasized that the admiral's work had saved hundreds of lives.[114] G. F. Chambers, a

112. J. R. Hind to W. H. White, March 28, 1840, Royal Meteorological Society Papers.

113. Fitzroy, *Weather Book*, 190, original emphasis.

114. "Meteorological Science," *English Mechanic*, June 23, 1865, 147–48.

well-established science journalist, deplored the summary action of the committee and blamed the "party who looked upon Fitzroy as a meteorological poacher."[115] Marine associations and insurance institutions and chambers of commerce in Liverpool, Manchester, Dundee, Glasgow, Edinburgh, and Leith also all protested the decision.[116] Scientific support for forecasts came usually from provincial circles, also seeing the decision as part of the program of a centralized scientific elite in London. Joseph Baxendell, a prominent member of the Manchester Literary and Philosophical Society, and astronomer to the city observatory, pinned the blame firmly on the Royal Society. The BAAS meeting the following year, at Dundee, became a focus for protests.[117] The main charge was led by Col. William Henry Sykes, the Liberal member of Parliament from Aberdeen (1857–72), a founding member of the Statistical Society, and a fellow of the Royal Society (1834). Echoing Baxendell and others, he called the reformed Meteorological Office a creature of the Royal Society, denouncing the "pedantic affectations" of that "coxcombry of science."[118] Such responses clearly recognized, and resisted, the claims of scientific elite to define science in an increasingly exclusive manner.

Under this barrage of protests, the Board of Trade in its turn began to pressure the Royal Society Meteorological Committee to restore signals.[119] By November, the committee agreed, but insisted on presenting a distinction between fact and interpretation—"we intend to telegraph actual *facts*, not *prophecies*," wrote Fitzroy's successor, Robert Scott.[120] It was a dubious compromise. Indeed these distinctions, intended to address concerns over judgment that had plagued Fitzroy, soon approached farce. The official memorandum issued to the public declared that the signal was "only intended to convey the information that there is an atmospheric disturbance somewhere which may possibly reach the place where the signal is hoisted."[121] In addition, information about

115. Chambers, "Fitzroy's Weather Forecasts," 265–66. Chambers repeated his denunciations decades later in *The Story of the Weather*.

116. See *Communications to the Board of Trade*, 185–205 (Parliamentary Papers); and Cooke, "Storm Signals and Forecasts."

117. The "exceedingly practical frame of mind" of Scottish philosophers (according to the local paper) was emboldened by the presidential address of the Duke of Buccleuch who strongly urged the resumption of storm warnings. Symons, "Meteorology at Dundee," 101; *Dundee Courier*, September 5, 1867, 3.

118. Symons, "Meteorology at Dundee," 101.

119. *Letters (in Continuation . . .)*, 393–94 (Parliamentary Papers).

120. Ibid., 413–14, original emphasis.

121. Ibid., 400–401.

the wind direction would no longer be issued, and the signals were to remain hoisted for a shorter period (thirty-six hours instead of Fitzroy's seventy-two) after the warning arrived. But this contrast between fact and prediction was manifestly weak. A warning in 1864 and one in 1868 sprang from the same kinds of observations: it was simply the way of representing the information that had changed. Moreover, since the committee used Fitzroy's signals and signal posts, the distinction between information and prophecy that the committee struggled to maintain was visually impossible. It must have been obscure to much of their audience, as Glaisher had anticipated at the Dundee meeting in 1867. "As to sending down the results of the meteorological observations to the coasts, [I] would say, for God's sake, do not do it. They would be read in a dozen different ways and would only lead to mischief."[122] Faced with the Royal Society committee's memo, the Edinburgh Chamber of Commerce agreed. Instead of "the intelligent readings and comparisons of barometrical disturbances and their application to practical use," the harbor now received "a confused record of high winds at various places[,] the whole signifying nothing. Can Meteorology do nothing more than this?" they asked.[123]

Conclusion

In 1853, as government meteorology was being organized, George Airy, the Astronomer Royal wrote that it was "totally unimportant to what Department of Government this is attached, its work will be so different from that of any official bureau that it will have no connexion with the bureau except in applications for money and authority, and in making reports—the *man* will be the only important selection."[124] This chapter has claimed that, in many respects, Airy was wrong. The development of forecasting in Britain undoubtedly owed much to a single powerful personality, Fitzroy, as the first head of the Meteorological Department. Yet, while it often chose to portray itself in terms of a history of individual intellects, in fact Victorian science struggled with the implications of collective science: science organized in groups, in official or semi-official bodies. In meteorology, formidable men such as Fitzroy, Glaisher, and Symons worked in clusters of institutions jostling one another for space, funds, and public attention. The leaders in the scientific community

122. This was a problem noted by *Scientific Opinion*, May 19, 1869, 543.

123. R. M. Smith to Charles Piazzi Smyth, March 6, 1868, Smyth Archive, A19/142.

124. George Airy to Capt. Henry James, October 13, 1853, Met. Office Papers, PRO BJ7/109, original emphasis.

as a whole explored a full range of roles: as voices of authority and progress under the banner of the BAAS meetings; as servants, issuing guidelines on scientific standards in response to government appeals; as pressure groups, pushing the government to take an interest in marine meteorological observations; as technicians, running the standard instruments department at Kew Observatory; as advisors and critics, serving on committees of inquiry. The variety of these positions was an expression not only of the richness of public life in the Victorian period, but also of an uncertainty about how to establish and maintain intellectual authority.

Meteorological networks seemed to confirm the difference between "real knowledge" of coordinated observations and the ill-founded pretence of individual speculations, such as those found in the almanacs of weather prophets. Collective science thus embodied the hierarchical ideals of science, distinguishing between mere observers and scientific thinkers. Yet shifting meteorology decisively into the public realm, as a consequence of the formation of a government department, brought out the challenges of collective science more openly. These challenges were not simply those of coordinating data among scattered groups. They included the struggle to distinguish fact or observation from interpretation, a distinction that shaped the social organization of meteorological networks. Forecasting muddied any sharp lines here, as those who wanted to preserve science from the uncertainties of weather prediction reframed the interpretation of facts as a matter of individual judgment rather than as a higher order of scientific reasoning. This move confirmed the tight links between personal character and intellectual authority that shaped scientific culture, as can be seen in the responses to Fitzroy's suicide. But at the same time, it opened up grounds for a critique of the scientific elite.

After 1867, the Royal Society committee, anxious to shape Fitzroy's department into a new model of state science, continued to wrestle with the public dimensions of meteorology. Quantitative knowledge and instrumental precision were the bricks of their new office. Yet applying these authoritative methods of modern science to the study of the weather was not straightforward. It led directly to further struggles with numbers and measurement, as seen, first, in the complicated status of a science of probabilities in the 1860s and 1870s and, second in debates over what kind of evidence quantitative meteorology neglected.

Precision and a Science of Probabilities

IN STRIKING contrast to the drama of the weather, meteorology required an overwhelming and monotonous registration of numbers. Reviewing the science in 1855, the Edinburgh natural philosopher David Brewster began with a portrait of weather as an emotional chart. "The scorching heat of summer, the biting cold of winter, the rain with its floods, the snow with its avalanches, the tempest with its thunder and its lightning—how many associations do they embosom, how many hours of joy, of disappointment, of grief, do they recall!" But he turned immediately to the painstaking work of observation, extolling an observer who carried out "hourly meteorological observations for *twelve years*, from 1828 to 1842."[1] What did such dedication to a quantitative account of nature mean? For a start, the contrast hid a fundamental similarity: regularity, routine, and detail were in themselves so praiseworthy that the grandeur of weather and a record of its minutiae inspired the same sort of awe. The slide from weather's passions to hourly observations in Brewster's account, then, would have been much less jarring to a contemporary audience than it is for the modern reader. But beyond the merit attached to self-discipline and archives, numbers summarized the ideals and methods of science. "Number, weight, and measure are the foundations of all exact science"; wrote Herschel in 1830, "neither can any branch of human knowledge be held advanced beyond its infancy which does not, in some way or other, frame its theories or correct its practice by reference to these elements."[2] Observatories, standardized

1. [Brewster], "The Weather and Its Prognostics," 173, 175.

2. Herschel, "[Review of Quetelet's] Letters on Probability," 41.

instruments, and trained observers together could manage a flood of particulars and would demonstrate that the weather followed simple, uniform, and universal laws. Numbers thus provided weather with a plot and gave the study of weather its setting, props, and actors. After Fitzroy's death, the reform of the Meteorological Office focused on accumulating accurate observations. A system of model observatories with standardized instruments and trained staff replaced the unwieldy network of unsupervised observers. Yet the program of precision observation that dominated meteorology after 1867 had the paradoxical effect of exposing its own weaknesses. Where we might expect to see the most confidence, the history of instruments and numbers in Victorian meteorology often shows men of science on the defensive, trying to define natural evidence on their terms in the face of public criticism.

The most obvious weakness centered on the nature of precision itself. Would exact observation forward the progress of meteorology, or was the science accumulating more numbers (at the taxpayer's expense) to molder in dusty volumes? These were not new concerns in meteorology. Writing to Maury in 1855, Fitzroy indicated the hazards of exact observation. "While we find the best standard barometers of Government—London, Greenwich, Kew, &c., scarcely agreeing with each other to the second place of decimals, can we conscientiously ask a seaman to be excessively scrupulous about the third place while observing a Marine Barometer—perhaps [while] pumping in a seaway?"[3] Fitzroy returned frequently to this complaint. In a popular article of 1860 he commented that to require "a perfect barometer" for weather observations "at a life boat station or fishing village might remind one of putting a racehorse in a cart, or using a razor to cut sticks."[4] Precision, Fitzroy thought, should be a relative, not an absolute, value. Fitzroy's references to seamen and racehorses emphasized a divide between popular and expert approaches to meteorology, but dissent about precision observations was equally acute within the scientific community. The reforms of 1867 drew on long-standing support for a program of expert observations: in British meteorology, that program was associated first with James Forbes's report on meteorology to the BAAS in 1832 and second, with the international project of physical observatories to record magnetic and meteorological data (the so-called magnetic crusade), which got under way in 1839.[5] Within scientific circles, the opposing position

3. Robert Fitzroy to Matthew Maury, September 19, 1855, Met. Office Papers, PRO BJ7/77.

4. Fitzroy, "On Weather Glasses," 329–30. Cf. Fitzroy, *Weather Book*, 7.

5. Forbes, "Report on Meteorology," *Report of the BAAS*, 1832, 196–207; Cawood, "The Magnetic Crusade."

can be represented best by a debate in the French Academy of Sciences, which Brewster's article summarized for his English readers shortly after the event in 1855. Pouring cold water on plans for an observatory in Algiers, a commission of academicians dismissed hourly observations as excessive and treated the photographic instruments pioneered at Kew as useless novelties. In the heated debate that followed, the mathematical philosopher Jean-Baptiste Biot referred to the fruits of recent meteorology as "large and very expensive quarto volumes filled with ciphers." Observatories, claimed Biot, cannot advance "fundamental questions of scientific meteorology"; still less do they address the practical demands of farmers.[6] This characterization of meteorology made it clear that systematic observations aroused as much distrust as zeal. In that sense, the observatory-centered reforms in Britain in the 1860s and 1870s only gave wider room for a critique of precision.[7]

This critique of course addressed in the first instance the classic problem of induction, or how to move from particulars to general laws. But embedded within this general problem were two other sorts of concerns: the public role of numbers and the support of quantitative forms of knowledge for a deterministic view of nature. Both of these concerns, as well as the general problem of particulars, developed most obviously in the context of a special form of exactness and numbers: statistics. Meteorology in Britain was peculiarly able to expose these unsettling aspects of statistical science. The office was manifestly a public one, developing a new relationship between the state and its scientific advisors. As a science of model observatories and precision instruments, it presented evident scope for debate over how facts would lead to natural law. And, last but not least, with weather as the epitome of chance and uncertainty, meteorology proved to be ideal grounds for arguments about determinism. Meteorological controversies threw light on the foundations and implications of statistical science, and statistics in turn explained why numbers and precision so readily overflowed into questions of authority, politics, and faith.

In order to explore the place of numbers in meteorology, then, it is necessary to look to changing ideas about statistics and probability. During the key years of the establishment of meteorological science in Britain, a science of probabilities was almost an oxymoron. Probability underwent a transformation in

6. [Brewster], "The Weather and Its Prognostics," 182–83.

7. For related discussions of precision in nineteenth-century science, see Gooday, "Precision Measurement"; Porter, *The Rise of Statistical Thinking* and *Trust in Numbers*; and Wise, ed., *Values of Precision*.

the first half of the nineteenth century, shifting from a measure of uncertainty to new associations with a measure of exactness. The key was the application of the mathematical calculation of chances to astronomical observations. This calculation, known variously as the astronomical error law, the method of means, or the method of least squares, grouped and measured variations in observations, allowing the astronomer to determine the most accurate position of the object under study. In the 1820s, a young Belgian astronomer studying in Paris, Adolphe Quetelet, became powerfully impressed with the error law. He translated the concept into a tool for tracing regularities in all sorts of phenomena, natural and social, when discrete individuals or records were examined collectively. That is, rather than studying variation to pinpoint the position of a star, he studied variation to detect the underlying patterns of order that were revealed. (By the end of the century, this law had acquired its modern names: normal distribution and standard deviation.) Quetelet's work laid the foundation for a new orthodoxy, in which statistics adopted the mathematics of probability not to measure chance but to detect order. Its governing principle, as the historian Theodore Porter has noted, was that "the greatest confusion at one level is not only consistent with but implies remarkable stability at another, which manifests itself in the form of statistical laws."[8] Quetelet thus supplied an answer to the problem of what to do with masses of particular facts.

The application of probability to statistical knowledge was one of the most critical intellectual developments of the nineteenth century. But its appeal depended on ignoring the ambiguity of the relationship between statistics and probability. In its new statistical uses, probability served to eliminate variation and promote exactness. But what about chance? Regularities, natural laws, and precision cut right across the older associations of probability with uncertainty. In the terms of mathematical philosophers of the previous century, probability was the reasonable calculus, a technique for developing rational judgments when our knowledge was incomplete and the outcome of events unknown. It only produced an expectation, or estimate. Probability in this sense developed as a self-conscious mathematical abstraction of common sense. James Clerk Maxwell reintroduced uncertainty into statistics only in the 1870s, arguing that because molecular movements were statistical, they were indeterminate and this indeterminacy characterized our knowledge of nature. Before the end of the nineteenth century, then, the broader awareness of statistics rested on

8. Porter, *The Rise of Statistical Thinking*, 70.

calculations of chance, on the one hand, and the novel regularities associated with Quetelet's social science, on the other. Meteorology exposed the tensions in Victorian statistics because the study of the weather fit squarely into both traditions. It was an ideal subject to tackle statistically, that is, to examine masses of individual events for evidence of an underlying order or regularity that would guide meteorologists to the natural laws. But weather also seemed to be exactly the kind of situation in which to apply disciplined reason, and thus protect oneself from doubt and confusion. The title page illustration to Augustus De Morgan's *Essay on Probabilities* (1838) recalled weather in just this way, picturing a woman seated in a harbor, with sextant and compass at her feet, looking over her shoulder at two ships on a stormy sea (fig. 4.1). The ships face the quintessential variability of wind and wave, but observations of science and the methods of probability, however indirect (like the woman's glance), will control our uncertainty.

The status of statistical knowledge mattered because it was inseparable from issues of public authority. Numbers were conceived of as a public form of science, and thus held all the appeal and liability of publicity. The general foundation for the connection between numbers and public authority was of course apparent in the etymology of "statistics": information for the state. More subtly, the use of statistics in science embodied a related conception of public and private authority. Precision observation was valuable because it suppressed private judgment and restrained speculation. Instruments that recorded mechanically, like a wind vane hooked up to an unrolling paper trace, were exact because they erased the presence of the observer. Similarly, when it was discovered that every individual astronomical observer made characteristic mistakes—called the personal error—the answer was to multiply observations and apply statistical calculations to the mass, in order to smooth out the error. A science of numbers, then, moved knowledge away from subjective perspective of an individual and into the realm of public judgment. Herschel, again, one of the most influential and thoughtful writers on scientific observation, nicely expressed this ideal in his article on statistics for the *Edinburgh Review*. "Publicity is the sine qua non of statistical science," he wrote, opening subjects to "the broad good sense of the thinking part of mankind" and clearing away "*professional* error and prejudice."[9] The objective, public qualities of the scientific enterprise—that is, its numerical data—defined

9. Herschel, "[Review of Quetelet's] Letters on Probability," 54, original emphasis. Cf. the studies of precision observation collected in Wise, ed., *Values of Precision*.

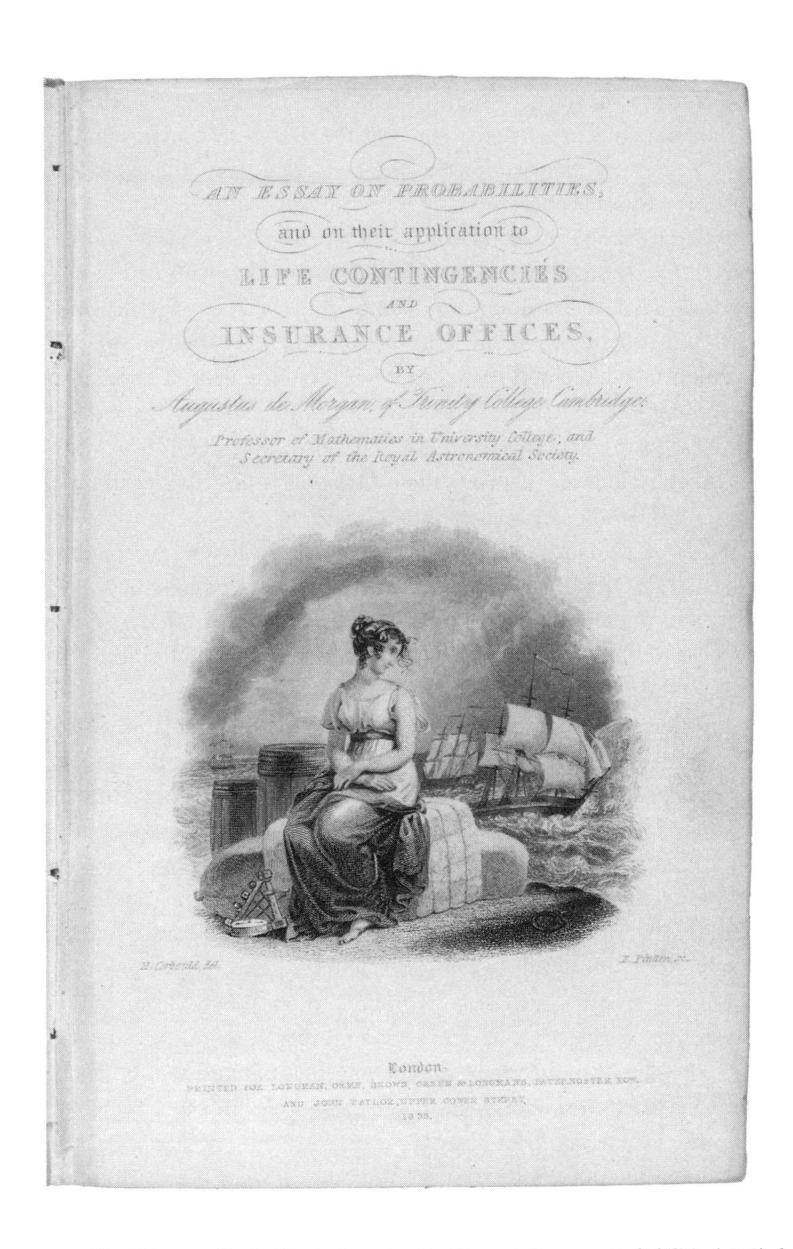
AN ESSAY ON PROBABILITIES,
and on their application to
LIFE CONTINGENCIES
AND
INSURANCE OFFICES,
BY
Augustus de Morgan, of Trinity College Cambridge:
Professor of Mathematics in University College; and
Secretary of the Royal Astronomical Society.

London:
PRINTED FOR LONGMAN, ORME, BROWN, GREEN & LONGMANS, PATERNOSTER ROW,
AND JOHN TAYLOR, UPPER GOWER STREET,
1838.

FIGURE 4.1 The title page illustration to Augustus De Morgan's *Essay on Probabilities* (1838) shows a young woman, seated on a bale of goods, with sextant and compass at her feet, looking over her shoulder at ships being tossed among the waves. The image suggested the indirect but scientific knowledge that probability would provide about the variations of nature and the economy. (Augustus De Morgan, *Essay on Probabilities and on Their Application to Life Contingencies and Insurance Offices* [London: Longman, 1838]; Thomas Fisher Rare Books Library, University of Toronto.)

its capacity to inform and guide the government. Statistics, then, showed the tight connections between scientific method and the larger claims of authority made by scientific culture.

The idea that the meteorological program after 1867 simply piled up useless facts was thus a profoundly damaging accusation that used but neatly reversed the values of exact science. Numbers offered the most authoritative kind of knowledge, but exposed science to independent scrutiny that might not be sympathetic. (When Herschel praised "the watchful inspection of the laity" that statistical science made possible, he was thinking of banking and insurance rather than the Royal Society as the objects of this attention.)[10] As meteorological statistics became increasingly identified with a hostility and resistance to practical knowledge such as forecasting, this split the connection between numbers and the idea of public value. The arguments about forecasting explicitly pitted the pursuit of strict observation against pursuit of utility. This was exactly the sort of public pressure and prospective loss of control that alarmed many in the scientific community. Similarly, meteorology's requirement of precise observations suggested to its critics a mindless, mechanical pursuit. As central as exact numbers were to the claims of scientific culture, the debates about meteorology showed how a focus on numbers could simultaneously undermine its dignity and expertise.

Statistics also notoriously identified scientific culture with a deterministic picture of nature and society. Here, too, meteorology was especially illuminating. Although meteorologists could and did identify regularities in their data with enthusiasm, they just as routinely acknowledged that the science was a long way from uncovering the natural laws of the atmosphere. Thus it was particularly difficult to convince critics that more observation would result in such laws, and particularly attractive to turn meteorology into a kind of last stand. Weather prediction, as usual, was a key part of such arguments. The unpredictability of weather could represent an order of natural event that resisted deterministic explanations. In addition, the weather offered a symbolically discrete event, like a single fact pulled from a bewildering succession of facts, or an extreme before it has been swallowed into an average. Both sides of the debate sought to characterize the relationship of small event to large patterns, while drawing opposite conclusions about inevitability and natural order. The debates about weather and natural law, therefore, nicely illustrate how in popular terms determinism merged with the problem of induction, the move from particular event to general laws.

10. Herschel, "[Review of Quetelet's] Letters on Probability," 54.

The zeal for quantitative observation and precision instruments appears to be one of the givens of nineteenth-century science. The history of meteorology in a sense bears this out amply. For both critics of the Meteorological Office and its supporters, precision observations and what to do with them was the key to the development of the science. However, meteorology also showed that skepticism about the value of numbers was equally common, and indeed that it arose from the tensions within the scientific ideal of quantification. This skepticism is particularly evident in an examination of the status of statistical science and the uneasy position of probability, both of which were clearly critical to the development of forecasting practices. It is not surprising, then, that arguments about instruments and numbers punctuated meteorological practice in the period 1860 to 1880.

Meteorologicum Odium: Instruments and Observations

In the wake of the 1866 report that had denounced Fitzroy's procedures, the department was renamed the Meteorological Office, and its management was assumed by a committee of the Royal Society.[11] Its reforms sought to transform the government meteorological department into a body that would exemplify the highest standards of scientific observation. The committee ended forecasts and shifted the responsibility for preparation of navigation charts based on the ocean data to the Hydrographic Office in the Admiralty. Instead of these practical aims, the committee concentrated on the establishment of six observatories to take land meteorological data, with Kew Observatory as the centerpiece of the system. In practice, the number of observatories soon climbed from six to eight: Kew and Greenwich plus six others, at Falmouth, Valencia, Aberdeen, Armagh, Glasgow, and Stonyhurst (fig. 4.2). The new budget reflected the priority of observations over any kind of practical work. The committee requested £4,250 for the six model observatories, plus an initial outlay of £2,500 for buildings and equipment. Their budget included a further £3,200 for ocean statistics and their analysis, designed to be completed in fifteen years, and £3,000 for "meteorological telegraphy" (telegraphic costs for the exchange of data, nationally and internationally, but not including predictions or warnings). The total of close to £11,000 annually was much

11. The initial committee was chaired by the Royal Society president, Edward Sabine, and also included Galton, Gassiot, Warren De La Rue, William Spottiswoode, Col. W. J. Smyth, and W. A. Miller. Charles Wheatstone joined the group after Miller's death in 1870, and Spottiswoode resigned in 1873.

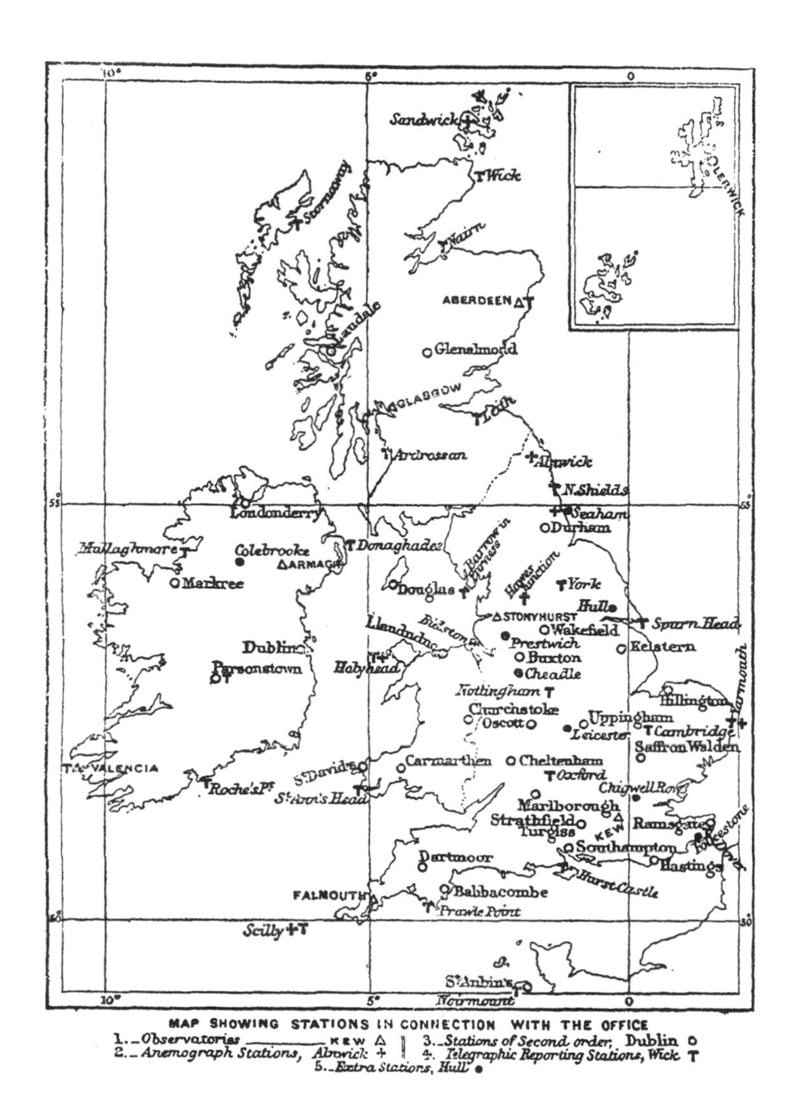

FIGURE 4.2 A map of 1880 shows the observation network of the reformed Meteorological Office. The seven main observatories, in Valencia, Armagh, Glasgow, Aberdeen, Stonyhurst, Kew, and Falmouth, are marked with a triangle. Stations that telegraphed reports to the central office are marked with a *T*. (*Report of the Meteorological Council for 1879–80*, 1881 (2741) XXXVII, 699 [Parliamentary Papers].)

higher than any of Fitzroy's budgets. But the office's scientific advisors insisted that exact observation was the necessary first step to any progress in the science. "The Committee are of the opinion that the £3,000 now annually spent on Telegraphy and Storm Warnings will be in a measure thrown away, on an unsound and practically useless system—unless the sum of £4,250 be incurred annually for placing it on a scientific and wholesome foundation."[12] The observatory network and its instruments would provide the pillars of a reformed meteorology.

These changes expressed two principles of modern science: a commitment to precision observation and its right to support from public funds. By any count, the Meteorological Office budget in the 1870s ranked high in the hierarchy of government spending on science, and if those monies were combined with the costs associated with meteorological work at Greenwich Observatory or the General Register Office, meteorology was outspent only by the large national and international surveying projects and by Kew Gardens (see table 4.1). The direction taken by meteorological research thus came to represent to critics the arrogance and expense of a centralized scientific elite. And indeed the Meteorological Office did become critical to scientific leaders as they claimed increasing political and cultural authority in the decade after Fitzroy's death. From its formation in 1854, the Meteorological Office had been a product of scientific lobbies. When Lord Aberdeen's government, preoccupied with developments in the Crimea, showed little interest in sending British representatives to the 1853 Brussels conference on ocean meteorology, a parliamentary committee in the BAAS exerted pressure that changed the decision. After the conference, Edward Sabine and Lord Wrottesley, both future presidents of the Royal Society, had a hand in setting up the Meteorological Department.[13] But in 1867, meteorology began to be perceived by the government as well as the public as "Royal Society business." Francis Galton in a private letter to Sabine explained that the Meteorological Office could be a kind of "*branch office* in London" for Kew Observatory and added that, "moreover, it would be a London office for the British Association which would be useful in many ways."[14] At the Board of Trade, Farrer remarked with satisfaction that "we . . .

12. Admiral G. H. Richards, "Memorandum on the Report of the Committee on the Meteorological Department of the Board of Trade," August 6, 1866, Met. Office Papers, PRO MT9/44/M2937.

13. Robert Fitzroy to Lord Wrottesley, February 3, 1858, Met. Office Papers, PRO BJ7/390; Burton, "Robert FitzRoy," 150–52; "Correspondence between the Board of Trade and the Royal Society."

14. Francis Galton to Edward Sabine, March 16, 1866, Royal Society of London, Sa 586. In the 1871 transfer of the observatory from the BAAS to the Royal Society a new Royal Society Kew Committee

TABLE 4.1 Government spending on science (in pounds sterling per annum)

Topographical Survey	32,000
Hydrographical Survey	121,000
Geological Survey	22,920
Botanic Gardens, Kew	17,572
Meteorological Office	10,000
Royal Observatory, Greenwich (astronomy)	5,642
Observatory, Cape of Good Hope	3,371
Royal Observatory, Greenwich (meteorological and magnetic department)	1,221
Royal Observatory, Edinburgh	805
General Register Office (meteorological observations)	150

Note: The table shows the relative position of the Meteorological Office: it received nearly £4,000 more than the Greenwich Observatory, but considerably less than Kew Gardens or the geological and topographical surveys. By comparison, in 1874, the budget for the national service in the United States, under the Army Signal Office, was $400,000 (equivalent to nearly £72,000). Data amassed by the Devonshire Commission in 1874–75 (*Royal Commission on Scientific Instruction and the Advancement of Science* [Devonshire Commission], *Fourth Report*, 1874 [884] XXII, 1; and *Eighth Report*, 1875 [1298] XXVIII, 417 [Parliamentary Papers]).

shall have nothing to do but to distribute the instruments, collect the logs and send them to Kew."[15] The new relationship reflected other developments in scientific policy. A series of government papers from 1871 to 1875, known as the Devonshire reports, had urged the reform of science education, recommended public support of research institutions, and proposed a separate Ministry of Science. Consciously or unconsciously, the administrative arrangements for the Meteorological Office in many ways echoed proposals for a Ministry of Science, or a Government Research Fund, both of which were beginning to be discussed by the Devonshire Commission.[16] Meteorological affairs in the 1870s

emerged to manage the observatory: it was essentially the Meteorological Committee under another name, as the membership was identical. Burton, "History of the British Meteorological Office to 1905," 102.

15. Farrer's note on Richards, "Memorandum on the Report of the Committee on the Meteorological Department of the Board of Trade," Met. Office Papers, PRO MT9/44/M2937. The Royal Statistical Society also saw things this way: drawing up a list of government support for science in 1870 it gave the Royal Society £11,000: 10,000 for the meteorology committee and 1,000 for the grant fund (Guy, "On the Claims of Science to Public Recognition and Support," 446).

16. The Government Fund, at £4,000 per annum, was administered through several committees of the Royal Society. For details, see MacLeod, "Support of Victorian Science," reprinted in *Public Science and Public Policy*.

therefore mapped out the movement that Victorians called "the endowment of research."[17] Under the circumstances it was not surprising that much of the animosity about the decision to cancel the storm warnings and forecasts in 1866 focused on the relationship between science and the government.

Two public inquiries into the Meteorological Office, in 1866 and in 1877, made the science a catalyst for discussion of scientific methods and values. Critics questioned the decision to build and equip new observatories, arguing that what the Greenwich and Kew observatories had not achieved in decades of observations was unlikely to be suddenly discovered in the next fifteen. Joseph Baxendell, for instance, wrote a public letter to the Royal Society after the cancellation of warnings late in 1866. Baxendell, secretary of the Manchester Literary and Philosophical Society (1861–85) and superintendent of a local astronomical and meteorological observatory at Fernley, described the Galton Report as "but an extended echo" of Royal Society opinions. He asserted that the society neglected the work done by many provincial observatories like his own and, most significantly, ignored the Royal Greenwich Observatory.[18] Similarly, the astronomer Richard Proctor, one of the principal opponents of public funding for science, argued that the committee perverted the meaning of utility when they vaguely spoke of "future" benefits while pouring money into the observatory network. At present, Proctor emphasized, the daily reports and charts were "utterly useless to the community . . . Our authorities, apparently unconscious of the inane absurdity of the proceedings, inform us placidly day after day of the weather of the day before, giving no hint whatever as to the weather probably approaching."[19] *Knowledge*, a journal Proctor edited, suggested "payment by results" for the Meteorological Office. It was a sarcastic allusion to the government's current method of funding scientific and technical education, with funds apportioned to schools based on exam results.[20]

As Baxendell indicated, the relationship between Kew and Greenwich became a significant part of the debate over meteorology. The acrimony in that

17. See Pattison, ed., *Essays on the Endowment of Research*; MacLeod, "Support of Victorian Science," in *Public Science and Public Policy*; Proctor, *Wages and Wants*.

18. *Manchester Courier and Lancashire General Advertiser*, March 26, 1867, 6. See also Baxendell, *On the Recent Suspension by the Board of Trade of Cautionary Storm Warnings*, a talk given to the Manchester Literary and Philosophical Society, December 11, 1866.

19. Proctor goes on to praise the contrasting system in the United States (Proctor, *Wages and Wants*, 87n).

20. Proctor, "Support for Science," 207. Cf. Appleton et al., *Essays on the Endowment of Research*, 96.

relationship—which one suffering mediator called the "odium meteorologicum"[21]—was revealing. It showed how standard instruments raised more questions about precision than they put to rest. Airy's quarrel began in 1862 as a disagreement about Sabine's magnetism work at Kew. In 1864 Airy pointed out that a set of extremely uniform results published by Sabine were likely to indicate "delusion on the part of observers." As Airy's own work at Greenwich had demonstrated, every observer generated a particular individual error that had to be calculated into the measurements.[22] Since Sabine's results showed none of this kind of variation, Airy implied that Sabine's too-perfect results indicated falsification. The accusation resurrected similar criticism by Airy about Sabine's work on terrestrial magnetism many years earlier. Airy tackled Sabine again in 1866, after Sabine published a summary of British meteorological work in the *Proceedings of the Royal Society* that was timed to reinforce the forthcoming report on the department, and especially to support the proposal to develop the network of model observatories centered on Kew Observatory. Airy took strenuous exception to Sabine's picture of meteorology as a science wasting from neglect of modern instrumentation when Greenwich had possessed self-recording instruments since the 1840s. This was, Airy thought, "wilful suppression" of the facts.[23] Airy and Sabine, it will be clear from these exchanges, had a history of bad feeling, but the quarrel was deeper than personalities. In attacking Sabine in 1866, Airy joined the critics of endowment of research: in 1871, he suggested that meteorological costs could be cut by using Greenwich observations *gratis* instead of paying Kew, and he refused to acknowledge the independence of the Royal Society committee, calling it "virtually a committee of the [Royal Society] Council."[24] But more significantly, Airy's complaints expressed his reservations about precision observation as an end in itself.

The antagonism between the Astronomer Royal and the Meteorological Office bubbled up a few years later. In 1877, a new government inquiry into

21. George Airy to Henry Smith, January 18, 1879, Airy Papers, RGO 6/704.

22. See Airy, *Autobiography*, 247–48; on precision instruments and characteristic error, see Schaffer, "Astronomers Mark Time."

23. George Airy to George Stokes, March 31, 1866, Royal Society of London, MC 7/317; George Airy to George Stokes, January 31, 1866, Airy Papers, RGO 6/393/37. Smith, "A National Observatory Transformed"; Chapman, "Private Research and Public Duty."

24. These matters are covered in Airy's correspondence (Airy Papers, RGO 6/394). See also a discussion of Airy's interference with the Meteorological Office in 1871–72 in Burton, "History of the British Meteorological Office to 1905," 113–16.

meteorological work was convened, prompted by public complaints about the expense and value of the Meteorological Office's work in the last decade. Specifically, the inquiry demanded whether "the appropriation of a large sum of public money in aid of meteorology is justified, bearing in mind that it is not the policy of the Government in this country to give direct assistance to the study of any science, except with a view to the more immediate application of scientific theories to practical purposes in which the Public rather than individuals have an interest?" In other words, were the Royal Society reforms in the direction of abstract science justified?[25] Galton and Farrer were members, as was General Strachey, and J. D. Hooker, director of Kew Botanic Gardens. Given the make-up of the committee, perhaps it was not a surprise that the report that emerged in 1877 smoothly turned into an endorsement of current policies. Despite the strong criticism of the observatory program by several witnesses, the report did not suggest reductions. Instead, it recommended a new £1,000 research fund.[26] In effect, the committee ignored the hostile implications of the official question prompting the inquiry. After 1877, a new Meteorological Council—different from the Royal Society Meteorological Committee only in name and the provision for payment of members—shifted ocean studies entirely over to the Admiralty and administered an increased grant with new provisions for research projects. Airy's opinion on these changes found voice in a new argument about Kew's standard instruments. The director of the Meteorological Office, Robert Scott, had suggested that the Royal Observatory thermometers were "sluggish" and described how Greenwich observations could be made to conform with Kew standards. Airy abruptly cut off Greenwich correspondence with the Meteorological Office and fumed in much the same terms as he had eleven years earlier. Scott had deliberately and publicly rebuked Greenwich data, Airy felt, and he dramatically claimed "there has never been such a transaction in the history of science."[27] In part, of course, Airy was again defending the importance of Greenwich, but Airy was also arguing that meteorology did not need more observations. Indeed, too much precision, as in Sabine's magnetic observations of 1864, was in itself suspicious. Far from solving meteorology's problems, self-recording instruments would simply provide overwhelming amounts of practically useless data.

25. *Report of the Treasury Committee*, 732 (Parliamentary Papers). The Treasury inquiry was announced in October 1875. The eighth and final report of the Devonshire Commission had appeared in June.

26. *Report of the Treasury Committee*, 739–40 (Parliamentary Papers).

27. George Airy to Henry Smith, January 18, 1879, Airy Papers, RGO 6/704/4–7.

Airy's complaints throw light on the detailed work of observations—on whose barometers were sluggish, the difference between credible and suspicious data, how the "delusions" of observers were managed. But these questions of scientific practice were subordinated to the larger question of whether to prioritize observations in meteorological work at all. Airy's criticism reflected tensions about different models of scientific work that, by the 1870s, seemed to be perfectly summarized in meteorology. The Royal Society reforms to meteorology had formally separated "practical" work (at best the chart making, at worst the forecasts) from precision observations. By making this distinction so clearly, the committee identified proper science with a certain approach to instruments and numbers. In the year immediately following the first reforms, Sabine; Balfour Stewart, the director of Kew; and Robert Scott, the new director of the Meteorological Office, all published work in the Royal Society's *Proceedings*. Modeling different styles of meteorological work, their papers showed how instruments created careful hierarchies that detached modern science from practical work and popular concerns.

First came Stewart, who published an analysis of errors in aneroid barometers, measuring the accuracy of this instrument under graduated variations in air pressure and temperature. Balfour Stewart was the director of Kew Observatory and became in 1867 the senior figure at the Meteorological Office, supervisor to Scott. He was well known as a scrupulous observer, fast developing the reputation that eventually led him into a professorship and influential laboratory at Manchester's Owens College in 1870. Stewart had been a student of James Forbes at Edinburgh, and then left the university at eighteen and spent a decade in marine commerce. He became an assistant to John Welsh at Kew Observatory in 1855, where he developed a thermometer for measuring extremes of temperature. He subsequently resumed his studies at Edinburgh and worked again with Forbes from 1856 to 1859 on the measurement of radiant heat. His own contributions to the study of heat earned him election to the Royal Society in 1862. From Scotland, Stewart continued to work with Welsh on a survey of terrestrial magnetism in 1857 and 1858. He was a natural successor when Welsh died the following year. At Kew, Stewart continued the self-recording magnetograph observations, helped establish a program for photographing the sun, and verified barometers, thermometers, and hydrometers. Stewart's associations with the scientific worlds and instruments of Edinburgh and Kew meant that he saw meteorology as part of a set of interrelated magnetic, electrical, and atmospheric phenomena—with connections so intricate that multiple, scrupulous observations and refinements of limits of error were the first principles of the work. Only then could science safely pinpoint what Stewart called the "residual phenomena" that would lead to novel insights.[28]

The investigations Stewart carried out for the 1868 paper were typical of Kew's responsibilities for standard instruments. It was the instrument under analysis that was most significant. The aneroid barometer was a relative newcomer to the scene of meteorological instruments, and there was as yet no strict consensus on its scientific value. The aneroid ("without liquid") barometer worked on mechanical principles, as opposed to the hydrostatic operation of the mercury barometer. A flexible diaphragm suspended over a vacuum in a small brass container responded to air pressure. The diaphragm was connected to a spring and via the spring to a lever that registered the change in pressure on the face of the barometer. This instrument had been devised by a Frenchman, Lucien Vidie, who patented the barometer in England first (in 1844) because of its weak reception in France. The English responded with greater interest to its novelty, and it earned a medal at the Great Exhibition of 1851. Because of its potential durability and portability compared to the glass-and-mercury barometer, the instrument was attractive to seamen and mountaineers. The London instrument makers Negretti and Zambra, for example, marketed aneroids as "especially convenient for Travellers."[29] But the accuracy of any particular diaphragm was limited—those expecting to encounter large changes in pressure might need to carry two or three aneroids with overlapping ranges. Stewart's investigations showed that the aneroid was *only* a popular instrument. Although the aneroid gave consistent observations, it could not be read as precisely as a standard barometer. Any measurements carried to the third decimal place were therefore less than reliable. Aneroids were not consistent across the barometric range and measured differently at low pressures, requiring large corrections. Such a variation was not in itself a source of inaccuracy, of course, if the correction was made, but did produce one more observation step or calculation. Small aneroids—precisely those most popular for the convenience of travelers—were the least trustworthy. The aneroid was priced as a popular instrument, at four pounds ten shillings cheaper than even an ordinary "fishery barometer & thermometer" made for the Board of Trade for public display at lifeboat stations ("a good sound working instrument,

28. Stewart, "An Account of Certain Experiments on Aneroid Barometers." On Stewart, see *Dictionary of Scientific Biography*; and Gooday, "Precision Measurement" [diss.], and "Precision Measurement" [*British Journal*].

29. Negretti and Zambra, *A Treatise on Meteorological Instruments*, not paginated.

admirably adapted for use in public institutions," according to the catalogue).[30] In short, Stewart's investigation laid out the distinction between a scientific and a lesser, popular instrument.

Sabine followed up Stewart's paper with an account of Kew's photographic self-recording instruments after their first year of operation in the new "central observatory of the British System of Meteorological Observations," as he called it. He presented tables of Kew's temperature, water vapor (measured as the difference between a regular thermometer and the wet bulb thermometer), and barometric pressure alongside those of two stations in the Russian system, at Nertchinsk and Barnoual, both in Siberia. The foreign comparison underlined Sabine's international connections and long familiarity with "very delicate and sufficient" observations of magnetism made at the same centers, work that "inferentially," as Sabine put it, established his authority to judge the quality of meteorological measurements in both Siberia and Britain. As Sabine described it, British meteorology was thoroughly focused on instruments and tables of data. The Kew organization monitored data from the central observatory and the seven outlying members of the network, then transmitted them to the Meteorological Office where the data was "formed into Tables" and "used" for "meteorological purposes," about which Sabine was vague. "The mode and extent in which the information thus obtained may be most suitably communicated to the public are not yet fully determined, but are receiving careful consideration," he ended.[31]

The recipient of Kew's scrupulous data and the link between meteorological science and its public was Scott. His paper, overleaf from Sabine's in the Royal Society's *Proceedings*, presented a strikingly different approach to that of Stewart and Sabine. Scott was a young man, recently married, who had spent the last decade casting about for scientific employment. A prize student of physical sciences at Dublin University in 1855, he subsequently spent two years studying in Germany, including stints with Liebig and the meteorologist Dove in Berlin. Scott was appointed both for his familiarity with meteorology (he had published a translation of Dove's *Law of Storms* in 1862) and his training in methods of observations and reliable reputation for record keeping. (Announcing his appointment, the *Athenaeum* favorably

30. Ibid. Cf. Fitzroy on the aneroid barometer, in Fitzroy, *Weather Book*, 7, 124.

31. Sabine, "Results of the First Year's Performance of the Photographical Self-Recording Meteorological Instruments," 5, 7.

noted his "habits of business.")[32] Scott's starting point was a recent study by Charles Meldrum, in charge of the magnetical and meteorological observatory in Mauritius, and a prominent member of the British Meteorological Society. Any Mauritian meteorology was bound to be concerned with the practical issues of storm prediction, because that island, an important port for trade to the Far East, was isolated in the Indian Ocean and equally famous for its sugar cane and hurricanes. Meldrum, accordingly, spoke with the authority of experience on the contentious issue of the origins of storms, arguing that cyclones in the Indian Ocean invariably emerged from two proximate but opposite streams of air, following Dove's theory on the conflict between polar and equatorial currents as the cause of cyclones. Crucial here, however, is not the influence of the theory itself, but the contrast between the styles of meteorological analysis that emerged. Triggered by Meldrum's speculations, Scott recalled a set of remarkable observations shortly after he took office, when there was a westerly gale at Portsmouth and an easterly one at Yarmouth, followed by a violent storm. He therefore tracked back through the previous two years of data, looking for examples of opposite currents and examining the incidence of storms. From a collection of fifty-seven cases, he concluded that when the currents were very close but the polar current was in a higher latitude (i.e., easterly winds are to the north of westerly winds) the atmosphere was considerably disrupted or stormy, whereas in the opposite case, when the equatorial current was in a higher latitude (winds westerly in the north and easterly in the south), the atmosphere was relatively calm.[33]

Even though all three individuals were operating within the same institutional circle and publishing in the same journal, the contrasts between the work of Stewart, Sabine, and Scott were sharp. Stewart and Sabine focused on instruments, which they subjected to experiment and comparisons, and produced tables of numbers that seemed comparatively—and deliberately—remote from both weather and public value. In contrast, Scott was concerned with storms. His procedures began with speculation, moved to an empirical rummaging through data that his narrative directly connected to destructive weather events. The information he studied was coupled to outcomes of clear public interest. In the writings of Sabine and Stewart, instruments and the observations seem detached and impregnable. If we read those accounts alongside Airy's correspondence, however, instruments and numbers seem

32. *Athenaeum*, February 16, 1867, 224. For further information on Scott, see *Oxford Dictionary of National Biography*.

33. Scott, "On the Connexion between Oppositely Disposed Currents of Air."

far less austere. In effect, numbers and instruments became isolated, but that isolation was not so much splendid as exposed. Airy, along with like-minded critics, insisted on remembering what precision was for, and in meteorology, a science he considered to be practical and public, he found exactness could be a handicap to its development.

Forecasting: Degrees of Precision or a Reasonable Guide to Life?

The rigorous observations after 1867 were promoted as a contrast to the inaccuracy of weather prediction. Weather prediction, the Royal Society committee concluded in 1866, was not the business of a "strictly scientific body."[34] The care and reluctance with which forecasting was resumed marked the priorities of scientific meteorology, which was Kew and the observatory program. Yet, like Fitzroy before him, Scott gradually embraced the more practical goals of meteorology. As early as May 1867, he told the committee he was prepared to resume warnings, assuring them that the extra work was negligible. Several months later, he proposed that observations be taken out of the hands of telegraph clerks, who by the terms of their appointment were not free to make unscheduled reports in case of sudden changes in barometric changes—the main sign of storms. By 1873, he proposed to issue twenty-four-hour forecasts. When the committee demurred, preferring a tentative, internal circulation of forecasts for a trial period, Scott replied, in terms remarkably reminiscent of his predecessor, that predictions must be done regularly and publicly. The predictions Scott was prepared to supply in 1873 in fact only resumed four years later.[35] A brief statement, covering the weather of the whole country, accompanied the weather chart that was provided to the newspapers starting in 1877. More systematic forecasts waited another two years, until April 1, 1879, when the office began to issue three daily forecasts, with the expense initially subsidized by the press.

But how inaccurate were the forecasts? Most of our information, significantly, comes from Fitzroy's time, a mark of the pressure to reckon up in numbers. However, we can start with a modern evaluation: Jim Burton has calculated the success of the storm warnings during two sample periods in

34. Minutes of the Meteorological Committee, April 21, 1869, Meteorological Office Library, Bracknell. Cf. *Royal Commission on Scientific Instruction and the Advancement of Science: Eighth Report*, 529 (Parliamentary Papers).

35. Minutes of the Meteorological Committee, November 17, 1873, Meteorological Office Library, Bracknell; Burton, "History of the British Meteorological Office to 1905," 150–55.

1863 grouping the coastal recipients by present-day sea areas. He concludes that Fitzroy's accuracy ranged from 76 to 90 percent—adequate to excellent by modern standards of weather prediction.[36] Fitzroy himself was reluctant to make extensive checks on the accuracy of storm signals, though he kept a log so that the fulfillment of prediction could be checked. He was aware of several factors that data would hide: that land observations did not indicate the strength of gales offshore, for instance, or that the absence of a storm warning was in itself a useful prediction, but would not be measured. Shortly after the storm warnings began, the Board of Trade asked the Coast Guard to keep an independent check of their success rate, solicited confidential views on their efficacy from the ports in February 1862, and pressed leading scientific men to give their opinion.[37] Two years later, the Wrecks Board, another department in the Board of Trade, carried out another check. For each day investigated, two diagrams were produced, one of wind force and one of wind direction, each carrying data for the forty-eight-hour prediction, the twenty-four-hour prediction, and the actual conditions. Both Tuesday's and Wednesday's prediction for Thursday, then, appeared on the same diagram (fig. 4.3). Not only were the predictions flawed, but there seemed to be little correspondence between Tuesday's and Wednesday's separate forecasts for Thursday. After Fitzroy's death, the Galton Report took all the available data and calculated the accuracy rate of warnings (ignoring all the difficulties that Fitzroy had pointed out) and extracted a percentage—22 percent correct predictions—which was widely quoted in the press. The gap between Galton's assessment of 22 percent and Burton's modern one of at least 76 percent, based on combining warnings into a geographical grouping of sea areas rather than particular coastal locations, stands as an eloquent testimonial to the different expectations about accuracy that governed nineteenth-century meteorology.

The series of investigations in the 1860s registered the difficulties of evaluating meteorology. The whole point of the controversy was that the numbers provided no sure answers. Meteorological forecasting was more or less right—or more or less wrong. The real question, then, was not, "how accurate?" but how to define and measure accuracy in the first place. Galton, one of the most numerical of Victorians, understood the problem clearly. Reviewing his 1866 report on meteorology for the Treasury inquiry in 1877, Galton thrice underscored the section that noted, "A clear understanding of the degree of Precision to be aimed at, lies at the root of all estimates of past and future

36. Burton, "History of the British Meteorological Office to 1905," 56–59, 326.

37. "Correspondence between the Board of Trade and the Royal Society."

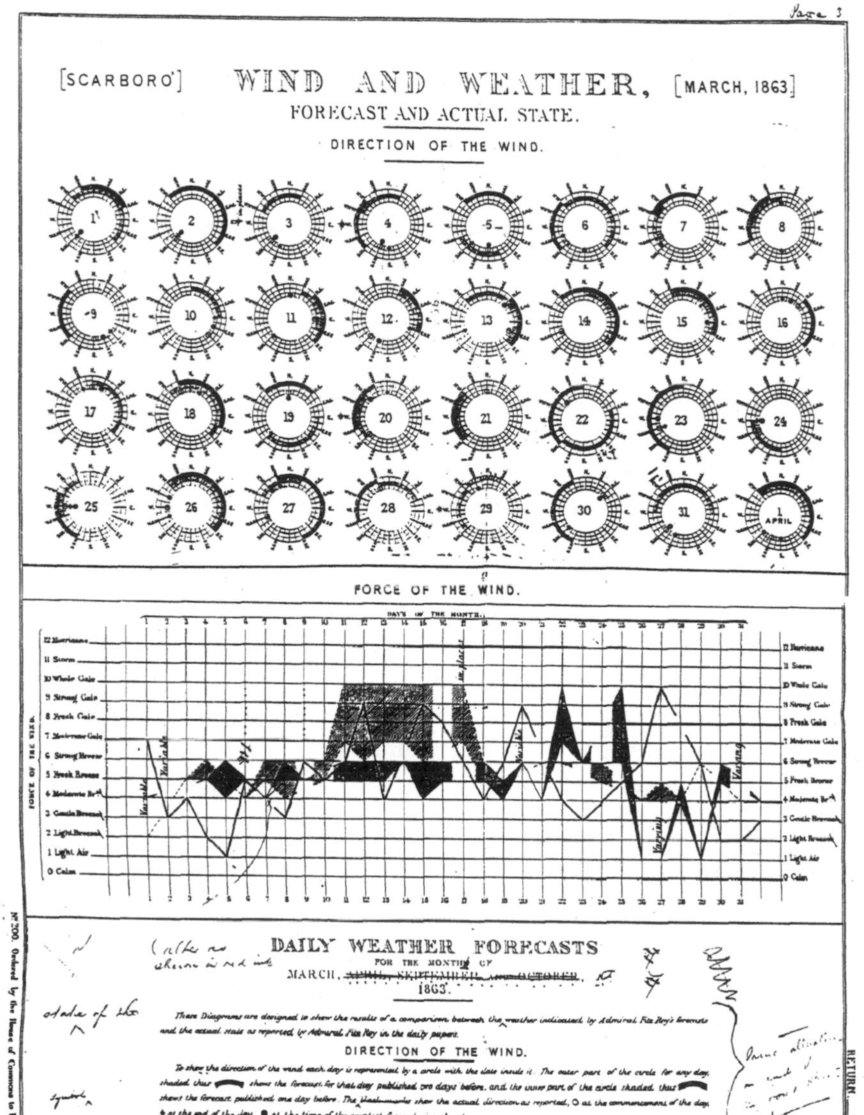

[SCARBORO'] WIND AND WEATHER, [MARCH, 1863]

FORECAST AND ACTUAL STATE.

DIRECTION OF THE WIND.

FORCE OF THE WIND.

DAILY WEATHER FORECASTS

FOR THE MONTH OF

MARCH, 1863.

These Diagrams are designed to show the results of a comparison between the weather indicated by Admiral Fitz Roy's forecasts and the actual state as reported by Admiral Fitz Roy in the daily papers.

DIRECTION OF THE WIND.

To show the direction of the wind each day is represented by a circle with the date inside it. The outer part of the circle for any day, shaded thus shows the forecast for that day published two days before, and the inner part of the circle shaded thus shows the forecast published one day before. The symbols show the actual direction as reported, ○ at the commencement of the day, ♦ at the end of the day, ● at the time of the greatest force during the day.

FORCE OF THE WIND.

On the Diagram for the force of the wind the thick black zig-zag line shows the maximum force reported during each day commencing at 8 a.m. The diagonal shaded parts show the forecasts—the part shaded thus being the forecast of two days before and the part shaded thus being the forecast of one day before.

The perpendicular lines of dots show that a storm warning signal was sent out for the date against which the dots are placed.

Board of Trade, April 1864.

RETURN. WEATHER FORECASTS.

No 200. Ordered by the House of Commons to be printed, 13 April, 1864.

FIGURE 4.3 Charts evaluating Fitzroy's weather forecasts were produced in 1864 at the request of the Board of Trade by the Wrecks Department. In this example, annotated at the Meteorological Office, we see the chart for Scarborough during the month of March 1863, comparing the forecast and actual direction of the wind and force of the wind over a two-day period. In the three rings of each diagram, the outer circle represents the wind direction forecast for two days earlier, the middle ring shows the forecast for one day earlier, and the circles and diamonds show the wind as it actually occurred: at the beginning, middle, or end of the day. The dark line shows actual force, while the shadings indicate predicted force one day earlier (lined) and two days earlier (hatched). The discrepancies dismayed critics. (Meteorological Office Papers, PRO MT 9/21/W595.)

work."[38] The struggle with "degree of Precision" must be disentangled before meteorological forecasting, and its critics like Galton, can be understood. Weather prediction brought the problems of numbers to a head.

The range of meanings that had accumulated around probability at the time of the forecasting controversy came from two directions, its enlightenment history as a model of rationality and its new and dramatic applications in statistics.[39] In the terms of eighteenth-century philosophers, probability was a calculus, a mathematical way of evaluating chances between different outcomes. The French mathematician and economist Marie-Jean-Antoine-Nicholas Caritat Condorcet, summarizing probability toward the end of the eighteenth century, had argued that every problem could be reduced to two models of urn and balls: if the proportions of different balls were known, then the chances of a particular outcome could be derived; if the contents were not known, their probable proportions could be calculated from the experience of a number of drawings.[40] The first case, direct probability, allowed for relatively straightforward calculations: if there were three black balls to every one white ball in an urn, for instance, the probability of drawing a black ball in the next instance was 75 percent. If we translate these abstract terms into weather conditions, the example would propose that if there are three rainy days to every clear day, the chances that tomorrow will be rainy is 75 percent. But direct probability offered a poor model of a natural system. There are not, of course, a priori proportions of every variety of possible weather. Condorcet's second category, known as inverse or a posteriori probability, was more relevant. It worked from observed instances backward to uncover the probable set of hidden conditions. The most difficult problem of inverse probabilities analyzed situations of unknown causes in order to estimate the likelihood of a particular later situation. (Imagine, for instance, taking observations of stormy weather, tracing those backward to an inference about the conditions that preceded storms, such as low barometric pressure or a westerly wind, and then speculating on the chances that a particular day will experience those conditions and be stormy.) Calculations that could be applied to a particular instance were indeed developed (separately, by English mathematician Thomas Bayes

38. Galton's annotated copy of *Report of a Committee Appointed to Consider Certain Questions Relating to the Meteorological Department of the Board of Trade*, 11, Galton Papers, UCL 118/1.

39. Daston, *Classical Probability*; Gigerenzer et al., *The Empire of Chance*; Hacking, *The Taming of Chance*; Klein, *Statistical Visions in Time*; Kruger, Daston, and Heidelberger, eds., *The Probabilistic Revolution*; MacKenzie, *Statistics in Britain*; Porter, *The Rise of Statistical Thinking*; Stigler, *The History of Statistics*.

40. Daston, *Classical Probability*, 230–32.

and by Laplace) but they were controversial. To students of probability in the nineteenth century, such calculations epitomized the weaknesses of probability as an arbitrary mathematical game, rather than anything approaching a model of nature.

The shift to events in the plural was the great transformation of probability into a tool for creating statistical knowledge. It emerged from astronomy, where probability calculations provided a method for determining errors of observations. In this technique, known as the error law or the principle of least squares, astronomers took a number of observations of an object. The most probable position (x) of the star is that for which the sum of the squares of the distance between x and the observations is a minimum.[41] In this formulation, we can see how "most probable" becomes the linguistic equivalent of "most accurate." The error law was an especially powerful example of probability as a method for evaluating accuracy of observations. Herschel called this the "first and most important application of the calculus of Probabilities (since it applies to all departments of science and affords a measure of the degree of precision attained in all numerical determination)."[42] The error law, a development primarily associated with Continental philosophers such as Gauss, Laplace, and Fourier, was established in astronomy by the end of the first decade of the nineteenth century. It assumed much larger significance the following decade, when Adolphe Quetelet, an astronomer studying in Paris, extended the error law to the study of anything that could be measured and counted. Quetelet found regularities in any collection of measurements, like that of mortality or incidence of crime or climate, and argued that these patterns pointed to the underlying laws of the phenomena involved. His study of the measurement of the human body, for instance, led him famously to the "average man," a type or ideal from which he argued all actual individuals deviated. Quetelet's "social physics" could be applied to everything that could be measured or counted, with the promise of finding the natural laws that produced these regularities in the same way that the accuracy discovered by the error law in astronomy displayed the laws of celestial mechanics. This extension of astronomical error law gave the study of society a new scientific foundation. Starting in the 1830s, the statistical study of society boomed. In Britain, the Board of Trade's statistical department was created in 1832, the statistical section of the BAAS appeared in 1833, statistical societies in Manchester and London sprang up in

41. See Kuhn, "The Function of Measurement in Modern Physical Science"; Schaffer, "Late Victorian Metrology and Its Instrumentation" and "Astronomers Mark Time."

42. Herschel, "[Review of Quetelet's] Letters on Probability," 17.

1833 and 1834, and the General Register Office emerged to collect mortality data in 1837.[43]

Quetelet was himself keenly interested in meteorology. In 1828, he was appointed director of the Royal Observatory in Brussels, and he presided over the first international conference on maritime meteorology in 1853. As he pointed out, meteorology was an obvious field in which to apply statistical methods. His own example in the 1838 *Letters on the Theory of Probability* of the use of statistics in disentangling complicated observations was a meteorological one. If your subject was the time of bloom for different species of plants, he suggested, you first collected observations of the time of bloom over some years, and then evaluated what meteorological data coincided with your bloom observations. If temperature range data matched the pattern, then it was a "relevant circumstance"—with accumulated evidence, it could become a theorem, or a law of nature. If anomalies in the combined observations remained (some years had temperatures as warm as normal but blossoms were later than normal) then further conditions—direction of the wind, amount of rainfall—could be added to the investigation. Probability entered into this process by distinguishing statistically significant variations from insignificant ones, and so could show whether the pattern was explained or whether anomalies remained.[44] As part of his monograph on Belgian meteorology in 1852, Quetelet applied probability calculation to the study of persistent rain, trying to determine the dependence of weather on previous states.[45] But if his work was among the most sophisticated, in its approach it was not unique or even fundamentally novel. At the beginning of the century, Laplace had studied daily variations in barometric pressure, proving that the oscillation was significant and not random. Alexander von Humboldt and Heinrich Dove had both pioneered the use of averages in meteorology as a spatial method, producing the curves called isobars and isotherms that linked points of the same average pressure and temperature. Studying weather averages over time rather than space also became standard practice in meteorology. In his *Climate of London*, for instance, Luke Howard argued for a weather cycle of eighteen years. Such cycles had obvious economic interest, and several British investigators, from

43. Hilts, "Aliis Extenderum; or, The Origins of the Statistical Society of London." Cf. Goldman et al., "The General Register Office."

44. Adolphe Quetelet, *Letters on the Theory of Probability*, 161–73; Ridder, *150 ans de météorologie en Belgique*.

45. Sheynin, "On the History of the Statistical Method in Meteorology."

George MacKenzie in 1829 to William Jevons in 1862, connected harvests or wheat prices with climate data to develop theories of climate and trade cycles, the beginnings of time series techniques. Meteorologists also repeatedly turned to statistics and probability to study the question of the moon's influence on the weather. In England, Glaisher conducted an elaborate analysis of Greenwich data to determine the influence of the moon's quarter on rainfall and direction of wind.[46] Such examples could be multiplied almost endlessly—the point is that the use of statistics took little different shape in meteorology than it did in dozens of other fields of inquiry, social and natural.

Fitzroy understood the direction of modern statistical usage, and he knew both Dove's work and Herschel's article on Quetelet. However, none of this helped him to forecast a particular day's weather, a practice that was simply incompatible with Victorian use of statistics. In "the doctrine of probability," as De Morgan put it in 1838, "there is prophecy, but not of particular events."[47] Fitzroy's own discussions of probability and the weather revealed the dilemma. "Forecasts," he asserted, "grow out of statistical facts," and he proposed to take "a broad general average or prevalence" and to consider "expectations for each district collectively in a group" in order to "*estimate* dynamic effects which may be anticipated." While maintaining that "meteorologic dynamics will soon be subjected to mathematical analysis and accurate formulas," Fitzroy simultaneously insisted that "it is extremely difficult to combine mathematical exactness with the results of experience." Forecasting was a judgment of "*probable* weather," not a prediction or prophecy, but strictly an "*opinion* [that] is the result of a scientific combination or calculation." He compared forecasting to a game of chess carried out in the head, without the assistance of a board and figures: his facts were "weighed and measured mentally."[48] Although he paid tribute to statistical methods, Fitzroy described probability in a way that emphasized interpretation and opinion.

His description, therefore, focused on the features of probability most alien to the current orthodoxy. The application of probable reasoning to a run of events, a series, was increasingly the only acceptable version of probability. Probability calculations of particular future events, as John Stuart Mill put it in his *System of Logic* (1843) were mistaken efforts, "a strange pretension"

46. Klein, *Statistical Visions in Time*. For an example of cycle theories, see Fullbrook, *The Wet and Dry Seasons of England*.

47. De Morgan, *Essay on Probabilities*, 113–14.

48. Fitzroy, *Weather Book*, 187, 193, 171, 190, 181.

emerging from the seduction of numbers that would "convert mere ignorance into dangerous error by clothing it in the garb of knowledge."[49] Fitzroy's description of forecasting also emphasized its subjectivity, a troublesome legacy of eighteenth-century ideas about probable knowledge. Drawing their examples from law and contracts, earlier philosophers had conceived of probability as a guide to action, a measurement of the degree of conviction that would prompt a rational being to make one decision or another. However, this rationality is a state of an individual mind. As Herschel put it, each calculation had "special and personal relevance . . . so that the same physical relation—the same future event—may have very different degrees of probability in the eyes of parties differently informed of the circumstances, the causes in action, the reputation for veracity of the testifying authors, or their opportunities for knowing the facts related."[50] Statistics, by contrast, offered objective knowledge that rose above the individual—typified in astronomy as the management of "personal error." By mid-century, the critique of the subjective foundations of probability was well established, led in Britain by mathematicians and philosophers such as Mill and, later, George Boole, and John Venn. In 1866, Venn's *Logic of Chance* roundly attacked the idea of probability as "the measurement of belief." This influential book, enlarged over several subsequent editions, argued that "actual belief at any given moment is one of the most fugitive and variable things possible." Venn compared mental processes to the swampy foundations of Rotterdam mansions. It was impossible to trace out and analyze the "substructure" of convictions so as to apply any precise value and calculate accordingly. Moreover, even if we could thus analyze our beliefs, Venn said, there is absolutely no reason to suppose that they properly correspond to actual situations. If that were so, lotteries would fold, since everyone could evaluate his or her miniscule chance properly. As Venn saw it, the notion of probability as reasonable belief allied measurement to the totally unreliable grounds of "natural tendencies" or "instincts." Logic and reason would disastrously collapse into the muddy depths of psychology.[51] Venn's arguments against probability as belief attacked the grounding claim of classical probability, the idea of the reasonable individual.

49. Mill, *Collected Works*, 8:545. Mill moderated his opinion after a correspondence with Herschel (see Porter, *The Rise of Statistical Thinking*, 82–83).

50. Herschel, "[Review of Quetelet's] Letters on Probability," 3. Cf. Cooper and Murphy, "The Death of the Author at the Birth of Social Science."

51. Venn, *Logic of Chance*, 119–78, quotations from 126, 127, 129. Venn was writing especially against the influence of Augustus De Morgan (Slamon, "John Venn's Logic of Chance").

Beyond philosophers like Venn, strict hostility to the subjective associations of probability was perhaps the exception rather than the rule. In Britain, the importance of religious beliefs to intellectual debate had shaped the reception of the epistemology of Continental philosophers like Laplace, resulting in a greater appreciation for the validity of subjective, or personal, knowledge that could leave room for conviction and faith.[52] Herschel, as the discussion above shows, was far more accommodating and, in that sense, perhaps more typical. He had a keen professional appreciation for the application of the error law, but also treated belief as the philosophical starting point for probability. "The reality, as an internal feeling, of the expectation that what has happened under given circumstances will happen again under precisely similar circumstances," Herschel noted, "is independent of metaphysical dispute and above it." This feeling belonged, as Herschel put it, to "contemplating minds."[53] By inference, disciplined reason could be applied to uncertain situations. And perhaps probability influenced the problems of weather prediction most of all on this simple level. If Venn saw here a slippery slide to irrational psychology, most of his contemporaries probably did not.

In any case the philosophical questions surrounding probability merged with questions of professional practice. If Fitzroy, in weighing and measuring his experiences mentally, showed a "contemplating mind," he was nevertheless not behaving like a Victorian man of science. Galton, by contrast, showed how one would analyze forecasting. In the report of 1866 that investigated Fitzroy's work, Galton was first concerned with managing the accuracy of the records. What he found appalled him. Not only was the method of transferring data from the registers of individual captains terribly inefficient, he noted, but it abandoned all definition of the quality of the information. When registers arrived, they were given a rating of quality, but then the notations were separately rewritten into logbooks divided up by subject and location: for example, wind records at a certain latitude. No attempt had been made to preserve the estimated quality of the data, so that in effect every observation was weighted equally, even when the register was known to be flawed. Looking beyond the problems of the data collection system, Galton also tried to evaluate the probability of the guidelines or maxims for weather prediction used in the department. If the information was managed more carefully, would the department's practices lead to reliable forecasts? Galton concluded not. He drew up a list of twenty-four different statements about the weather, such as that

52. Richards, "The Probable and the Possible."

53. Herschel, "[Review of Quetelet's] Letters on Probability," 1.

most storms move in a northeasterly direction, or that a southwest wind is light, warm, and moist, and then argued their indications should be calculated. As an example, he took a case where four of Fitzroy's rules or maxims came into play—a northeast current of air; a lowered barometer, a rising thermometer, and indications of a southwest wind; a gradual change in barometer overall; and a barometer lower on the continent, to the southeast, than in England. He argued that the probability of a correct forecast of stormy weather drawn from these conditions was the sum of the separate probabilities that each condition would be followed by a storm. Even if each had odds of 8 to 1, that is, were accurate in eight out of ten similar cases, the probability of an accurate forecast depending on the collection of four rules would be a much lower number, the sum of all the separate odds, or $8/10 \times 8/10 \times 8/10 = 500/1{,}000$. That is, "out of four such predictions, only two may be expected to succeed; or, the odds are equal that it will succeed or fail." This was hardly an impressive basis for authoritative predictions, Galton asserted. Moreover, in such a calculation, if any of the rules that connected conditions to particular types of weather were worthless (as he suspected), the problems mounted very quickly, because a false rule would completely erode "a chain of contingencies."[54] Government meteorology, in Galton's opinion, barely deserved the name of science.

Galton's frustration with the Meteorological Department methods and personnel may have biased his view. He did not find Fitzroy's loyal chief assistant, Thomas Babington, congenial or helpful (the fault here likely lay with Galton, who was notoriously tactless). Even before he encountered the ill-managed logbooks of his department, Galton probably disliked Fitzroy. Galton felt something approaching reverence for his cousin Charles Darwin, and Fitzroy's objections to Darwin's *Origin of Species*—moral outrage and biblical literalism—were of the kind most calculated to alienate Darwin's supporters. Moreover, Galton was increasingly unwell during the time of the inquiry. The meteorological report was published in May but he was ill by February of 1866, and his health broke down so completely by summer that he spent much of the next two years abroad recovering. All these factors may partially explain his rough treatment of meteorological statistics.[55]

Yet Galton's own evaluation of forecasts was fraught with difficulty. He had

54. See appendix 7 of *Report of a Committee Appointed to Consider Certain Questions Relating to the Meteorological Department of the Board of Trade* [Galton Report], xx–xxi (Parliamentary Papers).

55. On Galton, see Forrest, *Francis Galton*; Pearson, *Life, Letters, and Labours*; and Waller, "Gentlemenly Men of Science." Fancher, "Biography and Psychodynamic Theory," describes the breakdown Galton suffered from late 1865 to 1868.

treated each condition independently and ignored the possibility that the rules could be related to each other and a common cause, so that their multiplication would strengthen rather than reduce the chances of a correct forecast.[56] It was not a case of one weak link ruining a single "chain of contingencies," but rather that of several chains, possibly weak in themselves, combining to build a stronger one. Ironically, one of Galton's most famous contributions to statistics was his method for calculation the degree of interdependence of two or more variables, otherwise known as correlation. In 1866, of course, that achievement lay in the future. To suggest that Galton was operating with clumsy notions of probability himself is far from the point. Rather, Galton's response showed a definite understanding of what sort of thing a meteorologist should do, and his own difficulties can put to rest any suggestion that he used unusual or advanced standards. Indeed, the problem of related variations, though not its successful mathematical management, was the ordinary fare of observatory meteorologists, like those at Kew. Rainfall and wind direction and barometric pressure were all noted as separate statistics, but the phenomena were obviously not independent of one another. When, in 1888, Galton arrived at the technique of correlation, he was trying to connect the variation in length of arms with the variations in length of legs, a problem that he recognized as similar to the comparisons of variations in offspring and parent measurements in peas. Galton's studies of limbs and peas owed not a little to barometers and weather vanes, for the complex tangle of phenomena found in meteorological observations presented exactly this problem of dependent variations.[57]

Galton's condemnation of forecasting was based on a comparison of Fitzroy's practices with standard statistical research. Set against this background, Fitzroy's approach seemed to him painfully inadequate. But this was not the only context for his work. Where, after all, besides weather forecasting, did Victorians encounter probability? The most common encounter with probability was life insurance. By the 1860s, commercial life insurance had been established for about half a century, with its business concentrated among the prosperous classes. Life insurance fell within the reach of almost all earners when the government post office savings banks began to offer policies in 1864, following a plan promoted by William Farr, statistician at the General Register Office.[58] Although Farr built national life tables and sought natural laws, actuaries working for insurance companies took a less universal and

56. As pointed out by Burton, "History of the British Meteorological Office to 1905," 61–62.

57. See Porter, *The Rise of Statistical Thinking*, 270–96, esp. 292.

58. Eyler, *Victorian Social Medicine*, 84–87.

more pragmatic approach to statistics. As Thomas Sprague noted in 1892, in a critique of John Venn's *Logic and Chance*, "if insurance were a mere matter of chances, to be determined by the study of statistics, it would be a much simpler business than it is. There would be little scope for the exercise of judgment and skill by the company's officials."[59] Critics of life insurance could treat such a statement skeptically, as a sign of corrupt practices for which private insurance was notorious, in mid-century at least, though much less so at the time of Sprague's comment. However, Sprague's reservations about statistics also reflected the different goals of private actuaries. They examined the figures best adapted to their set of clients. Discussing the difference between census and mortality data on the one hand, and the records of insurance companies on the other, the actuary George King emphasized the superior accuracy of the latter. There we deal with individuals, he wrote, or "lives actually observed," instead of "lives in the aggregate . . . broad averages . . . in which there is ample room for error in the original facts." Life insurance was based, of course, on statistics, but, as King went on to show at length, there was ample room for disagreement about how to treat the data.[60] As practicing actuaries, both Sprague and King cautioned against idealizing statistics. Fitzroy resisted similar pressures in his approach to meteorology. Measurement was a relative matter, something to tailor to the demands of the work—Fitzroy always held in mind the sailor taking a barometer reading in rough seas.

Fitzroy, then, was caught in the middle of shifting ideas about probability and the authority of statistical information. As the man in charge of collecting meteorological statistics, he knew the purpose and value of quantitative observations. But as a weather forecaster, he appealed to probable knowledge in a different sense, as a reasonable judgment based on experience. For him, these treatments of probability were part of a spectrum, and he seemed genuinely unaware that contemporary philosophers so strongly rejected the application of probability to particular events. This disregard suggested that he had not grasped the general significance of probability within Victorian science. On the strength of its successes in the physical and social sciences, probability became a description of proper reasoning in science. Working out probabilities, Herschel wrote when introducing statistics to the readers of the *Edinburgh Review*, means to define "what are the really relevant circumstances on which events depend, and to analyze the complicated web of phenomena into a system of elementary and supposed uniformities, to which we assign the

59. Quoted in Alborn, "A Calculating Profession," 114.

60. King, "On the Construction of Mortality Tables," 225.

name of inductive theorems, or laws of nature."[61] Probability, in short, allowed the observer to "discuss" the data and move from numbers to knowledge.

Probability and statistics, therefore, were closely engaged with general convictions about the reach and security of natural knowledge. Moreover, as a man of Fitzroy's religious convictions was well aware, these convictions held intense moral and spiritual significance. Statistics was closely associated with determinism, a vision of nature and society governed by blind and inevitable forces. The connection between numbers and natural law was popularized by Thomas Buckle's *History of Civilization in England* (1857–61). Buckle introduced an account of the regularities in social events in order to ground its deterministic picture of history. The order that can be found in the most chaotic, seemingly arbitrary human events, such as murder or crime, proved that natural law governed society. Of all the nations, England, according to Buckle, best displayed the slow advance of liberalism that was the natural law of development—as a nation, it was the equivalent of Quetelet's average man. The limits on free will in this account of society disturbed many Victorians, and notorious statistics on crime and suicide drove home the link between determinism, immorality, and irreligion. In this context, the arguments about the value of meteorological statistics went beyond philosophy and politics to religion. Meteorology had become a signpost of the ambitions of the scientific leadership, and by the early 1870s, that leadership was identified with a strident materialist and determinist view of nature. Yet the study of the weather, even when armed with all its funding and instruments, hardly presented a picture of inevitability. Meteorology thus suggested the limits of the spreading claims of scientific culture. In the words of one critic, weather science became "an eligible battlefield" on which to define the extent of natural law.[62]

Measurement and Faith

The disputes over weather forecasting took place during a period of mounting tension between religious and scientific authority. Darwin's theory of natural selection appeared in 1859, the same year that Fitzroy outlined his proposal for storm warnings. Within a few months, a collection of essays by liberal theologians, *Essays and Reviews* (1860) challenged the authority of the established church by arguing against divine authorship of the Bible. These texts joined other evidence of apparent religious decline, such as the passage of

61. Herschel, "[Review of Quetelet's] Letters on Probability," 2.

62. MacLeod, *Scripture, Meteorology, and Modern Science*, 55–56.

the Divorce Act of 1857 and the poor record of church attendance revealed by the first national census of 1851. In this atmosphere, the 1860 BAAS meeting was expected to draw sparks, and did: Samuel Wilberforce, Bishop of Oxford, condemned Darwin's theories in front of an audience that included Thomas Henry Huxley, a young naturalist whose fierce response began to earn him his reputation as "Darwin's bulldog." Also in the audience was the man who had sailed Darwin around the world, Robert Fitzroy—he joined in against Huxley and for the bishop. That BAAS meeting has been treated ever since as a symbolic encounter of science and religion, but an equally symbolic issue at the same meeting was meteorological—specifically, rainfall, and whether to pray for it.[63]

A few months before Britain's men of science descended on his Oxford diocese for the annual meeting, Bishop Wilberforce had formally requested his clergy to begin prayers for fine weather during harvest. His action triggered a long exchange on prayer and natural law. Modern accounts of the prayer controversy show how Wilberforce's harvest prayers were connected with other prayer campaigns, such as those against cattle plague in 1865 and in support of the typhoid-stricken Prince of Wales in 1871. The broader context for any treatment of prayer was the question of miracles and natural law. Could providential actions interrupt the unrolling of natural events? Both prayer and miracles, then, presented ways of considering authority and evidence.[64] As a Church of England theologian summarized in 1865, "It soon appears that the two sides have no common criterion of good evidence and bad—that what is strong evidence to one man is weak to another; what is sufficient to one is defective to another."[65] Evidence, however, is a general category, and it is possible to indicate what was at stake much more clearly by noting that this controversy centered on a particular form of evidence: numbers. The peak of the prayer controversies came in 1872 and 1873, over the measurement of prayer for the sick, using hospital data—from this point contemporaries

63. For the encounter between Huxley and Wilberforce, see Desmond, *Huxley*, 276–83.

64. Mullin, "Science, Miracles, and the Prayer-Gauge Debate"; Turner, "Rainfall, Plagues, and the Prince of Wales."

65. Mozley, *Eight Lectures on Miracles*, xxii; Temple, *The Relations between Religion and Science*; Chalmers, *The Efficacy of Prayer Consistent with the Uniformity of Nature*; Fowler, *Mozley and Tyndall on Miracles*. For modern discussions of miracles and natural evidence, see Cannon, "The Problem of Miracles in the 1830s"; Chadwick, *The Victorian Church*; Tweyman, ed., *Hume on Miracles*; and Daston, "Marvellous Facts and Miraculous Evidence."

dubbed it the Prayer Gauge Debate. A focus on the quantitative values under examination helps explain why the prayer controversy began with rain.

Wilberforce's orders were an accepted, if occasional, function of clerical authorities. Public Days of Prayer had been called previously for protection against cholera in 1833 and 1849, and for harvest thanksgiving as recently as 1854, as well as during the Crimean War in 1854 and the Indian Mutiny in 1857, in the form of national humiliation and fasting.[66] The 1860 call for prayer for fine weather, however, moved prayer away from human political events. Even cholera, though clearly a natural event, could be interpreted as an extraordinary crisis, not part of the normal order of things (for Britain at least). The weather, however extreme, was different; its variations were manifestly natural. Charles Kingsley, a leading broad churchman and keen naturalist, responded to Wilberforce with a sermon insisting that the prayers were misguided and presumptuous. The call for prayers misunderstood the power and extent of natural law, Kingsley thought: "Shall I presume, because it has been raining too long here, to ask God to alter the tides of the ocean, the form of the continents, the pace at which the earth turns around, the force, the light, the speed of the sun and the moon?" The unity of natural law transformed any particular, individual desire to an enormous scale, he suggested. "For all this and no less, I shall ask, if I ask Him to alter the skies, even for a single day."[67]

John Tyndall echoed Kingsley's remarks in 1861, pointing out that in this sense there could be no distinction between a small and a great miracle. The dispersal of a mist or the slightest fall of rain, attributable to intercession with God, must be considered as much a disruption of the laws of nature as cancellation of an eclipse, or the reversal of the Niagara Falls.[68] Increasingly, following the rainfall debate, treatments by men of science on the question of natural versus spiritual laws included careful references to meteorology, as a science that was peculiarly susceptible to misinterpretation.[69]

In his challenge to scientific authority, Wilberforce's choice of meteorology was not idle. Earlier writers had often used meteorology to trace God's hand

66. See Turner, "Rainfall, Plagues, and the Prince of Wales."

67. Kingsley, *Charles Kingsley*, 2:112.

68. Tyndall, "Reflections on Prayer and Natural Law," 1–5.

69. Newman, *Controversy about Prayer*, 5–6. Cf. earlier discussion of weather as the epitome of (apparent) irregularity, Chambers, *Vestiges*, 328; and Prout, *Chemistry, Meteorology, and the Function of Digestion*.

in the design of the natural world. In the canonical series on natural theology commissioned by the will of the Eighth Earl of Bridgewater in 1829, meteorology was considered in two different treatises. William Prout's volume on the atmosphere and the weather extolled the suitability of the earth's climate for human habitation, and the benevolence of the cycle of rainfall and evaporation. William Whewell's preparations for his contribution to the series, *Astronomy and General Physics*, left him "hugely delighted" with the "most admirable set of Bridgewaterisms" he found in meteorology. "I do not know any subject which is at present in so instructive a condition," he enthused to a friend.[70] Wilberforce was quite prepared to take meteorology from this "instructive" natural theology to a world of active divine intervention, where the natural order served to judge human society. His timing was deliberately confrontational, coinciding as it did with efforts to establish the science of meteorology as well as with the BAAS meeting in Oxford. Fitzroy had proposed a system of storm signals in late 1859 and won approval from the Board of Trade six months later. The initial network of thirteen stations, issuing daily reports by telegraph to London, was in place by September 1, 1860. A direct appeal to God made a nice contrast with these laborious, uncertain experiments of the government's meteorological department.

A few years later, a Scottish clergyman gave a much more explicit description of the attraction of meteorology for critics of scientific culture like Wilberforce. Alexander MacLeod of the Scottish Presbyterian Relief Church used the weather to illustrate "the supremacy of Biblical teaching regarding physical forces and the laws of God, and to show the entire subordination of all these to His direction and control."[71] MacLeod came from an evangelical and millenarian strain of Scottish Presbyterianism: raised in Glasgow, he attended Glasgow University in the late 1830s and was ordained in 1844. His crucial ministry, from 1855 to 1865, was in the large John Street Church, Glasgow, as colleague to William Anderson, who was himself a contemporary and protégé of Thomas Chalmers, the influential Scottish evangelical of the 1820s and 1830s.[72] In *Scripture Meteorology*, MacLeod argued that the science was particularly effective for a demonstration of God's laws because it epitomized

70. William Whewell to Rev. R. Jones, July 23, 1831, in Todhunter, *William Whewell*, 2:124. On natural theology and the Bridgewater treaties, see Brooke, *Science and Religion*, 192–225; and Topham, "Beyond the 'Common Context'" and "Science and Popular Education in the 1830s." The two volumes that dealt directly with meteorology were Prout, *Chemistry, Meteorology, and the Function of Digestion*; and Whewell, *Astronomy and General Physics*.

71. MacLeod, *Scripture, Meteorology, and Modern Science*, 194.

72. MacLeod, *A Man's Gift and Other Sermons*; *Dictionary of National Biography*.

both providential care and infinite variability. The difficulty of predicting the weather made it only too evident that the atmosphere was not "a regularly moving machine," and, therefore, MacLeod triumphantly concluded that "If indeed the atmosphere . . . had been constituted for the special purposes of confuting infidelity, and exploding the chimera of a paramount, insensate law, it could not have been more adapted than it is now for such an end . . . we must reckon it the most eligible battle-field whereon to meet and readily discomfit the impugners of a divine and active Providence."[73]

When the prayer question resurfaced in 1872, it comes as no surprise to find quantitative evidence at the center of the debate. The trigger in this case was the successful recovery of the Prince of Wales from an attack of typhoid. Official rejoicing emphasized his recovery as proof that a national day of prayer had been effective. In *Contemporary Review*, a leading monthly journal, an anonymous article (later identified as the work of a well-known London doctor, Henry Thompson) suggested that praying for sick could be investigated scientifically, resulting in an "absolute calculable value" for appeal to divine intervention. A hospital ward, filled with patients suffering from diseases whose causes and treatments were well understood, could become the focus of prayer "by the whole body of the faithful."[74] After a trial of three to five years, Thompson proposed, the rates of mortality in the ward's patients could be compared to the national statistics of mortality for the given condition. The difference between the two mortality rates would accurately define the power of prayer. The article garnered the desired attention and set off a debate that flourished for many months. One of the supporting articles, also in the *Contemporary Review*, written by William Knight, extended the argument to "the economy of the universe." The experiment, Knight insisted would show "the absolute fixity of natural law" and the physical causation of all phenomena. Like Tyndall, Knight saw clearly that meteorology could become the thin end of a wedge. The "capricious" reputation of weather, he claimed, had mistakenly made it a target for the "crude notion" that the winds "may be sent forth on special errands in response to human entreaty." Whereas proponents of the efficacy of prayer supposed that some super-physical force can subtly influence some "local incidence," Knight insisted that no such event can isolate itself from the whole "chain of physical sequence." "The fluctuations of weather between

73. MacLeod, *Scripture, Meteorology, and Modern Science*, 5, 55–56, 53–54. Cf. the evangelical William Scoresby's ideas about terrestrial magnetism and mesmeric phenomena as evidence of divine will permeating the universe. Winter, "'Compasses All Awry,'" 81–82.

74. [Thompson], "The 'Prayer for the Sick,'" 205–6.

two seconds of time are as rigorously determined by law as are the larger successions of the seasons," he concluded.[75]

Knight's picture of physical causation overwhelmingly recalled statistical regularities, and the science of serial observations. It comes as no surprise, therefore, to find Galton joining the debate. Galton in 1872 was also one of the most active members of the Royal Society committee that supervised the Meteorological Office, and he was following studies of heredity. He had recently published his first genealogical and statistical inquiry on talent, *Hereditary Genius* (1869), and was full of figures about different classes and professions. Notoriously, he elaborated on Thompson's original proposition by suggesting that, since congregations prayed weekly for the king or queen, one could compare the longevity of the sovereign against the national average. Since British sovereigns lived much shorter lives than their most affluent class of subjects, the efficacy of prayer, according to Galton, was dubious indeed. (He made similar arguments about the "praying professions," clergy and missionaries.) With a fine dualistic flourish, Galton argued that investigations into prayer could be resolved into a "simple statistical question"—"are prayers answered or are they not?" Two lines of research were possible: "the study of isolated incidents" or the examination of "large classes of cases, and to be guided by broad averages." The latter "promises most trustworthy results," according to Galton, whereas the former potentially leads an investigator into real or perceived errors of personal judgment.[76] His comments, including the reference to statistical regularity as a means of avoiding the dangers of personal judgment, paralleled his critique of forecasting several years earlier. The undertones of these arguments would have been quite clear to Victorian readers. The alternative to good science (following out the regularities of natural laws) was willfulness—an ill-discipline that was both morally and scientifically reprehensible. Isolated incidents, like illnesses or stormy weather, were a distraction to the true investigator.

These accounts of prayer and weather, on both sides, drew attention to the connection between natural law and individual event. As the study of regularities and averages, statistical science treated the individual event as a potential expression of error, or idiosyncrasy. In the minds of Wilberforce or MacLeod, on the other hand, a particular event was treated as a possible

75. Knight, "The Function of Prayer in the Economy of the Universe," 185, 186–87. Turner, "Rainfall, Plagues, and the Prince of Wales," has a more complete list of articles on the debate published during this period.

76. Galton, "Statistical Inquiries into the Efficacy of Prayer," 126.

signal of God's presence: the interruption rather than the regular course held all the significance. Divorced from theology, this contrast appeared in other contexts as well. Engineers, for instance, vigorously questioned the privileging of regularity at the expense of singularity. From the point of view of men concerned with the water supply of towns and the control of rivers, the mean was far less significant than deviations from it—the isolated incidence of droughts and floods, rather than rainfall averages. Although they did not question the value of averages, they objected to observations that were exclusively or predominantly directed to the calculation of the mean. However, the tone of these objections also implied that the leveling powers of statistical science were inadequate as a description of the natural world. Frederic Pratt, a meteorologist who wrote for *English Mechanic*, expressed this point of view in 1864 when he insisted that meteorology should be a science of extremes. "At present it is almost wholly that of 'means' whereby the science is being deprived of its fullness and grandeur, transmitting only the dry bones of statistics to puzzle and embarrass future generations."[77] Alexander Thom made a similar point two decades earlier in a discussion of storm theories. In meteorology, he admitted, results come "from a multiplicity of events" rather than "separate accuracy" of any individual event. Yet "fractional minuteness" and "rigorous exactitude," though desirable and necessary, interrupt "a clear view of the whole." The routine of making and perusing logbooks "has a tendency to unfit the mind to break through formality and deal with more stirring events as they deserve."[78] Instead of managing the problem of subjective judgment, statistics might prevent any kind of judgment at all.

Conclusion

One of the most memorable accounts of meteorology's problems with numbers was a poem published in *Nature* in 1877 dedicated to the Council of the Royal Society. "Meteorology of the Future: A Vision" recorded the moans of a drowned sailor in 1881, denouncing "the board that telegraphs the weather." That board, "with facts and figures stored," had failed poor Jack: it was "jobbed," or corruptly stuffed with men of science whose concerns were remote

77. Pratt, "Modern Meteorology," 32. Related concerns about the relative value of synoptic versus mean charts were raised at the 1868 BAAS meeting (*Report of the BAAS*, 1868, 146).

78. Thom, *An Inquiry into the Nature and Course of Storms*, vi, 11. This emphasis on the "human record" (ibid., 110) echoed earlier tensions between an instrument-based narrative of the regular course of nature and accounts that pursued the "uncommon" or "extraordinary" event (see Jankovic, *Reading the Skies*, 78–102).

from meteorology. One, for example, was "a mind of many sides, well filled with *a* and *b* / And *x* and *y*, and likewise *z*— / But he didn't know the sea." This was an allusion to the mathematician Henry Smith, and other verses gave equally barbed references to the rest of the Meteorological Council (including Sabine's "Divine . . . god-like presence"). The list culminated with Galton whose recent statistical studies of inheritance had led him to measure the variations among size of peas. Galton "knew the sea" for he was "the rover of the crew" (a reference to Galton's explorations and 1868 publication of *The Art of Travel*), but unfortunately he could give no attention to the problems of a sailor—he was "busy shelling peas."[79] The preoccupation with statistics and the regularity of nature seemed to draw meteorologists away from the weather and its impact on human existence.

Although the rhetoric of precision and numbers often seemed to reign unchallenged in Victorian discussions of scientific method, there was in reality a broad range of responses to the role of instruments and numbers. Airy's quarrels with Sabine, Galton's impatience with Fitzroy, or his outright hostility to religious arguments like those advanced by Wilberforce or Macleod, give a sense of the contests shaping Victorian scientific culture. Those contests were intimately connected with statistics and probability, as it spread from the error law to the study of fluctuating, complex phenomena like social behavior—or like meteorology. Because nineteenth-century statistics promised to manage the move from small fact to general law, it dealt with how knowledge was made. Yet the meteorological controversies, pointing to the accumulation of endless details without result, questioned that promise and insisted moreover that the problem was not simply how knowledge was made, but how it was used. The notorious connection of statistics with determinism was in that sense perhaps a red herring—the really vital issues that meteorological statistics raised were funding, the isolation of research from practical goals, and the growing authority of scientific institutions. Weather prediction exposed the often jumbled philosophical bones of statistical knowledge, about whether observations and experience could give scientific knowledge about future particular events or whether this idea was, in Mill's terms, a strange delusion. However, weather prediction also focused attention on the social and political claims of modern science to funding, control of research, and increasing public authority.

The meteorological controversies show how the commitment to precision instruments and statistical forms of knowledge defined Victorian science. Galton, Sabine, and others in the Kew Observatory and Royal Society insisted that

79. "The Meteorology of the Future," *Nature*, July 5, 1877, 193.

practiced observers and precision instruments were paramount; Fitzroy and his supporters argued that a wider collection base was more important than minute accuracy. Yet the focus on instruments and their quantitative records gave critics of scientific culture their arguments as well. Faced with problems in constructing meteorological knowledge from the weight of precision observation, meteorologists turned their attention toward kinds of knowledge that stood outside conventional methods and instruments, however extensively situated, however precise and continuous. The reputation of popular weather wisdom explained how meteorology persistently remained a key site for attacks on the dogmatism of scientific culture, and it forced meteorologists to consider the problem of evidence that seemed to escape the forms of number, weight, and measure.

CHAPTER FIVE

Maps, Instruments, and Weather Wisdom

MAPS were one of the most striking innovations of nineteenth-century meteorology, an attempt to give shape and structure to the invisible forces of the atmosphere. The best single instance of the power of two-dimensional representation were the isobars, the lines linking points of equal barometric pressure, which as one meteorologist said in 1887, "entirely alter the attitude of mind with which we regard weather changes."[1] Yet even as they incorporated the numerical data into visual form, maps pointed to the limitations of numbers and the observation programs that produced them. Indeed, by the end of the century, as the meteorologist Napier Shaw remarked in the 1920s, maps had created "a curious alienation" of the "theoretical or experimental physicist," interested in precision observations themselves, from the meteorologist, interested in exploiting the visual suggestions of forces provided by mapping such observations.[2] As well as participating in the development of conventions of precision observation, then, meteorological maps could represent a critique of such conventions. The history of this critique extends beyond the professional scientific community to the world of popular knowledge about the weather. Applying a distinctive phrase—"meaning at a glance"—to both weather wisdom and maps, Victorians recognized the visual sensibility as well as qualities of accuracy and speed that connected the two. Any analysis of

1. Abercromby, *Weather*, 10. Dove's prized isothermic maps, plotting over 500 global observations of average temperatures, give another example of Victorian responses to maps of this type. See chapter 3, pages 88–89.

2. Shaw, *Manual of Meteorology*, 1:154.

the weather map draws on the standard history of visual representations in nineteenth-century science, but it must also include another story, that of the instruments, techniques, and problems of observation represented by popular models of knowledge.

One of the oddest of these instruments was built for the Great Exhibition of 1851, by a physician from Whitby, Yorkshire, appropriately named George Merryweather. The thousands who thronged to the London exhibition could visit a section of the vast hall dedicated to scientific instruments and see the self-recording barometers and delicate thermometers of the elite instrument makers, as well as an extensive selection of telegraphic equipment. Down the aisle was Merryweather's "Atmospheric, Electro-magnetic Telegraph, conducted by Animal Instinct," or the "Tempest Prognosticator," for short (fig. 5.1). It consisted of a circle of pint bottles, each containing a leech. When disturbed by atmospheric conditions, the leeches would crawl into the top of the bottles, passing over a whale bone lever that would trigger a signal to the central dome. As Dr. Merryweather noted, the device could be connected easily to a central telegraph network. His account of the invention invited the reader to consider that with his "Whitby pygmy Temples" he "could cause a little Leech, governed by its instinct, to ring Saint Paul's great bell in London as a signal for an approaching storm."[3] The dome of his ornamental instrument was a St. Paul's in miniature.

Eccentric as it may appear, the Tempest Prognosticator embodied widely shared assumptions about forms of knowledge, instruments, and meteorological science.[4] Its plausibility was based on two key perceptions: first, the precision and infallibility of sensations; and second, the importance of instruments to modern knowledge. Both are crucial for understanding the relationship of weather wisdom and scientific meteorology. The complex instinctive behavior of some "lower" forms of life modeled a natural form of automatic precision.

3. Browne, *The Story of Whitby Museum*, 134; Merryweather, *An Essay Explanatory of the Tempest Prognosticator*, 48.

4. For a selection of recent discussions about instruments in the history of science, see Anderson, Bennet, and Ryan, eds., *Making Instruments Count*; Bud and Cozzens, eds., *Invisible Connections*; Hackmann, "Scientific Instruments"; Hankins and Silverman, *Instruments and the Imagination*; Hankins and Van Helden, "Instruments in the History of Science"; Maas, "An Instrument Can Make a Science"; Turner, *Scientific Instruments, 1500–1900*; and Warner, "What Is a Scientific Instrument?" On particular instruments, see Bud and Warner, eds., *Instruments of Science*; Middleton, *The History of the Barometer* and *Invention of the Meteorological Instruments*; Negretti and Zambra, *A Treatise on Meteorological Instruments* and *Negretti and Zambra's Encyclopaedic Illustrated and Descriptive Catalogue*.

FIGURE 5.1 The "Atmospheric, Electro-magnetic Telegraph, conducted by Animal Instinct" proposed to connect the sensibility of leeches to a national telegraphic forecast. Eccentric in itself, the object nevertheless presented a conventional understanding of the relationship between instrumental knowledge and the natural sensibilities associated with weather wisdom. (George Merryweather, *An Essay Explanatory of the Tempest Prognosticator* [London: John Churchill, 1851], frontispiece; Whitby Museum.)

Merryweather's device incorporated this superior precision by turning the instinct of leeches into a telegraphic process. Sensibility, in other words, could be converted into rational judgments about nature. Merryweather deliberately alluded to the collective process of reason in his instrument by making the number of his leeches twelve, as in "a Jury of Philosophical Counsellors," and by arranging "his little comrades" in a circle, so that they "might see one another and not endure the affliction of solitary confinement."[5] The Tempest Prognosticator was a collaborative instrument in a science in which, as we have seen, Victorians considered that practical and philosophical progress depended on collaboration and exchange. Increasingly, as the conventional forms of that collaboration bore little fruit, the science also seemed to depend on the exploitation of popular knowledge about the weather, represented in terms of an unusual and primitive sensibility.

The critique of conventional observation developed in association with these ideas about weather wisdom and the subtle knowledge it represented. For Merryweather, the leech was the agent of weather wisdom; more commonly, Victorians associated it with the sailor or shepherd. The history of weather wisdom and of "meaning at a glance," then, traces an engagement with popular knowledge that can too readily disappear from pictures of scientific work. As George Harvey wrote in the *Encyclopaedia Metropolitana*, "it is humiliating to those who have been most occupied in cultivating the science of meteorology, to see an agriculturist or a waterman, who has neither instruments nor theory, foretell the weather . . . with a precision which the philosopher, aided by all the resources of science, would be unable to attain."[6] An idiosyncratic instrument like Merryweather's indicated the kind of concerns and assumptions that shaped the widespread visual experimentation with weather charts and maps. The Tempest Prognosticator was a hybrid product, a combination of telegraphy and meteorological instrument that appeared to produce a special kind of instant sensibility. So was the synoptic weather map. Indeed, if exhibition visitors viewed the Tempest Prognosticator between August and October 1851, they could have also seen a daily weather map published by the Electric Telegraph Company, which collected the observations produced by such instruments and conveyed them to the site by telegraph.[7] Maps and some kinds of instruments potentially rescued scientific meteorology, capturing inaccessible phenomena

5. Merryweather, *An Essay Explanatory of the Tempest Prognosticator*, 48.

6. See Butler, *Philosophy of the Weather*, xiii.

7. Marriott, "An Account of the Bequest of George James Symons," 258.

in ways that offered to replicate the speed, accuracy, and sensitivity of weather wisdom.

However, because they marked out the intersection of different approaches to the study of the weather, maps and their counterparts also offer an alternative history of precision, showing how Victorian scientific men and their audiences wrestled with the limitations of a statistical approach to the weather. Experimenting with an analysis of European weather maps in the early 1860s, Francis Galton expressed a common point of view when he argued that maps would transform meteorology. "When observations are printed in line and column, they are in too crude a state for employment in weather investigations." Sorting the observations into charts helps matters, Galton felt, "but it requires meteorographic Maps to make their meanings apparent at a glance."[8] He concluded that "the ultimate condensation of bulky reports into small intelligible maps is the aim we must keep stedfastly [*sic*] in view."[9] Galton's very characteristic Victorian interest in mapping must be placed within a web of other practices and instruments, from telegraphic codes and the study of clouds, to the camera and the spectroscope. But before we turn to these, we need to account for the ways Victorians defined the skills of the "weather-beaten weather-wise old sailor," to see how he "might prove a more certain authority . . . than *all* the Savans with *all* their *meters*."[10]

Weather Wisdom as Natural Precision

Some of the most evocative descriptions of weather wisdom appear in the Wessex novels of Thomas Hardy. In *Far from the Madding Crowd* (1874) and *The Mayor of Casterbridge* (1886), Hardy gave weather the part of a player in the drama. The former novel, set in the town of Weatherbury, portrayed a monumental electrical storm threatening the harvest on the heroine Bathsheba's farm. The shepherd, Gabriel Oak, holds the key to foreknowledge of the storm. For this observer, the natural world was full of signs. "Every voice in Nature was unanimous in bespeaking change," the narrator recorded. A "heated breeze," "dashes of buoyant cloud . . . sailing at right angles to another stratum," and the "lurid metallic look" of the moon, the rooks' flight, the "timidity and caution" of the horses, and the huddling of fearful sheep—all told Oak to expect a thunderstorm, while the appearance of toads, and spiders from the thatched

8. Galton, *Meteorographica*, 3.

9. Ibid., 5.

10. John Locke to James Tennant, November 30, 1859, Met. Office Papers, PRO BJ7/726.

roof, and slugs crawling indoors, indicated that the storm would be followed by cold continuous rain.[11]

Hardy's narrator recounted the shepherd Oak's deductions flatly, as a catalogue of observed facts. His list was typical of the data associated with weather wisdom—subtle, unmeasured indications from the sky, animal behavior, the built environment, even one's own body. For Victorians, weather wisdom was classically associated with men like Oak, the shepherd or seaman. The rules of the so-called "Shepherd of Banbury" published by John Claridge in 1740 were perhaps the most systematic collection. In weather maxims with accompanying notes, it set out the wisdom of forty years' experience of a shepherd for whom (as an introduction to the 1827 edition recounted) "every thing in Time becomes . . . a Sort of Weather-Gage" and "Instruments of real Knowledge."[12] An equally popular set of verses, variously attributed to Dr. Edward Jenner, the inventor of the smallpox vaccine, and to Dr. Erasmus Darwin, gave a summary of weather indications: the appearance of clouds, rheumatism and the creaking of floors, or the activity of large animals and tiny insects.[13]

Hardy also gave a brilliant account of weather wisdom and its connection to ideas about modern scientific precision. In the *Mayor of Casterbridge*, Hardy contrasted weather wisdom with more "mechanical" learning. The comparison is part of a broader theme, in which the rough ways, ready passions, and "rule of thumb" approach of one character, Henchard, must yield to newcomers. For instance, a young intelligent assistant Farfrae takes over Henchard's business, and "the scales and steelyards began to be busy where guess-work had formerly been the rule." Although the novel acknowledges Farfrae's success and Henchard's failings, Henchard, with his constantly responsive or impetuous temperament, is ultimately a far more sympathetic character. In the middle of the novel, as Henchard prepares for a visit to the village weather prophet, Mr. Fall, the narrator comments on the weather wisdom of the community as a whole. Because wheat and weather shape the life of the farmer, "in person he became a sort of flesh-barometer, with feelers always directed to the sky and wind around him."[14] The agriculturist, in other words, possesses a human version of the infallible senses of insects, with his behavior and moods ("elated," "sobered," and "stupefied" by various weathers) directly responding

11. Hardy, *Far from the Madding Crowd*, 174.

12. Claridge, *The Country Calendar or the Shepherd of Banbury's Rules*, 18. For a discussion of its "folk epistemology" in the eighteenth century, see Jankovic, *Reading the Skies*, 133–35.

13. For the verses and attribution see Drewitt, *Edward Jenner.*

14. Hardy, *The Mayor of Casterbridge*, 257, 295.

to the atmosphere. And this attention is exclusively local: "the local atmosphere was everything to him; the atmosphere of other countries a matter of indifference." Hardy's narrator relates this to an unmodernized landscape, comparing the farmer's sensitivity to the old roads, "steep in their gradient, reflecting in their phases the local conditions, without engineering, levelling or averages." In contrast to modern engineering—efficient, mathematical but less faithful—local knowledge is like a simple and natural instrument. It is a "flesh-barometer," or, in Hardy's similarly evocative description of the town of Casterbridge, "the bell-board on which all the adjacent villages and hamlets sounded their notes."[15] Hardy's images subverted the conventional estimation of instruments, tying their authority to inarticulate field workers instead of national observatories and scientific men.

The significance of weather wisdom, as Hardy indicated, hinged on its relationship to instruments. Because it was perceived as a form of automatic operation, weather wisdom was akin to the tools of modern science. From this perspective, the uncertainty of prediction was a consequence of the limits of the human ability to trace such natural information. Moreover, the highest levels of ability were associated with the roughest sorts of men and women—those without formal education, living close to the land or sea. Weather wisdom suggested the instability of the hierarchy of intelligence and sensibility—a hierarchy in which more learning was not necessarily a source of superiority. Sensibility and trained intelligence, then, existed in an inverse relationship: leeches or horses, rural laborers or seamen, women and other races, were all perceived as more sensible than the urban European male.[16] Andrew Steinmetz, a London writer of a number of works on popular science, discussed this phenomenon in a work on meteorology published in 1867, at the height of the forecasting controversy. "In our social life," Steinmetz noted, "there are all sorts of concerns that prevent us from hearing the voice of nature, and which render us deaf and blind in the presence of the most evident signs." While animals can "with certainty" predict changes in the weather from their sensations, man must "rely on those instruments which his intellect has devised, and which are, so to speak, substitutes for the instinct of the former."[17] Insensibility was the handicap of a rational existence.

Hardy's picture then drew upon an ambivalent account of instruments

15. Ibid., 257.

16. Some examples of the range of discussions about sensibility are Anderson, "Instincts and Instruments"; Cox, "Sensibility as Argument"; Romanes, *Animal Intelligence*; and Winter, *Mesmerized*.

17. Steinmetz, *Sunshine and Showers*, 5–6.

within scientific culture. The history of the barometer, moreover, suggests that meteorological instruments had particular significance within these accounts. The barometer developed into an instrument for understanding the atmosphere shortly after its invention by Torricelli in 1644, and by the middle of the following century, barometers were part of the regular commercial stock of clock making establishments, opticians, and mathematical instrument makers. An emblem of enlightenment rationality, barometers provided an instrumental guide to the weather, but it was not an instrument of precision. Rather, its fluctuations, as Jan Golinski has noted, were read alongside traditional prognostics. The measurements of the mercury column in a barometer "called for qualitative interpretation and judgement; they offered the opportunity to compare numerical results with subjective impressions and feelings."[18] Indeed mercury, or quicksilver, itself had associations with animation, vital fluids, and the circulation system of the body that reinforced such parallels between subjective and objective observations of the atmosphere. Familiar representations of the human passions as a graduated instrument depended on these parallels: the barometer of gullibility that we saw in Cruikshank's satire of the Murphy almanac craze; or the thermometer as gauge of sexuality, from modesty to abandon; or, more soberly, the evangelical barometer of spirituality and Bentham's gauge of happiness.[19]

Instruments were thus a tribute to the power of science, but they could also be considered imperfect solutions to the challenge of natural observation. As John Herschel put it, "None of our senses . . . gives us direct information for the exact comparison of quantity." As a consequence of "this emergency," he continued, "we are obliged to have recourse to instrumental aids, that is, to contrivances which shall substitute for the vague impressions of sense the precise one of number, and [to] reduce all measurement to counting."[20] If instruments were responses to the "emergency" of the state of sensational knowledge, then weather wisdom represented that state of knowledge in workable form, finely tuned and effective (though not numerical). Better instruments left this problem unchanged, as can be seen in a much later account of instruments and sensations given by the noted entomologist, John Lubbock. Writing on the nervous systems of insects in 1888, Lubbock concluded that microscopes had reached the limit of their usefulness for exploring minute structures.

18. Golinski, "Barometers of Change," 91.

19. Castle, "The Female Thermometer"; Sivasundaram, "The Periodical as Barometer"; Bentham, *Principles of the Civil Code*, 304.

20. Herschel, *Preliminary Discourse*, 124–25.

Magnification had already reached beyond the point at which the properties of light began to interfere and obscure technologically assisted vision. Lubbock argued, therefore, that the complexity of these systems defies our ability to fully distinguish between the sensibilities of living beings. We cannot then impose our judgment about any hierarchy of sensations, Lubbock claimed. "The smallest sphere of organic matter which could be clearly defined with our most powerful microscopes may be, in reality, very complex; . . . there may be an almost infinite number of structural characters in organic tissues which we can at present foresee no mode of examining." Lubbock went on from this point to draw an eloquent conclusion about the range of human senses and those of insects. "We cannot measure the infinite by our own narrow limitations," he insisted, and speculated on "fifty other senses," as variable as our own. "To [insects], it [the world] may be full of music which we cannot hear, of colour which we cannot see, of sensation we cannot conceive."[21] Lubbock's account of an insect nervous system vividly described perceptions that lay beyond the grasp of human reason and (to use Herschel's term) "contrivances." Lubbock's tribute to the potential of insect sensibility, like Merryweather's Tempest Prognosticator, gives another version of the authority of sensation.

Just as it raised questions about the nature and limits of precision, weather wisdom also pointed out the problems of local knowledge. Both literally and figuratively, it was not knowledge that traveled well; it could not be readily translated across space or among unlike minds. As one clerical observer wrote to the Meteorological Office, "it is curious that my weather wisdom always leaves me when I am from home."[22] In this respect, weather wisdom epitomized the problem facing meteorology in moving from local to general understandings of the atmosphere. Meteorologists understood clearly that their task was to replicate a profoundly local experience, as national and international networks of instruments tried to pull a universal science out of particular details. Observations for a scientific calendar, for instance, required minute attention "year after year, in the same place" and even "on the same individual plants." Trees of the same species, rooted in the same spot, with branches intertwined, one observer warned, may nevertheless "vary as much as a fortnight in their times of leafing or flowering."[23] This close study of weather indications,

21. Lubbock, *On the Senses, Instincts and Intelligence of Animals*, 191–92.

22. Rev. William Worsley to Robert Fitzroy, November 27, 1863, Met. Office Papers, PRO BJ7/754.

23. Lowe, "A Calendar of Nature," 152.

given the name "phenomenology" by some nineteenth-century observers, was the heir to the tradition of studying weather as an attribute of place.[24]

The inarticulate nature of weather wisdom was thus interpreted as a question of locality as well as dumbness (of animals, or simple folk). These qualities combined to lend meteorologists a noticeable preoccupation with language and communication. Scott and Fitzroy, the first two directors of the Meteorological Office, shared the difficulty of bringing local, extra-instrumental knowledge into the scientific exchange. Each thought of it as a problem of language. In 1863, Robert Fitzroy wrote to John Herschel, in terms remarkably reminiscent of Thomas Carlyle's *Sartor Resartus*, on the limits of an articulate description of nature. "Language seems but a dress—or many dresses—surrounding a form of perfect character—and to realize the truthfulness and beauty of such a form, the rags heaped upon it by ignorant (more or less) translators or copyists must be looked through."[25] Fitzroy had begun to apply the word "forecast" to his activities in order to shift his predictions away from the distasteful connotations of "prophecy" or "prognostications," and he repeatedly expressed his anxiety about forecasts by reference to the nursery parable of the boy who cried wolf.[26] In his newspaper publication of forecasts, where his words reached their widest audience, Fitzroy's concern was most acute. He explained his dilemma: "As newspaper space is very limited and as some words are used in different sense by different persons, extreme care is taken in selecting those for such brief, general and yet sufficiently definite sentences as will suit the purpose satisfactorily."[27] Scott expressed similar concerns when he took over from Fitzroy. In 1867, Scott compared "the local experts" with "an expert sitting at Whitehall," like himself, and admitted frankly that the former made better judges of weather.[28] "Over and over again," he wrote somewhat plaintively, "the remark has been made by me or to me, when our office has failed to issue timely warning of a gale, 'If only our reporters at the out-stations

24. Jankovic, "The Place of Nature and the Nature of Place." Cf. Alberti, "Amateurs and Professionals in One County." The classic example within England of this tradition of minute local observation was White, *Natural History of Selbourne*. For an example of such details in an observer's record, see Ormerod, ed., *The Cobham Journals*.

25. Robert Fitzroy to John Herschel, January 15, 1863, Royal Society of London, HS 7/262.

26. Robert Fitzroy to George Airy, June 12, 1860, Airy Papers, RGO 6/702.

27. Fitzroy, *Weather Book*, 185–86. Galton denounced the term "forecast" as sloppy (*Report of a Committee Appointed to Consider Certain Questions Relating to the Meteorological Department of the Board of Trade* [Galton Report], 17n [Parliamentary Papers]).

28. Minutes of the Meteorological Committee, February 25, 1867, Meteorological Office Library, Bracknell.

had told us the appearance of the sky and landscape, in addition to sending up their instrumental readings.'"[29]

As Scott's remarks about the "the appearance of the sky" suggested, the study of clouds was an especially significant instance of these problems of scientific language and communication.[30] Cloud observation is accordingly one of the best places to trace meteorologists' awareness of popular expertise about the weather. Revealingly, an American point of view on cloud study was quite different from accounts in Britain, reflecting the greater confidence about the authority of weather prediction services in the United States. The American meteorologist Cleveland Abbe regarded cloud study as a field in which a trained meteorologist could display the expertise that turned observations into science. In 1869 Abbe had collected cloud information from a network of observers he initiated at Cincinnati Observatory, and continued the practice in Washington as the civilian scientific director of the national network, which was from 1871 to 1891 under the auspices of the U.S. Signal Corps, a division of the Army. (In 1891 this organization became the civilian U.S. Weather Bureau.) Benefiting from an extensive network, funding, and trained technicians, Abbe had been drawing up cloud maps three times a day since 1872 as part of the preparation for forecasting.[31] In recognition of American leadership in meteorology, the editors of *Encyclopaedia Britannica* commissioned Abbe to write the article on meteorology in 1890. There Abbe contrasted the "casual glance" of the rude sailor with the rigor of the scientific gaze. When "properly understood and interpreted," clouds would reveal "the most important features" of the atmospheric processes. Abbe treated the conclusion as obvious: "If the farmer and the sailor can correctly judge of the weather several hours in advance by a casual glance at the clouds, what may not the professional meteorologist hope to do by a more careful study?"[32] Buoyed by two decades of pioneering work and expanding resources, Abbe was certain that clouds could be analyzed. However, his attitude was not typical. Across the Atlantic, where weather forecasting was attended with less conspicuous success, the response was more muted. Like Abbe, British meteorologists identified clouds

29. Scott, "Forecasting the Weather," 452.

30. Hankins and Silverman, *Instruments and the Imagination*, 140–47, describes the relationship between visual records and language in graphing.

31. Abbe was director of the Cincinnati Observatory from 1868 to 1870 and then headed the Weather Service of the Signal Corps starting in 1871 (*Dictionary of Scientific Biography*; Fleming, *Meteorology in America*).

32. Abbe, "Meteorology," 272.

as the special province of the weather-wise farmer or sailor, but were far less confident of the capacity of scientific observation to supersede such knowledge. John Drew, a popular science writer and early member of the Scottish Meteorological Society, noted simply that "Clouds are that page of nature's book which only [sailors and farmers] are competent to read."[33]

The scientific study of clouds could be quite literally considered a question of language, because its central problem appeared to be a suitable and accurate nomenclature. The first compelling classification was that of Luke Howard, set out in his *Essay on the Modifications of Clouds* of 1803.[34] Howard defined four basic types, which he significantly called "Structures" in order to disclaim any description of "precise form or magnitude." These structures were the now-familiar cirrus, cumulus, stratus, and nimbus. Howard quite clearly saw his nomenclature as an attempt to deal with incommunicable knowledge and compared the instrument-bound philosopher, working on his records without knowledge of clouds, to the doctor who "only attends to the pulse" instead of gathering a more complete history.[35]

Yet the categories, as important as they were in founding a systematic basis for observation, and as persistent as they were in the literature, did not satisfy all meteorologists. By the time an international classification was established at the end of the century, meteorologists recognized the ubiquity of Howard's terms—the stratus, cirrus, cumulus, and nimbus—but they insisted that even if the nomenclature was universal, its usage was not. More rigorous classification was called for, meteorologists argued, because the same label was used for quite different natural objects, depending on the observer's inclination, nationality, and training.[36] Accordingly, more elaborate nomenclatures had appeared regularly throughout the century. Most of the later nomenclature built on Howard's terms, either by linking terms to create new hybrid names for intermediate forms, or by setting their terms alongside his. Both forms of influence appeared

33. Drew, *Practical Meteorology*, 177.

34. Hamblyn, *The Invention of Clouds*; Forster, *Researches about Atmospheric Phenomena*, gives an account of the reception of Howard's work. For a brief survey of the history of classification, see Kingston, "A Century of Cloud Classification," and "A Historical Review of Cloud Study."

35. Howard, *Essay on the Modifications of Clouds*, 1–2. Howard's comparison with medical knowledge was significant, for debates over "experience" and "instrument-based" knowledge similar to those in meteorology occurred in nineteenth-century medicine (Lawrence, "Incommunicable Knowledge").

36. See, e.g., Abercromby, "Modern Developments of Cloud Knowledge," 3. The complaint was still echoing decades later in Clayden, *Cloud Studies*. For further discussion of clouds and objectivity see Anderson, "Looking at the Sky."

in the cloud terminology that Frederic Gaster, a fellow of the Royal Meteorological Society, published in the society's journal for 1893. Gaster sought to "enlarge" rather than to "violate" Howard's terms, singly or linked, developing two basic classes, cumuliform and stratiform, which were modified by several prefixes (fracto, mammato, detached, turreted, furrowed, and, confusingly, cirriform) and then classified by height (surface, lower medium, higher medium, and highest). Clement Ley, another English meteorologist, built his system on the formation process of clouds, relating Howard's terminology to new categories called radiation, interfret, inversion, and inclination. In yet another complex reorganization of the nomenclature, the director of the Havana Observatory, André Poey, insisted on the need for multiple and interchangeable forms of cloud description that would include Latin, vernacular, and symbolic languages; an extensive Latin nomenclature based on Howard; a convenient symbol form for efficient record in a logbook; and corresponding vernacular nomenclatures in French, German, English, Italian, and Spanish to ensure that observers could match their local experience with the categories of scientific observation.[37]

With these efforts to develop a stable and accurate nomenclature, meteorologists claimed that the chief obstacle in cloud studies was a matter of finding correct labels for cloud forms that could be used in a uniform manner. Yet framing the problem of cloud study in terms of its nomenclature simultaneously allowed for a far less positive approach. It also expressed the limited extent to which scientific meteorology could hope to mimic the weather-wise reading of the sky. In an article on clouds in 1888, Ralph Abercromby, a prominent member of the British Meteorological Society, undermined the possibilities of uniform or universal translation. "Clouds always tell a true story, but one which is hard to read; and the language of Scotland is not the language of Borneo. Form alone is equivocal, for the true import must be gathered from the surroundings, just as the meaning of many words can only be judged by the context."[38] This comment pointed to the direct experience of the observer, and to the inadequacy of words. Similar concerns emerged in the work of Henry Francis Blanford, the director of the meteorological observatory in India from 1875 to 1888. An observer who traveled widely on inspection work across the many cultures and dialects as well as climates of his posting, he too spoke of the difficulties of accurately translating the eloquence of the sky into the "language of physical sciences."

37. Poey, *Comment on observe*, 39; Ley, *Cloudland*.

38. Abercromby, "Modern Developments of Cloud Knowledge," 18.

> To an experienced eye the forms and movements of the clouds and the general aspect of the sky are eloquent . . . but save seamen, and among intelligent and observant nations, farmers and herdsmen, there are few who have learned their language and can rightly interpret it; while such knowledge as those possess is for the most part empirical and little capable of being harmonised with the facts and translated into the language of physical science.[39]

This knowledge could not, Blanford argued, "be learned from books" because the terminology can never be sufficiently flexible and precise, and the match between cloud masses and "mere" words is too difficult. "The first difficulty . . . is the want of sufficiently exact language to designate the protean forms in which cloud masses present themselves to the eye . . . and second the difficulty of identifying the particular kind of cloud intended from a mere verbal description."[40] The appearance of the sky was difficult to transpose into a scientific form. It remained the province of unlearned sailors, herdsmen, and farmers.

Accounts of weather wisdom therefore expressed intense concern with scientific communication as well as accuracy and the process of generalization from local knowledge. These problems of communication were summed up in a cartoon, "Weather Wise or Otherwise," published in 1867 in *Judy* (a satirical weekly and an imitator of *Punch*; see fig. 5.2). In the foreground of the picture, two local mariners squint at the sky, while elegant tourists take their seaside strolls in the background. "What do you think of the weather?" says one. The other replies, "Why, I think as you thinks." It seems that no one—neither readers, nor tourists, nor his companion—can be the wiser for this reply. Is this the joke? Both local mariners are fools, and neither admits it. Perhaps. But if we recall the contemporary reputation of sailors, who are "wiser than all the Savans with all their meters," the joke changes. Perhaps the mariners are sharing a genuine, if inarticulate, understanding. In that case, it is those who cannot participate in this cryptic but almost effortless exchange—the genteel tourists, and by extension, the *Judy*'s readers—who are both ignorant and oblivious.

Victorians characterized weather wisdom in several ways, all of which raised fundamental questions about observation, precision, and scientific exchange. Such knowledge drew from an intimate direct experience of the natural world, quintessentially associated with those who spent their lives in closest contact with the elements, the sailor or shepherd. The variety of observations

39. Blanford, *A Practical Guide to the Climate and Weather*, 56.

40. Ibid.

WEATHER WISE OR OTHERWISE.

1st Boatman. WHAT D'YE THINK OF THE WEATHER, BILL?
2nd Ditto. WHY, I THINKS AS YOU THINKS, JACK.
1st Ditto. WELL, AND I THINKS THE SAME.

FIGURE 5.2 A cartoon showing the ambivalent response to weather wisdom. Weather wisdom, associated with unlearned and inarticulate seamen or shepherds, could attract ridicule but was difficult to dismiss entirely. In this exchange, Bill and Jack may be two ignorant seaman, trading empty bluffs—or they could be sharing a genuine, albeit unintelligible knowledge. In that case, the elegant and fair-weather tourists in the background, one with a telescope to her eye, are the ignorant parties. ("Weather Wise or Otherwise," *Judy*, June 23, 1867, 272.)

falling under the rubric of weather wisdom—the appearance of the clouds, behavior of insects and animals, the transparency of the air—were in detail less important than the seductive model of knowledge that they represented. Above all, weather wisdom seemed to offer speed. In contrast to the labor of gathering observations from instruments, or pouring over comparisons of recorded numbers, weather wisdom provided immediate insight. Speed and certainty was the ideal of weather forecasters, and it was this ideal that traditional authorities—the "weather-beaten, weather-wise old sailors"—represented. As knowledge of sensation, it was associated with qualities of exactness and precision. That quality in one respect seemed to fit it for appropriation by scientific instruments, in the way that Merryweather thought that the instinct of leeches could be absorbed into the electric telegraph, one kind of precision blending into another. But in other ways, weather wisdom underlined the difference between the artificial sensation of instruments, and the superior precision of natural knowledge. The latter was deeply embedded in a particular locality, and notoriously difficult to express outside that context.

Weather wisdom was, of course, not an isolated case of popular knowledge in negotiation with emerging scientific disciplines. Herbalists, midwifes, and other healers found their practices systematically attacked by the nineteenth-century medical reformers who, with their control of education and medical licensing, strove to assure the preeminence of their version of medical practice. A similarly broad confrontation with traditional authority developed in geology as scientific men sought to reconcile the teachings of the Bible with their interpretations of strata and fossils.[41] Exchanges between popular practices and scientific elites, far more equivocal at the time than they appear in retrospect, share common elements with weather wisdom, especially the question of local knowledge and the preoccupation with methods of transferring knowledge between different localities and approaches.[42] Yet knowledge about the weather appeared to be particularly slippery. Because (as in Hardy's plots) weather wisdom appeared to be so effective as a means of prediction, meteorologists often had to concede its significance, and to face the task of defending the value of their instruments and theory. In recognition of its reputation, the term "weather wisdom" was often, as we have seen, an ambivalent acknowledgment of popular insight.

One characteristic response of meteorologists to weather wisdom was that seen in Merryweather's Tempest Prognosticator: to incorporate weather wisdom into the forms and practices of science. The reputation of weather wisdom encouraged meteorologists to turn to instruments and techniques that modeled a sensation-based knowledge and that engaged with problems of local knowledge and scientific communication. The two most critical of these were the telegraph and the weather map. The telegraph offered instantaneous communication; the map offered a visual form of knowledge that replicated the direct intuitive grasp of weather wisdom. The two techniques reinforced each

41. For general accounts of the significance of contested knowledge in Victorian science, see Winter, "The Construction of Orthodoxies and Heterodoxies"; Cooter and Pumfrey, "Separate Spheres and Public Places"; and Desmond, "Redefining the X Axis." On particular examples in medicine, geology, and paleontology, see Brooke, *Science and Religion*, 226–38, 248–63; Rudwick, *The Meaning of Fossils*; Desmond, *Archetypes and Ancestors* and *The Politics of Evolution*; and Donnison, *Midwives and Medical Men*.

42. Secord, "Science in the Pub" and "'Be What You Would Seem to Be.'"

other. Together, they seemed to liberate the individual observer and literally open new vistas, enabling the meteorologist to see beyond the narrow confines of his own columns of data. Both the telegraph and the map represented a form of sensibility, the answer of modern science to the intuition and experience of the local authority.

Maps, Instruments, and Codes: A Visual Model of Scientific Knowledge

In extending the leeches' sensibility to a telegraphic apparatus in his Tempest Prognosticator, Merryweather was drawing on a parallel in contemporary perceptions of nerves and telegraphy. As physiologists developed an anatomically based understanding of the nature of sensations and thought, they evoked an analogy with the transmission of information along the lines of the electric telegraph system. In his *Principles of Mental Physiology*, for instance, the influential English physiologist William Carpenter produced an extended comparison of the nervous system with the telegraph. The production of the electrical current by chemical reaction in telegraphy, he explained, was like the reaction that takes place between the blood and the central nerve cells or peripheral nerve fibers, and the "axis cylinders" of neighboring nerve fibers are protected with insulator, "just as are the numerous wires . . . which are bound up together in the aerial cable of the District Telegraph."[43] Across the channel, Hermann Helmholtz's studies of the process of perception in the late 1840s and 1850s also relied on analogies with the electric telegraph. For Helmholtz, indeed, the electrical signals of the telegraph literally modeled sensation, as he carried out his experiments with apparatus supplied by the telegraphy industry.[44] By the 1850s, the analogy between the nervous system and the telegraph was sufficiently established that the comparison operated as readily in the other direction: mental acts were like telegraphy, but telegraphy was also like mental acts. In 1858, when the first Atlantic cable briefly connected two remote continents, a London newspaper pictured the telegraph as a giant nervous system, "a complete network of electric filaments will overspread every civilised land in the world, and converge to great ganglionic cables which shall stretch . . . across every ocean and sea." The resulting dramatic increase

43. Carpenter, *Principles of Mental Physiology*, 38, 35.

44. Lenoir, "Helmholtz and the Materialities of Communication." Lenoir notes the discussion of this analogy in Helmholtz's *Die Lehre von den Tonempfindungen als physiologische Grundlage für die Theorie der Musik*, which was not translated into English until 1885.

in human perception would transform knowledge: according to this vision, the telegraph was in fact the ultimate scientific instrument.[45]

Although the telegraph thus defined and made possible the modern forms of weather prediction, it simultaneously provided a framework that connected the work of official meteorological observatories with popular knowledge. Elsewhere in scientific work, instruments represented a boundary between elite and popular practices.[46] Yet the instrumental apparatus of weather prediction by contrast was equivocal. Although clearly an elaborate and expensive system that privileged the government meteorologist, the combination of instruments and telegraphy also served to suggest the affinity of popular and elite practices. As he repeatedly spoke of the advantages that this combination offered, Fitzroy illustrated the connection between sensibility and telegraphy. The telegraphic network, he wrote, offered a "means of *feeling*—indeed one may say *mentally seeing*—successive simultaneous states of the atmosphere."[47]

> While no man had the means of knowing anything about the weather beyond his sight, or the *feeling* of his instruments, it was scarcely possible to foretell changes of importance . . . but now the case is exceedingly different. A daily glance at the published 'Weather Reports,' a recollection of their previous features during the few previous days, a look at the glasses at home, and an eye turned occasionally to the heavens, may enable anyone . . . [to predict] the principal changes.[48]

Telegraphy thus symbolically extended the sensibilities of the central office and so offered a way for the modern scientific investigator to compete with the experience of the local experts.

As Fitzroy's remarks indicated, the new sensibility that the telegraph offered was portrayed primarily in terms of vision. Although touch or hearing might seem equally suggestive sensations to associate with the tap of the telegraph, Fitzroy had singled out the "daily glance," the "eye to the heavens," and his "mental vision" of the atmosphere. The powerful appeal of visual methods

45. *The Illustrated News of the World*, August 28, 1858, 1. For an extended discussion of electricity, see Morus, *Frankenstein's Children*.

46. For a typical account that stressed the sophistication and precision of the instruments, the "most vigilant care" required for observations, and the "mass of calculations" afterward, see Glaisher, "The Magnetical and Meteorological Royal Observatory, Just Completed at Greenwich," *Illustrated London News*, March 16, 1844.

47. Fitzroy, *Weather Book*, 170, original emphasis.

48. *Report of the Meteorological Department*, 1862, 448–49, original emphasis (Parliamentary Papers).

in meteorology thus had two sources. One was general, a commonplace of Victorian science: the proliferation of visual evidence in an era of new instruments and printing techniques.[49] Given this context, it was not surprising how often meteorologists contrasted the completeness of a visual result with the partial insight derived from other observations. A second source, however, was philosophical, part of contemporary discussion of scientific method. Visual evidence had a special appeal as a summary of collective instances. Herschel had noted this aspect of visual evidence in the *Preliminary Discourse* (1830) in his description of "the parabolic form assumed by a jet of water spouted from a round hole." It was "a *collective instance* of the velocities and directions of all the motions of all the particles which compose it *seen at once*, and which thus leads us, without trouble, to recognize the laws of motion of the a projectile."[50] Visual evidence, a summary of particulars, accordingly seemed the natural resource of a science depending on collective observation.

This sort of appeal came across vividly in a response to one of Fitzroy's innovations, the wind star. One of the first productions of his department, the wind stars were revised versions of Matthew Fontaine Maury's compass figures for indicating the prevalent winds in a particular area of the ocean. They consisted of a central circle, marked with latitude and longitude, surrounded by differently sized "rays" spreading out in narrow triangles (fig. 5.3). The longest and broadest of the rays corresponded to the greatest number of observations of wind in that direction, that is, giving the most common or reliable wind that a mariner could expect. (In fact, the wind stars were flawed. Since ships took observations at regular intervals, slow progress in a certain direction would weight the star in that direction; a strong wind might result in fewer records for the same ocean area.)[51] The astronomer Charles Piazzi Smyth congratulated Fitzroy, noting that "a moment's glance, as when looking at a physiognomy, may give instant and true judgement: not so the looking at a page of numbers and figures."[52] Rather than simply acting as intermediate

49. Examples of the extensive work in this field are Anderson, *The Printed Image*; Baigrie, ed., *Picturing Knowledge*; Blum, *Picturing Nature*; Brennan and Jay, eds., *Vision in Context*; Crary, *Techniques of the Observer*; Galison, Jones, and Slaton, eds., *Picturing Science, Producing Art*; Hentschel, *Mapping the Spectrum*; Pang, "Visual Representation and Post-constructivist History of Science"; Rothermel, "Images of the Sun"; Rudwick, "The Emergence of a Visual Language"; Schaaf, *Out of the Shadows*; Smith, *Victorian Photography, Painting, and Poetry*; and Tucker, "Photography as Witness, Detective, and Imposter."

50. Herschel, *Preliminary Discourse*, 185, original emphasis.

51. Pinsel, "Wind and Current Chart Series"; Pearson, *Life, Letters, and Labours*, 2:55–56.

52. Charles Piazzi Smyth to Robert Fitzroy, November 5, 1855, Met. Office Papers, PRO BJ7/530.

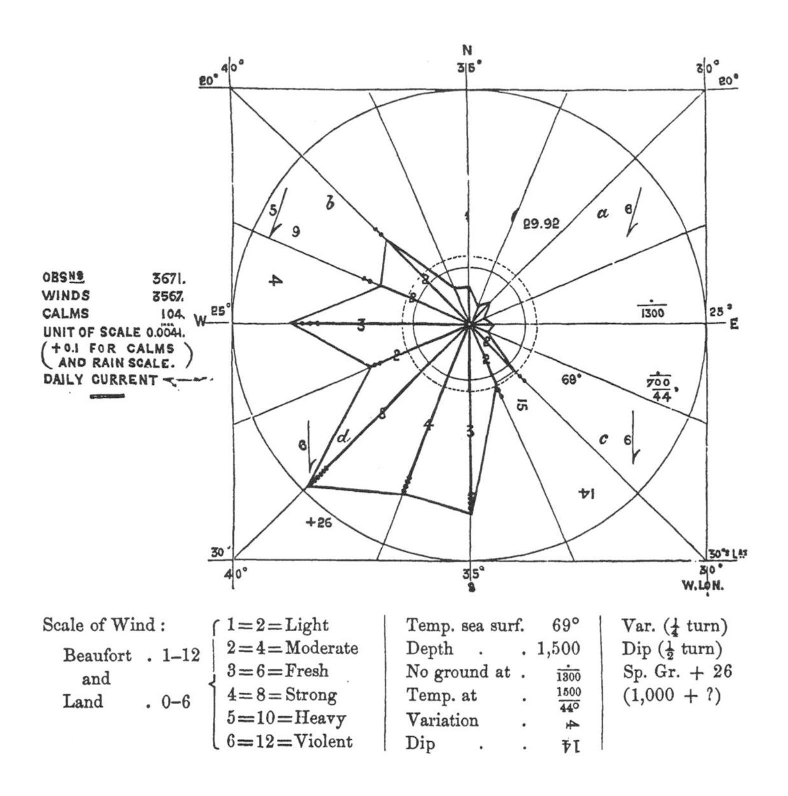

FIGURE 5.3 Fitzroy's "wind stars," which the astronomer Charles Piazzi Smyth thought gave "instant and true judgment," were an example of the department's effort to compress multiple observations into an instantly accessible visual record. The charts, a modification of nautical wind roses developed by Matthew Fontaine Maury and others, sought to represent at a glance the direction and strength of the winds a ship could expect most often to encounter within a given region and season (in this case, an area of the South Atlantic near Rio de Janeiro, January to March). The numbers within the star indicate wind force according to the Beaufort scale; other notations, some distinguished by their rotation (placed sideways or upside down), give magnetic variations, barometric pressure, sea temperature, and sea depth. (Robert Fitzroy, *Weather Book*, 2nd ed. [London: Longman, 1863], 414.)

steps, visual representations seemed to represent an ideal grasp of the subject at hand—in Smyth's words, "instant and true judgement."

Rapid insight into collective observation was embodied above all in the weather map. The origins of the weather map can be traced to Heinrich Wilhelm Brandes, a professor at the University of Breslau, who referred to the possibility of laying out atmospheric records on a map as early as 1816 and published an example in 1826. Brandes drew on data gathered by Societas Meteorologica Palatina, Mannheim, an early meteorological society that brought together observations across Europe in the decade 1781 to 1792. He almost certainly developed his suggestion in imitation of the maps laying out lines of equal magnetic force (isogons) and of trade winds that had been published by Edmund Halley a century earlier, in 1688.[53] Around the same time as Brandes, Alexander von Humboldt experimented with maps linking points of equal temperature (isotherms) and showing the currents of the trade winds. By 1850, these and similar physical maps had become the modern standard (fig. 5.4), available to serious students in the influential *Physical Atlas* of Hermann Berghaus (1836), and reworked for a British audience by Alexander Keith Johnston of Edinburgh (1849).[54] With their new prestige, maps became an obvious resource in the storm theory debates. William Redfield mapped the West Indies hurricanes of 1835, James Espy used maps to establish the direction of wind movement during storms, and Elias Loomis more systematically explored the value of synoptic maps of storms. After Loomis, the isobaric lines showing deviations from normal pressure joined the clustered wind arrows as typical features of meteorological maps (fig. 5.5).[55]

Maps gave a two-dimensional presence to what were often invisible forces or undefined space like the atmosphere or ocean. Fitzroy, following Maury's example, insisted meteorological work would make the ocean recognizable land-like space rather than merely points of latitude and longitude, and on occasion he called the isobars "soundings," self-consciously linking such maps with his own hydrographic career.[56] Meteorological mapping held a special interest for Fitzroy beginning in 1857, when he first began to collate observations on the north Atlantic and transfer them to a series of charts. By 1860, he and his assistant had produced hundreds of these charts, designed to express

53. Halley's map of the trade winds, some form of which must have been reprinted in every reputable nineteenth-century account of meteorology, was published in 1688. Monmonier, *Air Apparent*, 24–26, summarizes the evidence on cartographic "firsts." Cf. Robinson and Wallis, "Humboldt's Map of Isothermal Lines" and *Cartographical Innovations*; and Robinson, *Early Thematic Cartography*.

54. Camerini, "The Physical Atlas"; Robinson, *Early Thematic Cartography*.

55. On Espy and Loomis, see Fleming, *Meteorology in America*; Kutzbach, *The Thermal Theory of Cyclones*, 22–28; and Garber, "Thermodynamics and Meteorology." On Loomis, see Hellmann, *Meteorologische karten*. On the general characterization of mapping in nineteenth-century science, see Cannon, *Science in Culture*; and Dettelbach, "Humboldtian Science."

56. *Report of the Meteorological Department*, 1857, 321 (Parliamentary Papers); Fitzroy, "On British Storms," *Report of the BAAS*, 1860, 60. On Maury, in addition to his influential *Physical Geography*, see Bruce, *The Launching of Modern American Science*, 171–86; Deacon, *Scientists and the Sea*; Dick, "Centralizing Navigational Technology"; Pinsel, "Wind and Current Chart Series"; Smith, "Marine Meteorology"; and Williams, *Matthew Fontaine Maury*.

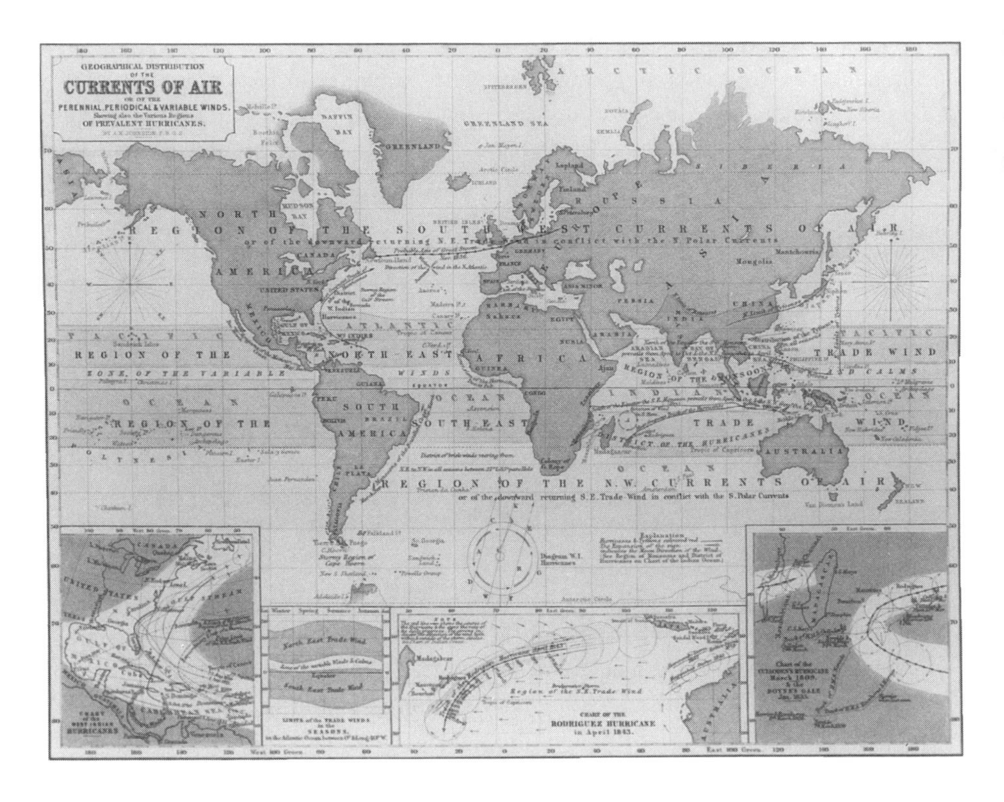

FIGURE 5.4 An example of the standard meteorological representations in the physical atlases of the mid-nineteenth century. "Currents of Air" marked the prevailing winds across the globe but also supplied maps (inset) giving the path of hurricanes, with the revolving cyclonic movement indicated by successive circles, in the West Indies, South Pacific, and Indian Ocean. (Alexander Keith Johnston, *The Physical Atlas* [Edinburgh: Blackwood, 1850], opp. 54; Special Collections, York University Library.)

"consecutive simultaneous states of the atmosphere—as if an eye in space looked down on the whole North Atlantic" at regular intervals.[57] His map of the *Royal Charter* storm in 1859 gave such a view (fig. 5.6). Because this dramatic disturbance passed over the middle of England, affecting many densely populated areas, Fitzroy's data was especially extensive. This map included isobar and isothermal lines, wind vectors that showed the direction of the wind and indicated its force by their length (longer line, stronger wind), and symbols for cloud cover and different forms of precipitation. Fitzroy concluded that many

57. Fitzroy, *Weather Book*, 103.

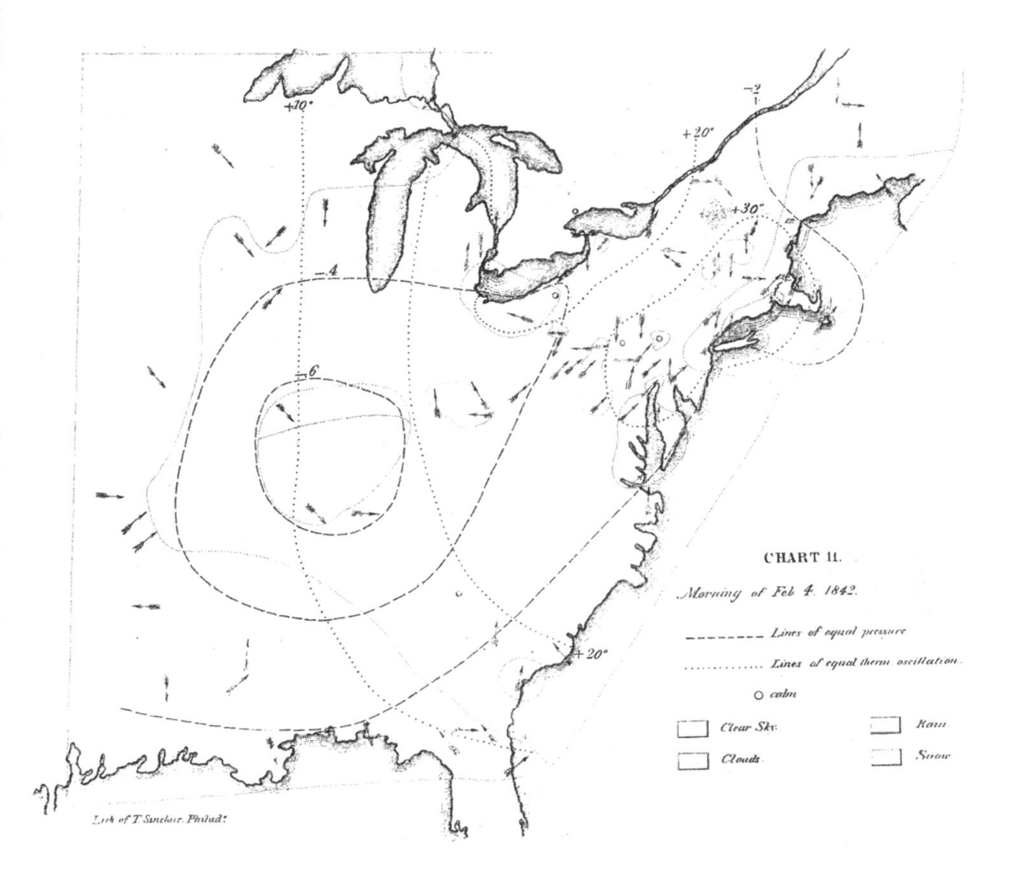

FIGURE 5.5 The storm maps of Elias Loomis analyzing a storm of 1842 brought together the main features of what became the standard representation of the weather: isothermal lines, here connecting points of equal deviation from average temperature; isobars, here connecting points of equal deviation from average barometric pressure; and arrows giving wind direction. Loomis presented his analysis to the American Philosophical Society in Philadelphia in 1843, in an effort to resolve the rival theories of wind movement during storms—whether winds whirled around a center, or blew in toward the center from all directions. In his maps, though wind directions remained visually confusing, the regions of air under different pressures literally took on shape. To properly appreciate the impact of such weather maps, we need to consider the contrast between the startlingly empty spaces of this image, against which the shapes stand out so sharply, with the raw records, densely printed columns of numbers and remarks. Loomis did not close the controversy about storms, but he made clear the overwhelming advantage of transferring massed observations to synoptic charts. (Elias Loomis, "On Two Storms Which Were Experienced Throughout the United States in the Month of February 1842," *Transactions of the American Philosophical Society* 9 [1846]: 161.)

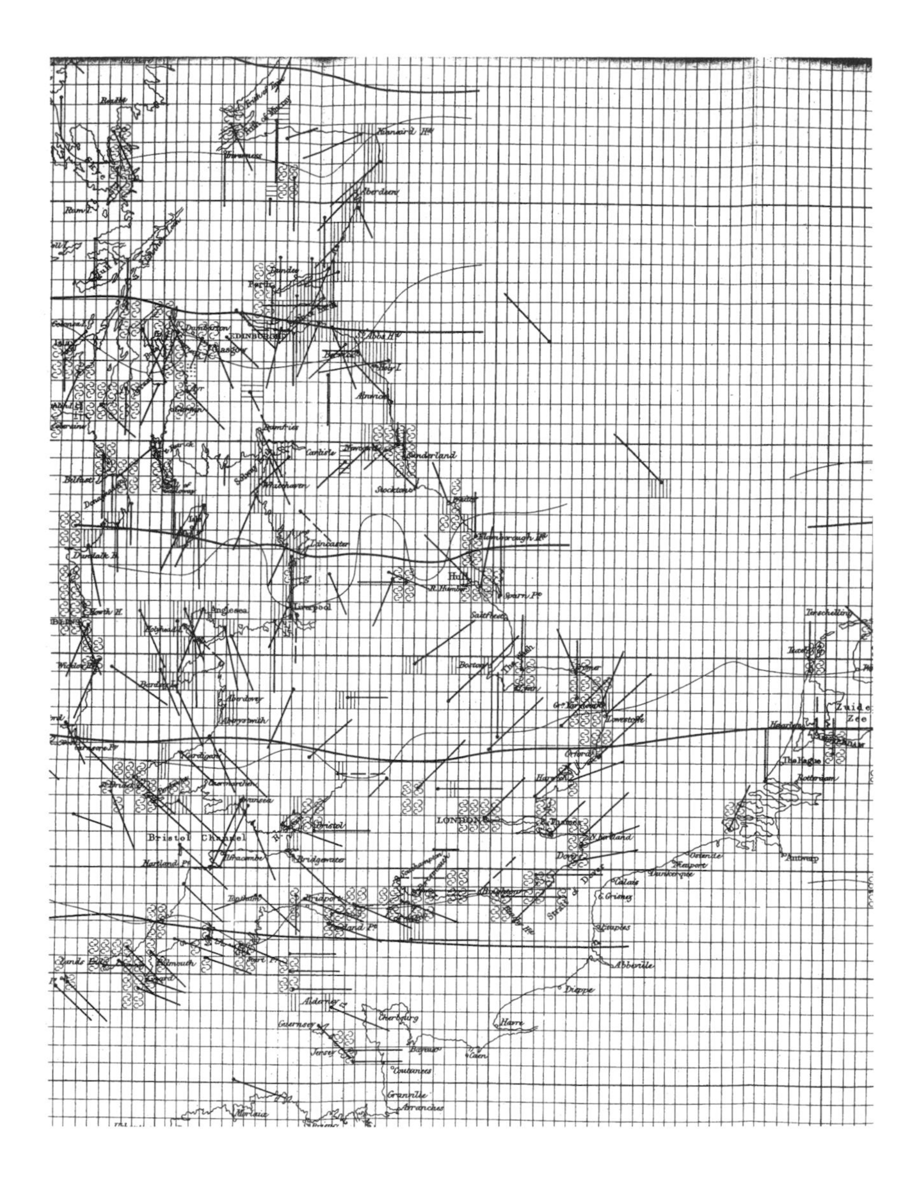

FIGURE 5.6 Detail of Fitzroy's synoptic map of the *Royal Charter* storm at 9 a.m., October 26, 1860. Mapping the gale's progress convinced him that storms could be predicted, although the visual complexity of the end result suggests why in many different cases of nineteenth-century weather maps, the experience of producing a map inspired more conviction than that of encountering it in completed form. Also, in comparison with the continental scale of the map in figure 5.5, we can see the geographical disadvantages of British weather study: Fitzroy's observations produced no clear isobaric shapes, although the whirling counterclockwise movement of air did emerge from the placement of wind "arrows" (here represented as simple lines, with their length correlated to force). Different shadings indicated cloud, hail, snow, rain, or fog. (Robert Fitzroy, *Weather Book*, 2nd ed. [London: Longman, 1863], plate xiii.)

telling indications preceded the disturbance, such as a very low barometric pressure, a clear atmosphere, and magnetic disturbances affecting telegraphic wires.

His maps also convinced him to view the atmosphere in terms of contesting currents of warm and cold air in which local eddies, carried along in larger bodies of moving air, produced stormy conditions. It was a version of the convective or thermal theories of cyclones, similar to other theories that were emerging among U.S. and European meteorologists, and it was particularly indebted to Dove's insistence that atmospheric changes depended on the alteration of polar and equatorial currents. Fitzroy's contribution to debates about cyclone theory was modest at best. Yet maps shaped the essential feature of his meteorological career: his conviction that observations could be carried into the practical arena of forecasting. In that sense, the *Royal Charter* storm map was insignificant in comparison to the hundreds of other maps he and his assistant compiled. These masses of charts, showing "the ordinary course of nature," founded his confidence that weather prediction was not only possible, but "a daily public duty."[58]

Maps and telegraphs also both promised to address problems of intelligibility and communication. This is an obvious point to make about maps, for one of their advantages of course was that they did away with the differences of scales that hindered easy exchange of data among European meteorologists. It is a less obvious point to make about telegraphy, and the key lies with understanding the telegraphic codes as a symbolic representation that transcended the details of numbers, in the same way that the symbols on a map or chart did. As a kind of meta-instrument, the telegraph was a technological recourse for the obstacles of communication that appeared to hinder the development of meteorology. Indeed it provided an elaborate physical model of the stages and labor involved in moving from individual observation to public conclusion. In Fitzroy's network, observers recorded measurements from their instruments six days a week at 8 a.m. and sent to London rainfall, temperature, and air pressure readings, as well as the direction and force of wind. The cost of transmitting messages using Britain's privately owned telegraph companies led Fitzroy in 1861 to develop a complex code that maximized the amount of information he could receive at least cost. (Telegraph companies charged messages by the word, and in 1860 each figure, even a decimal point, was charged in England as one word; on the Continent, by contrast, five figures

58. Ibid., 108.

were considered one word.)[59] An instruction manual of 1860 described a set of four "words" of five numerals each, which conveyed measurement of the barometer, the air temperature, and a humidity reading as well as indications of wind direction, the amount of cloud or the appearance of the sky, and an estimate of wind force.[60] Fitzroy claimed triumphantly that he could send full data from six observation points to Paris in twenty of these "words," including addresses.[61] Nor was this encoding the end of the process. The telegraphed words would in the London office be combined with other reports, national and international, studied, and then retransmitted as descriptions of probable weather, in written rather than numerical versions, to newspapers. Finally, warnings would be wired to the coastal stations, where they would be translated again into the other, more famous code developed by Fitzroy, the cone and ball signals hoisted on docks or headlands for all the marine community to see (see figs. 3.6, 3.7). The impressive quality of this mixture of electric transmission, codes, and symbols may be hard to appreciate now; but for Victorians it was dramatic and startling.

Fitzroy's enthusiasm for telegraphic codes and storm signals also indicated how British meteorology participated in a general discussion of the merit of symbolic representation as a way of connecting scientific advance and popular education. Such discussions provide further evidence of the affinities between telegraph and visual products like the signals, charts, and maps. This approach to symbols can be traced in a text that was widely known in the maritime community. Lieutenant Henry Raper's *Practice of Navigation and Nautical Astronomy* was a virtual textbook in naval and merchant marine training for much of the nineteenth century. Raper had worked on the *Adventure* during 1815–20 under William Smyth, an eminent hydrographer and astronomer. He left active service in the 1820s, and instead put his energies into the emerging scientific societies, the Geological and Astronomical Societies.[62] His *Practice of Navigation* was first published in 1840, when it was awarded a gold medal by the Royal Geographical Society and it became part of the required library of naval vessels. By 1870 it had reached ten editions. In the work, Raper discussed questions of numerical accuracy in connection with a wider debate about what was valuable information for the practical navigator.

59. Ibid., 355; for a contemporary discussion of the costs, see Anderson, "On the Statistics of Telegraphy."

60. Robert Fitzroy to George Airy, September 22, 1860, Airy Papers, RGO 6/702.

61. Robert Fitzroy to George Airy, September 19, 1860, Airy Papers, RGO 6/702.

62. See *Dictionary of National Biography*.

Commenting in the first edition on the "very indistinct and erroneous notions" about computation "among practical persons," Raper argued against a kind of "useless precision" that lent false authority to a computation. He asserted that calculations made without consideration of the quality of the information only obscure the subject. They tend to persuade people "that a result is always as correct as the computer chooses to make it."[63] In later editions, it became clear that Raper's alternative to the problem of long calculations and false precision was the same as his answer to the problem of long descriptions—whittle them down to practices that could be performed quickly with reasonable accuracy under the often trying circumstances of a seafaring life. For these purposes, he held, there was nothing like the substitution of symbols that could speedily give all necessary information, expediently and (genuinely) accurately. Just as "every exertion should be made to abridge computation," so should symbolic notations replace long descriptions. "Suitable and expressive symbols . . . [are] not merely a convenience," Raper argued, but have the "still greater consequence, namely, certainty."[64] In his second (1841) edition, Raper accordingly developed "a new system of Symbols." His example was a symbol for anchorages, a small anchor, surrounded by a box, to the right of which were two numbers arranged vertically, the higher giving the depth at high tide and lower at low tide. With such symbols, Raper claimed, the reader's work "was done at once; he seizes in an instant the information given and his mind is altogether unembarrassed by circumstances of narration."[65]

Raper clearly saw this work as part of contemporary changes in navigation and science and linked it to developments in geographical mapping. Citing Berghaus's *Physical Atlas*, he noted that "a growing tendency to the use of symbols manifests itself on all sides." While "form[s] addressed to the eye" might a few years ago have been dismissed as "rash innovation," symbols were clearly now the way of the future. They "arrest and fix the attention," powerfully compress information in the smallest possible space, and contrast sharply with the "monotonous aspect of alphabet writing."[66] Raper's advocacy of symbols in the 1840s was followed in the next decade by reforms to the signaling code of the British marine. Here the parallel treatment of description (the monotonous alphabet) and numbers (the excessive and misleading computation) can be seen once more. In 1857, an old code that had been

63. Raper, *Practice of Navigation*, 1st ed., vi–vii.

64. Raper, *Practice of Navigation*, 2nd ed., xiv.

65. Ibid., 383.

66. Ibid., vi–vii.

"perpetuated on the vicious principle of numerals" was rejected as "complicated and inadequate" by an investigating committee of the Board of Trade. The older numerical signals had been confined to the digits 0 to 9, and the smaller number of possible combinations had led to a complicated system of substitute flags and index flags. The latter gave identical signals different meanings, depending on the division of signals indicated by the index flag. The new signal book was issued with a total of 18 flags or symbols. These were alphabetic and would be hoisted in groups of either three or four, giving a "power" of 78,682 separate signals "complete in themselves" (i.e., hauled up at one place and moment). The former category of three flags was designed to be international and independent of English—"a Universal Language of Signals, equally available to all nations."[67] Symbols were both more accessible and offered a greater depth of information.

Both these navigation examples, Raper's system and the Signal Code of 1857, spoke of symbols as modern responses to new orders of complexity and professionalism. (A ship's identity code could be transferred directly as telegraphic code, for instance, for production of the shipping lists.) The necessity for "genuine" accuracy, and for methods that reached and defined a widening community, motivated the developments. In this, there were clear parallels with other kinds of scientific work and, especially, a ready connection with meteorology, whose findings were above all directed to the maritime community. Meteorology's interest in maps and charts thus took place in a context of general enthusiasm for the powers of symbolic representations among the group that was apparently the natural beneficiary of, and perhaps natural audience for, the science. Fitzroy himself built his telegraphic codes and his charts on a careful reading of Raper. His directions for compiling a meteorological register, with diagrams of barometer and thermometer readings, stressed their superiority to "numerical figures." Their compactness and accuracy, he claimed, would "shew the principal atmospheric changes—at a glance—in the least space of paper . . . under all circumstances" (fig. 5.7).[68] In these remarks, Fitzroy paid a characteristic tribute to the way symbols met a blend of practical and scientific needs.

Yet precisely because of the reputation and purpose of accessibility represented in these forms of visual knowledge, they raised questions about expertise. How could the notorious difficulty of meteorology be squared with

67. "The New Code of Signals." Cf. the editorial on the old and new system, *Times*, April 10, 1857.

68. Fitzroy, *Weather Book*, illustration XVI to appendix Q, "Explanation of Diagrams" (not paginated).

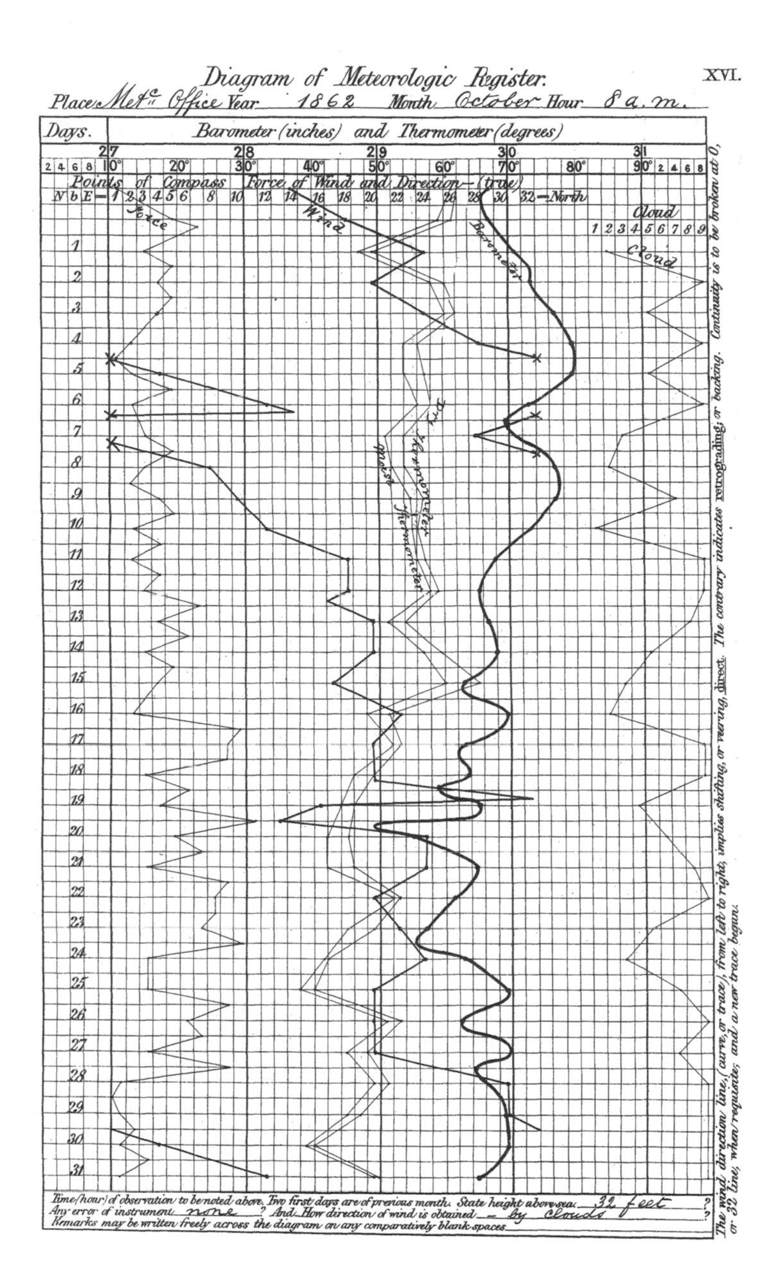

FIGURE 5.7 Another instance of Fitzroy's enthusiasm for visual records. (Robert Fitzroy, *Weather Book*, 2nd ed. [London: Longman, 1863], plate xvi.)

meaning so transparent that "anyone" could see it? And if anyone could glance at the sky and the weather maps, what happened to the role of the scientific expert? Fitzroy, in the remarks cited earlier, had emphasized that telegraphic weather data opened forecasting up to anyone. With a newspaper in hand, a local barometer, and a look outside, any individual could become a kind of weather center himself. But of course, this glossed over the telegraphic network centered on the department and Fitzroy in Parliament Street. In fact such a network was so expensive and required so much coordination that private equivalents were really not an option. This was one of the main arguments for government leadership—rapid, international collection of meteorological data could not be undertaken by private scientific enterprise. The intervention of an expensive technology like the telegraph then evoked privileged access and control as well as unlimited participation. One enthusiast who visited Fitzroy with his theories complained of the latter's great "want of courtesy," which arose, the writer concluded, "from an expressed conviction on his part that no one *could* know more about the weather and registrations than he did, he having so great a correspondence with observers."[69] The new technique of telegraphy, distributed in the press, may have allowed any observer to "see" beyond the confines of his or her own horizon, but it could also give Fitzroy, as the direct recipient of nationwide observations, a form of omniscience not available to anyone else. With these ambiguities in mind, then, we need also to consider how mapping complicated boundaries between practitioners and audiences of meteorological science.

A series of European synoptic maps, *Meteorographica*, produced by Francis Galton in 1863, provides a starting point. He himself came to meteorology from an interest in maps, cultivated during his explorations of southwest Africa from 1850 to 1852. An early publication on geography in 1855 included a plea that maps should strive for the most intensive information possible, in order to rise above "an abstraction, or a ghost of . . . vivid recollection." Strikingly, Galton contrasted the potential of the visual sense, impoverished by the "meagre information" in ordinary maps, with another sense, that of touch. In present inadequate maps, Galton claimed, "a blind man fingering a model could learn as much from his sense alone, as they convey to our eyes."[70] Mapmakers needed to find better ways to exploit the power of vision. By 1865, Galton developed one fairly successful and enduring solution, a method for stereoscopic maps,

69. Stephen Saxby to George Airy, December 28, 1860, Airy Papers, RGO 6/700, original emphasis.

70. Quoted in Forrest, *Francis Galton*, 78–79.

photographing mountainous landscapes with a double-view or stereoscopic camera.[71] Two years earlier, however, he had pursued the same goal of accurate, intensive visual realization in meteorology with much less such success.

Meteorographica presented a series of ninety-three maps of European weather in December 1861, three per day for the month. Galton had solicited observation from across Europe, but in the end his data from about eighty stations came principally from Belgium, Holland, Austria, and Prussia—the national observatory in France was uncooperative, Italian data failed to arrive, and few observations were available from Norway, Denmark, or Ireland. Yet even with partial coverage, his maps of the wind and barometric pressure in Europe produced at least one important result. They revealed that high pressure areas were calms and surrounded by clockwise winds—a condition he termed "anti-cyclone" in reference to the "cyclone" pattern of low pressure, calm areas surrounded by counterclockwise winds.[72] Galton felt optimistic. The work showed, he argued, that meteorological maps could transform observations into useful scientific data. "A few judicious sweeps and shadings of a draughtsman's pen may embody the simultaneous observations of hundreds of meteorologists."[73]

Despite the anti-cyclone and Galton's confident tone, *Meteorographica* was nevertheless not a conspicuous success. One critic compared the symbols to "Chinese spelling"—a synonym for "unintelligible"—and even his own student and biographer, Karl Pearson, referred regretfully to the "feeling of repugnance which the crudely hatched masses of red and black on his charts excite in our minds" (fig. 5.8).[74] Although maps seemed to promise rapid, accessible, and transparent knowledge, they were just as likely to raise problems of interpretation and call attention to a hierarchy of technological and analytical expertise. Although Galton had emphasized the smooth development of knowledge from scattered observations to the panoramic viewpoint of centralized meteorology, maps also underlined the interpretive work of that process, and the *absence* of definite, transparent meaning. The meteorological maps of the mid-Victorian period often seemed to represent a movement from fact to opinion, rather than from fact to scientific knowledge.

71. Galton, "Notes on Modern Geography" and "On Stereoscopic Maps"; Pearson, *Life, Letters, and Labours*, 2:21–35.

72. Galton, *Meteorographica*; Pearson, *Life, Letters, and Labours*, 2:38–39.

73. Galton, *Meteorographica*, 3, emphasis added.

74. Pearson, *Life, Letters, and Labours*, 2:41.

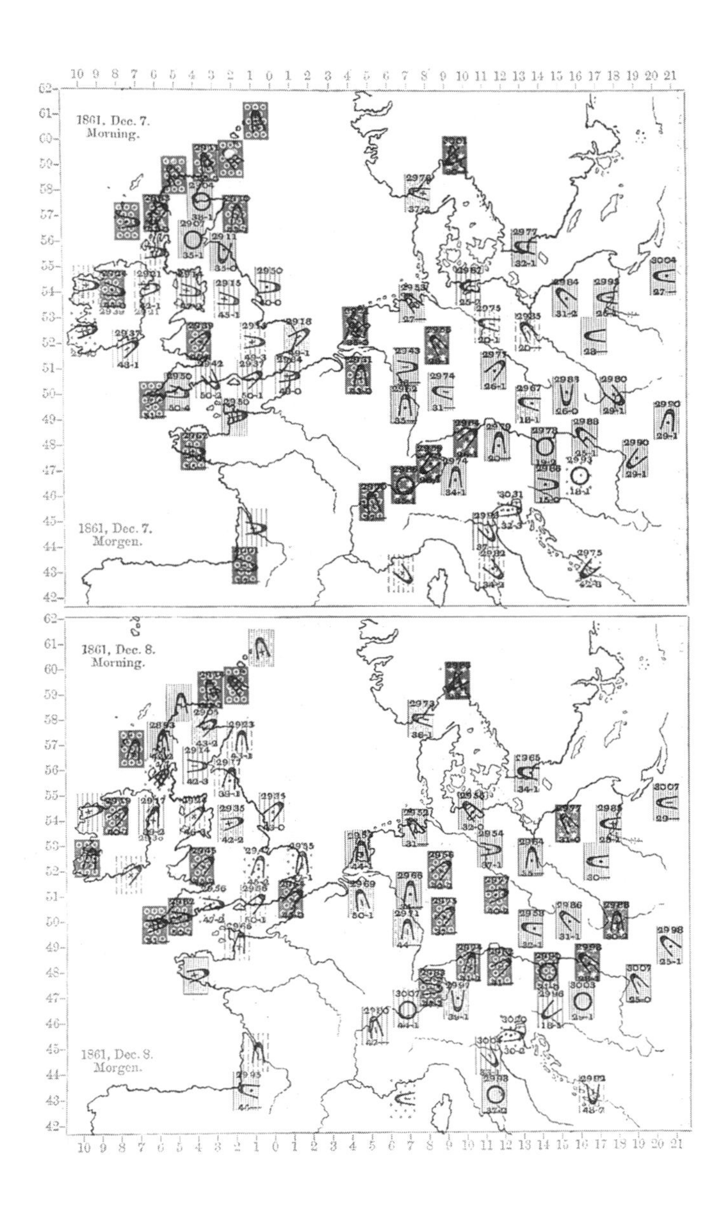

FIGURE 5.8 Galton's weather maps in *Meteorographica*, 1863. They were strikingly unlike the synoptic charts that were becoming standard: Galton situated an observation on a geographical space, but did not submerge separate records or eliminate numbers in the way that isolines did. ("December 7 and 8, 1861," in Francis Galton, *Meteorographica* [London: Macmillan, 1863].)

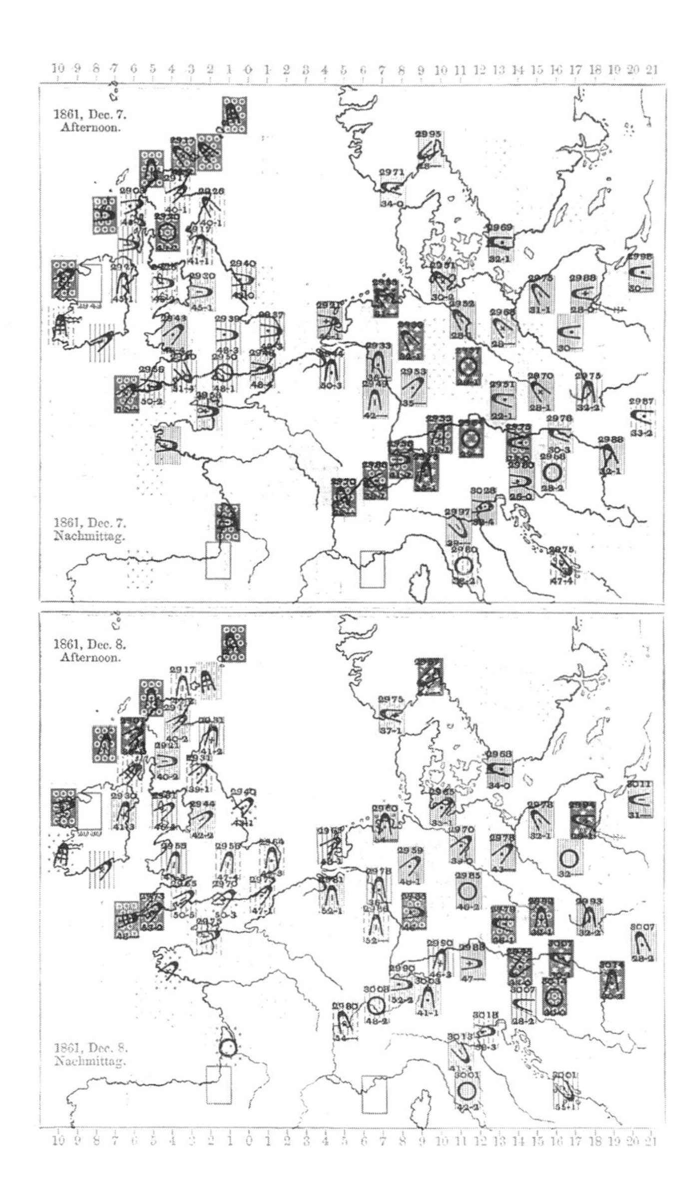
1861, Dec. 7.
Afternoon.
1861, Dec. 7.
Nachmittag.
1861, Dec. 8.
Afternoon.
1861, Dec. 8.
Nachmittag.

In 1863, the year *Meteorographica* was published, Galton sat on the editorial board of a new London scientific publication, the weekly *Reader*, an unsuccessful predecessor of *Nature*. He likely authored a review of LeVerrier's introduction of maps in his *Bulletin de l'observatoire imperial de Paris*, which appeared in the *Reader*. The review concluded that, despite the "audacious dogmatism too common among meteorologists," LeVerrier's publications showed that maps were not a powerful tool for prediction. "No living meteorologists who did not know them by heart could place them back again into their proper order."[75] A year earlier, Fitzroy in his *Weather Book* had similarly acknowledged the dilemma of mapping: "as even the earnest and experienced student of meteorology is often perplexed, if not in error, while drawing hasty conclusions from the first view of facts: how much risk of incorrect decision must there be when persons insufficiently acquainted with the subjects are obliged to decide."[76] In his own work, Galton had uneasily recognized the "wide room for fancy" when observations were inadequate. "Exercising the right of occasional suppression and slight modification, it is absurd to see how plastic a limited number of observations become, in the hands of men with preconceived ideas."[77] The answer to the problem was expertise—"patience and training of the eye"[78]—but given the lack of agreement among meteorologists about which maps were most useful, this counsel was less than convincing.

This plasticity seemed obvious to contemporaries, and it became a feature of the popular response to the charts and maps of meteorological science. In the novel *Tono-bungay*, H. G. Wells satirized the illusory appeal of the weather chart when Mr. Ponderevo, a chemist, succumbs to his own heady vision of judicious sweeps and shadings. Taking the idea from his nephew's meteorological graphs, he develops "an innocent intellectual recreation . . . called stock market meteorology." As he excitedly explained, "It's as plain as can be . . . See, here's one system of waves and here's another . . . It's absolutely scientific. It's verifiable. Well, and apply it! You buy in the hollow and sell in the crest, and—there you are!"[79] Naturally, his exercises led the family to bankruptcy. Wells published the novel in 1906, but he may well have been familiar with some of the earlier satirical commentary on the weather. In October 1875, a few months after the resumption of daily weather observations in the daily papers, *Punch*

75. *Reader*, December 19, 1863, 730, a review that George Airy kept (Airy Papers, RGO 6/703).

76. *Report of the Meteorological Department*, 1862, 457–58 (Parliamentary Papers).

77. Galton, *Meteorographica*, 5.

78. Ibid.

79. Wells, *Tono-bungay*, 66–67.

published a letter and chart from "the Meteorologist" entitled a "Weather-Eye-Opener" (fig. 5.9, *left*). Requesting allowances for a "first attempt" by a "young man who . . . apparently knows nothing about drawing and still less about Meteorology," the *Punch* cartoon featured a blank central office surrounded by chaotic and incomprehensible symbols. Wind arrows pointed every which way and alphabetic codes sprouted randomly alongside (a = awful, b = beastly, e = everlasting, etc.). The accompanying remarks referred to breakdowns in the instruments: the barometer tapped too hard with an umbrella and the thermometer's mercury tube destroyed by "a youthful and scientific member of the family," who placed it in the fire. Not willing to leave a good joke quickly, *Punch* later published "Our Whether We Like It or Not Chart," here converting the wind arrows to signs of illness (fig. 5.9, *right*). Hayfever and Summer Catarrh swept in from the north as Mumps settled over Ireland, Croup descended on London, and finally, Appetite rose across the Channel in France. As these spoofs made abundantly clear, the interpretation of the image was far more difficult to control than its production.

The engagement with ideas about weather wisdom in mapping persisted well beyond the first experiments with official weather prediction in the 1860s. It can be seen for instance in Ralph Abercromby's *Weather: A Popular Exposition* (1887), a volume in the well-known transatlantic International Scientific Series, which presented authoritative summaries of modern science for the lay reader in the last quarter of the century. Listing seven fundamental shapes of isobars in a synoptic map, Abercromby maintained that all prognostics could be matched to these forms. Accordingly, his diagram of a cyclone incorporated descriptions of the sky, merging the observations associated with weather wisdom into the maps of "modern meteorology." An idealized cyclone moving northeast showed muggy weather in the front, cool weather in the rear, with distinct cloud forms accompanying its passage. Besides precipitation and clouds, Abercromby noted "Restless animals," "Corns," and "Watery Sun," and he combined on the map scientific cloud nomenclature such as "Detached Strato-Cumulus" with more general descriptive terms such as "Dirty Sky" or "Hard Sky" (fig. 5.10).[80]

Although weather maps gave shape to the forces of the atmosphere, then, they were ambivalent resources. Because maps made instruments and observers, collectors of the data, invisible, they could be inadequate to convey distinctions between respectable and unrespectable weather science, or between various levels of expertise. Weather maps often seemed unintelligible,

80. Abercromby, *Weather*, 18, 28.

PUNCH'S WEATHER-EYE-OPENER.

Mr. Punch, resolved not to be behind his contemporaries in enterprise, has determined to publish henceforth a Meteorological Chart. Allowances are requested for his first attempt, as the young man who was engaged (on his own representations) for the purpose apparently knows nothing about drawing, and still less about Meteorology.

WEATHER CHART.

Explanation.—The *arrows* fly with the wind, in fact, *bow* before it. *a*, awful; *b*, beastly; *c*, cursed; *d*, never mind; *e*, everlasting; *f*, frantic.

Rainfall.—Wet through twice a day, and through the ceilings of the attics.

Barometrical Readings.—None. Somebody having smashed the barometer by tapping it too hard with his umbrella.

Thermometrical Readings.—None. A youthful and scientific member of the family having robbed the thermometer of its practical utility by putting it in the fire to see the mercury run up the tube.

General Remarks.—Shan't make any, for fear of using bad language. Dreadful cold in my head.

The Meteorologist to Mr. Punch.

OUR WHETHER WE LIKE IT OR NOT CHART.

See what we're in for :—

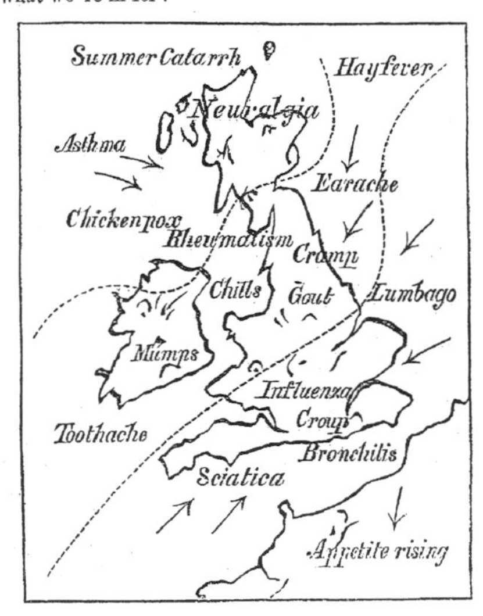

FIGURE 5.9 *Punch*'s spoof of weather charts. The first cartoon shows the continuing reputation of weather charts as unintelligible, with a blank central office surrounded by randomly scattered arrows and letters (*a* = awful; *b* = beastly, *c* = cursed, *d* = never mind, *e* = everlasting, and *f* = frantic). "Our Whether We Like It or Not Chart" mirrored the standard chart, showing the familiar isolines and arrows, but marked weather conditions with human ailments and disease. Note the "appetite rising" on the Continent. (*Top*: *Punch*, September 10, 1881, 119. *Bottom*: *Punch*, October 30, 1875, 182.)

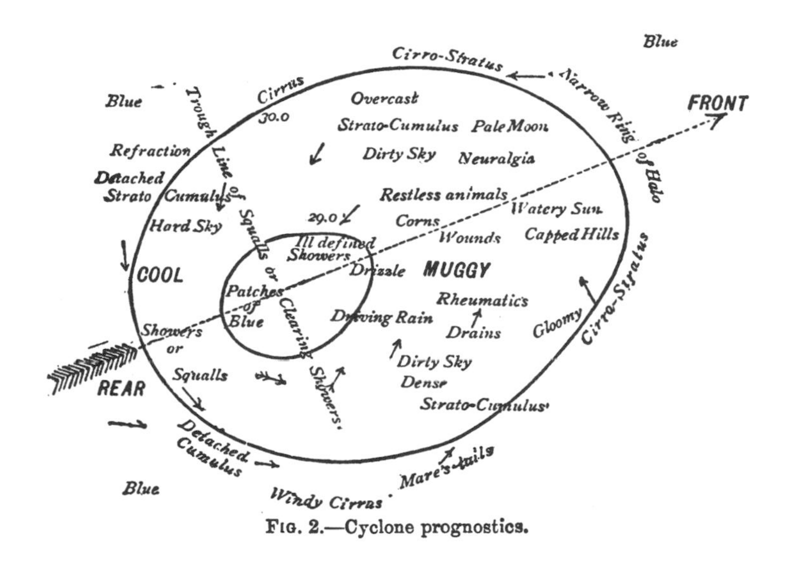

FIGURE 5.10 Ralph Abercromby's map of cyclone prognostics (1887) blended the distinctive isobaric shape of a scientific weather map with a catalogue of features associated with weather wisdom, like animal behavior, bodily sensations, and the appearance of the sky. Note the mixture of descriptions of clouds, from "Strato-Cumulus" to "Mare's-tails" or "Dirty Sky." (Ralph Abercromby, *Weather: A Popular Exposition of the Nature of Weather Changes from Day to Day* [London: Kegan Paul, Trench, 1887], 28.)

leaving their creator unable to express his insights, just like the weather-wise but inarticulate seamen who could not express theirs. The problems with mapping stand out clearly, and perhaps unfairly. It should not be forgotten that their background was a series of very successful objects and techniques, like Humboldt's air currents maps published in the Berghaus atlas. Nevertheless, at the same time as it displayed the power of visual analysis, the weather map certainly illustrated the vulnerability of meteorology.

A final example will indicate how clearly meteorologists understood this vulnerability. In 1875, as we have seen, the Meteorological Office did resume weather predictions. Earlier weather predictions had been accompanied by a table of data; but on April 1st, the Meteorological Office began to put out daily weather charts and weekly barographs that were published in the *Times* (fig. 5.11). Descriptions of this innovation indicated a guarded understanding of the popular interest in meteorology. "It is hardly necessary," wrote *Nature*, "to allude to the value of such charts . . . as a means of leading the public to gain some idea of the laws which govern our weather changes. As soon as they appear in our afternoon papers we may hope for a more intelligent comprehension of the difficulties which beset any attempt to foretell the weather of these islands for the space of even twenty-four hours."[81] In other words, instead of using maps to show the weather at a glance, these maps were to emphasize the opposite: how difficult it was to understand and predict the weather. In an equally significant move, the account turned its attention instead to the technology of the central office: the process of map production that supported the exclusive expertise of government meteorologists. The bulk of the article was devoted to describing this impressive sequence of steps that resulted in the image, from the collection of telegraphic data, to the pantographs that reduced the outline map, to the engraving block that could be engraved rapidly "without blurring or chipping," to the templates that would provide a uniform typeface of the characters.[82] The description emphasized the pressure and speed of the process and the sophisticated printing technology that the map required—highlighting the technological foundations of the privileges of the central office. No other observatory or society, let alone an individual meteorologist, had the resources to make these daily charts. By shifting readers' attention to the *process* of image production, rather than the predictions themselves, the new charts reasserted the powerful position of the central office and cautioned the public against easy interpretations of modern scientific data.

The reputation of weather wisdom made the management of visual experience in meteorological science particularly acute. Its local, unlearned insights seemed inaccessible to conventional instruments and difficult to exchange across geographical space and between individuals with different experiences. In this way it was the antithesis of universal science. Yet, paradoxically, weather wisdom unsettled conventional science because as knowledge based on sensation, it seemed to be automatic and exact—a natural model of the kind of precision that scientific instruments tried to replicate. Given the important role that precision instruments played in building the authority of science, this comparison was potentially troublesome. Did impressive technologies like the telegraph, or parallel representational techniques, like maps, provide adequate substitutes for weather wisdom? As we have seen, meteorologists were hopeful, but not assured. Often, their instruments and techniques led them into further doubts about the relative authority of popular insight versus

81. "The *Times* Weather Chart," *Nature*, April 15, 1875, 474. Cf. Scott, "Weather Charts in the Newspapers."

82. Pearson, *Life, Letters, and Labours*, 2:44–47.

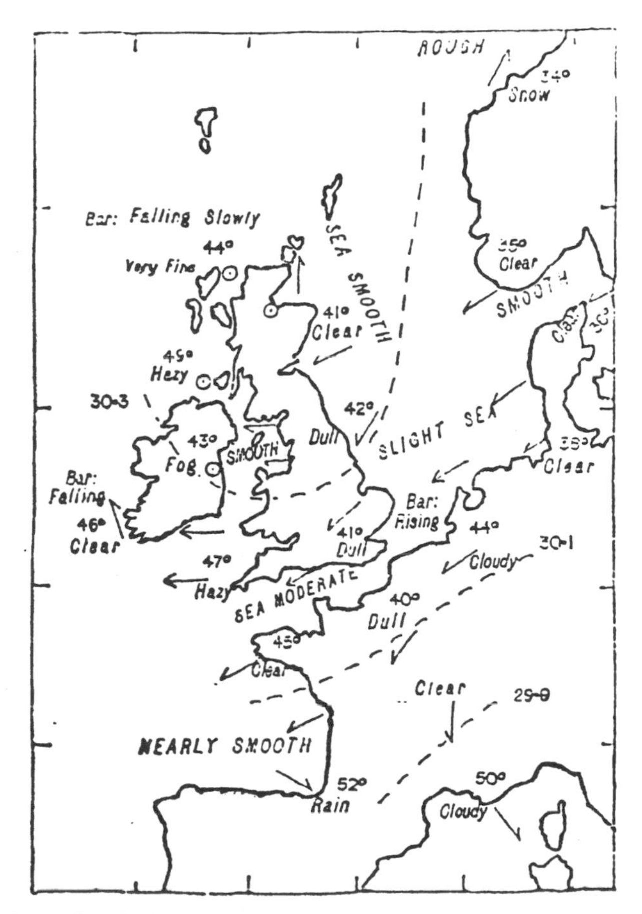

The dotted lines indicate the gradations of barometrical pressure, the figures at the end showing the height, with the words " Rising," " Falling," &c., as required. The temperature at the principal stations is marked by figures, the state of the sea and sky by words. The direction and force of the wind are shown by arrows, barbed and feathered according to its force. ⊙ denotes calm.

FIGURE 5.11 The *Times* weather chart, first appearing on April 1, 1875. An article discussing the innovation in *Nature* stressed that it would lead to "a more intelligent comprehension of the difficulties" of forecasting, a remark that entirely rejected meaning at a glance. (*Nature*, April 15, 1875, 473.)

official expertise, contributing to criticisms of "mere numbers" and raising questions about interpretation.

The history of Victorian weather maps thus traces a pattern of eagerness and irresolution in meteorological science, as the boundless promise of the technique returned meteorologists to established problems with interpretation and expertise. The same pattern emerges in the case of the camera and the spectroscope. These other instruments showed how meteorologists repeatedly sought to establish boundaries for the consideration of visual evidence. After the initial experimentation with maps and weather telegraphy in the 1860s, ideas about weather wisdom continued to play a significant role in developing the critique of conventional observation in meteorological science, merging with concerns about aesthetics and language. The best place to analyze these developments is the work of Charles Piazzi Smyth.

As a young astronomer's assistant at the Cape of Good Hope in the late 1830s, Smyth had taken notes and submitted observations on meteorology; at the other end of his career, in the 1880s and 1890s, he turned to cloud study and spectroscopic analysis of the atmosphere. In the middle, as Astronomer Royal of Scotland, Smyth's Calton Hill Observatory was responsible for analyzing the data collected by the Scottish Meteorological Society and submitting the results to the registrar general. Smyth's varied involvement with meteorology over several decades suggested that although his instruments advanced the cause of precision, they also had other effects. He argued that they broadened the view of nature rather than simply honing its exactness. Images, for instance, could highlight aesthetic responses to nature. The very success of his instruments, moreover, in Smyth's view, served to emphasize the limits of "the human hand, eye and memory."[83] Like the clouds, whose shape dissolves "in soft formless vapour when we come close to it," nature's meaning recedes before the scientific observer. Smyth's example shows that at least for some observers, the exploration of meteorology and the nature of popular knowledge was an exploration of the reach of science. What could be communicated? Can all observation translate itself into the languages of science? And could the limits of scientific observation control the dogmatism of centralized scientific communities? With varying degrees of confidence and optimism, Smyth addressed versions of these questions in his scattered meteorological investigations from the 1870s to the 1890s. The most important of these experiments concentrated on two instruments: a weather spectroscope and a cloud camera.

83. Smyth, *Cloud Forms That Have Been*, 1:3 (held at Royal Society of London; see note 124 below).

An Integrating Glance: The Weather-Wise Instruments of Charles Piazzi Smyth

Smyth's experiments with the spectroscope and cloud camera were typical of his intensely visual approach to scientific work.[84] He was a fine watercolorist, filling his diaries with ink, pencil, and watercolor sketches. One of his earliest publications was a discussion of astronomical drawing, in the Royal Astronomical Society's journal (1843), based on sketches made at the Cape during the visit of Halley's comet at the end of 1835 and beginning of 1836.[85] Family practice encouraged him to see such productions as an ordinary part of scientific observations. His father, William Henry Smyth, was one of the best hydrographers of his generation, and his pilot charts of the Mediterranean included his own sketches of headlands and features of the coastline.[86] One of the best examples of his son's immersion in a visual culture was his interest in color. His publications and private notebooks show many instances, from his relish for colors observed with the spectroscope to a study of the changing effects of light on a cloudscape during sunset, to the vivid observation on a journey to Tenerife that "chromatically, [we] were in another hemisphere."[87] In 1879, reporting research on the color of double stars, he urged that observers should develop their sense of color, suggesting that some "400 practical tints" could be detected and recorded. Equipped with these sensibilities, he was powerfully impressed by colored ranks that appeared in the spectroscope. Turning to table-top observations of gases in the early 1880s, Smyth paid eloquent tribute to the ordered and beautiful view thus obtained. Here resolved in their "ultimate condition," the lines of the spectrum could be seen "beautifully ordered . . . perfect in structure . . . indexed ready to one's hand by colour."[88]

As one of the most ardent enthusiasts for the spectroscope in astronomy, Smyth trained the lens of that instrument in many directions. It was quite characteristic that he came to champion a vest-pocket-sized instrument that

84. Warner, *Charles Piazzi Smyth*; Hentschel, *Mapping the Spectrum*.

85. Smyth, "A Memoir on Astronomical Drawing." Cf. Hughes and Stott, "Two Piazzi Smyth Comet Paintings"; Pang, "Victorian Observing Practices"; Rothermel, "Images of the Sun"; and Schaffer, "On Astronomical Drawing."

86. For an example, see Smyth, "Taormina, Sicily." On the elder Smyth's work and charting in general, see Ritchie, *The Admiralty Chart*; De Luna Martins, "Mapping Tropical Waters."

87. Smyth, *Teneriffe*, 28.

88. Smyth, "Colour in Practical Astronomy," 791; "Gaseous Spectra in Vacuum Tubes," 96. Cf. the discussion of Smyth's spectroscopy in Hentschel, *Mapping the Spectrum*.

could be whipped out anywhere, on a train journey or a walk in the hills, as well as in the observatory. Now a rather obscure footnote in the history of meteorological instruments, the rainband spectroscope seemed to Smyth in the 1870s a scientific tool of boundless promise, as revolutionary in meteorology as other versions of the spectroscope had been in physical astronomy and chemistry.[89] Smyth first became interested in the meteorological possibilities of spectrum observation on a summer visit to Sicily in 1872, when, as usual, he could give more attention to meteorological subjects in his break from observatory duties. En route, Smyth had noted the marked difference between observations of the spectrum of the atmosphere taken before and after a sirocco, the drying wind off the Saharan desert. Once more off to the Continent in 1875, Smyth again undertook regular spectroscopic observations of the atmosphere, and at this time his interest was further stimulated by fresh attention to the challenges of forecasting. In Paris, Smyth and his wife paid a visit to Urbain LeVerrier, the autocratic director of the Paris Observatory. LeVerrier was less than ideally hospitable to his visiting Scottish colleague, and in his journal afterward, Smyth took malicious pleasure in contrasting the prediction of dry weather that LeVerrier's institution had made earlier that day with the evening's "great storm of lightning and rain"—LeVerrier's discomfiture was clearly some solace to Smyth as he and his wife "were left to find our way home without any refreshments or assistance."[90] That storm was part of a long bout of wet weather. Both the French and British countryside suffered from excessive rainfall in the spring and summer of 1875, and experienced considerable flooding. On April 1 of the same year, the *Times* had begun producing daily weather maps for the first time, and a new government investigation into the working of the Meteorological Office was announced in October. Thus, after a lull of several years following the Fitzroy controversy, weather prediction was again prominent in public affairs.

In the observations of his 1875 journey, Smyth noticed a distinctive band in the solar spectrum, a cluster of lines to the left of the red D band, which

89. Austin, "A Forgotten Meteorological Instrument." I am grateful to Mary Brück for this reference. There is a summary of contemporary pamphlets on the rainband spectroscope in Peterson, "The Zealous Marketing." On spectroscopy in general, see Gucken, *Nineteenth Century Spectroscopy*; James, "The Establishment of Spectro-Chemical Analysis"; Sutton, "Herschel and the Development of Spectroscopy"; and (a near-contemporary textbook) Watts, *An Introduction to the Study of Spectrum Analysis*.

90. Charles Piazzi Smyth, *Journal*, July 7 and 9, 1875, Smyth Archive. See also the account in Brück and Brück, *Peripatetic Astronomer*, 182–83. The archive is described in Brück, "The Piazzi Smyth Collection."

reliably appeared before a rainfall. In France again the following year, he visited a fellow worker in spectroscopy, Pierre Janssen. He proved to be a sympathetic colleague. Sharing Smyth's interest in mountain-top astronomical observations, Janssen was also one of the new leaders in the developing field of spectroscopy, and had just moved to his own physical astronomy observatory in Meudon.[91] Comparing his results with Janssen's investigations of the spectrum of water vapor in high pressure steam pipes, Smyth concluded readily that the rainband, as he called it, measured the water vapor in the atmosphere. He and his wife then undertook systematic observations of its predictive value. Rainband notes with diagrams of the spectrum began in his diary in July 1875 (fig. 5.12), and by 1877, Jessie Smyth was "undertaking a regular investigation of the influence of rainclouds on absorption bands of the spectrum," according to Smyth's correspondence with a friend.[92] Over the next several years Smyth published several enthusiastic descriptions of his meteorological experimentation with the spectroscope.[93] He described his instrument in *Nature* in 1880: a tube a mere four inches long, three-quarters inch in diameter, with a fine slit at one end and an eyepiece with a compound prism at the other, available for only two pounds from Adam Hilger, the London instrument maker, and able to fit nicely in a vest pocket.[94] To observe, one looked through the instrument ten to twenty degrees above the horizon (to maximize the thickness of the atmosphere in the view), with the sun at a distant angle, or better yet, behind a cloud. The rainband would appear to the red side of the solar D line, and must then be judged greater or lesser, darker or lighter, than its usual appearance (fig. 5.13).

Reading these accounts, it is evident that for Smyth rainband spectroscopy captured the knowledge that contemporaries associated with the weatherwise. The instrument gathered the meaning of the sky at a simple glance and transformed that glance into scientific data. The humble nature of the instrument, which Smyth invariably emphasized, initially established this set of associations. "Meteorological Spectroscopy in the Small and Rough," Smyth called an account published in 1877, contrasting the "very small affair . . . [purchased]

91. Smyth, *Journal*, June 5, 1876, Smyth Archive. On Janssen, see *Dictionary of Scientific Biography*.

92. Thomas William Webb to Charles Piazzi Smyth, April 10, 1877, Smyth Archive. In *Madeira Meteorologic*, 25, Smyth referred to Jessie Smyth's "long-continued enthusiasm" for the observations of the rainband spectrometer.

93. Smyth published numerous articles on the rainband in the late 1870s and 1880s in scientific and popular journals. See titles in the following notes and in the bibliography.

94. Smyth, "Rainband Spectroscopy," 90–91. Cory, *How to Foretell the Weather*, 86, gives prices of £1 12s. 6d. for a spectroscope from Browning (an "ordinary") and £3 8s. 6d. for one from Graces.

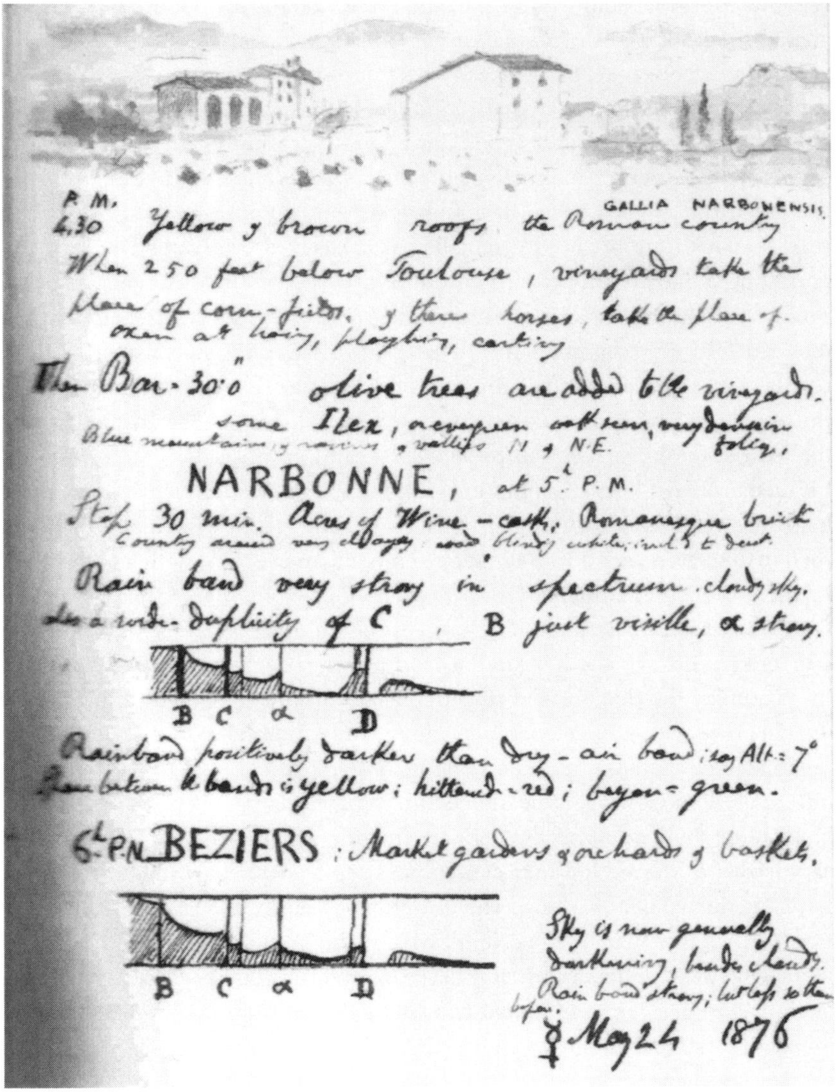

FIGURE 5.12 The combination of the new charts and forecasts and a wet summer returned public attention to meteorological science in 1875. Charles Piazzi Smyth, traveling in France during a period of widespread flooding, took up spectroscopic observations of the sky, identifying a "rainband" that foretold wet weather. His diary of the following summer showed his observations on another journey, traveling by train south to the Mediterranean. At Narbonne, the rainband is "very strong in spectrum, cloudy sky," and at Béziers, "rainband strong, but less so than before." The diary noted details of the landscape as well as barometer and stereoscope readings, recording Smyth's characteristic range of interest in visual observations. (Photographed by permission of the Royal Society of Edinburgh from the RSE Piazzi Smyth Archive, held on deposit at the Royal Observatory, Edinburgh.)

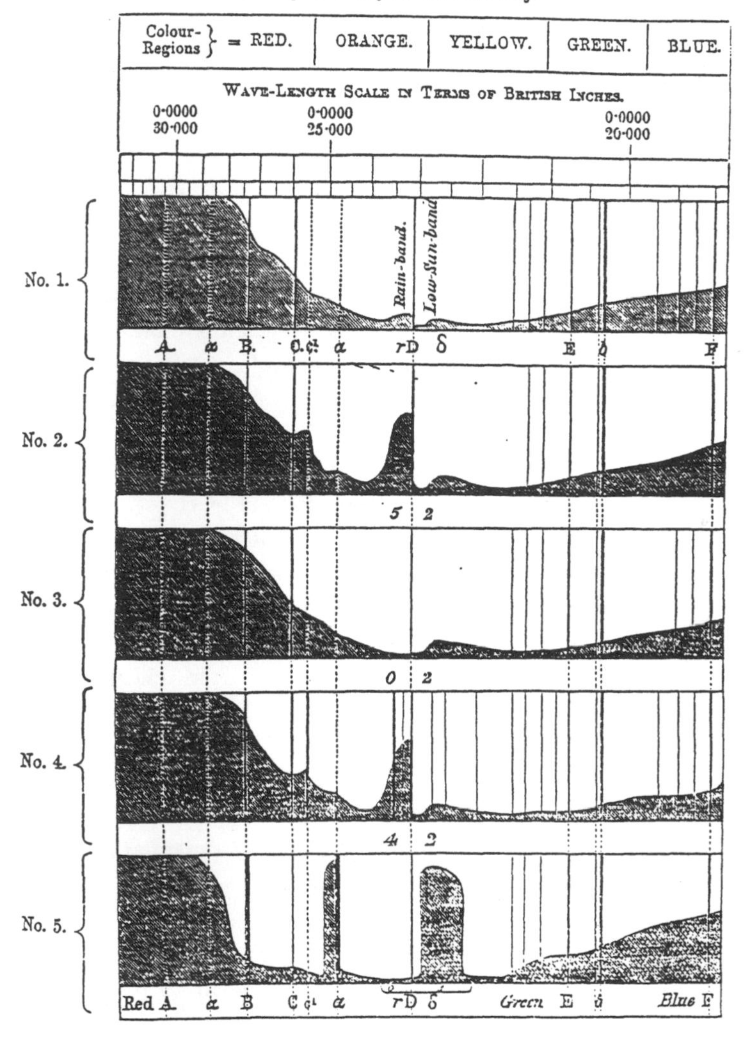

FIGURE 5.13 A diagram showing examples of the rainband, just left of line D in the spectrum. Although these black and white diagrams became the standard representation for spectroscopic records, they of course failed to represent the intense colors shown in the spectroscope, and Smyth continued to insist on the necessity of the observer's sense of color. (Charles Piazzi Smyth, "Rainband Spectroscopy," *Journal of the Scottish Meteorological Society* 5 [1880]: 95.)

at my private cost" with the more elaborate spectroscopes he used in his solar observations ("Solar Spectroscopy in the Large").[95] Smyth repeatedly emphasized the rapidity of the observation, calling it, in one memorable phrase, the "instantaneous, integrating glance."[96] The phrase evoked comparisons to weather wisdom, as did the manner in which the accounts juggled references to certainty and experience. On the one hand, the use of the instrument readily provided "infallible" information; on the other, the instrument required steady observation, and familiarity with the appearance of the spectrum at different air temperatures. The parallels with weather wisdom emerged further with Smyth's argument that his new method captured insights that conventional meteorological instruments did not. For instance, it was especially good for "quiet" rains not indicated by barometric change. Smyth announced in 1878 that "No meteorological instrument is so portable, so easily and quickly observed with, and so certain as to the genuine character of its phenomena . . . as the spectroscope when used, as it may be now, by everyone for a *certain class* of rain prediction."[97]

The reaction to Smyth's advocacy of the rainband spectroscope left no doubt that contemporaries understood the way in which such descriptions linked the instrument to disputes over weather wisdom. One enthusiast called it "the missing link in forecasting" that would "annihilate its difficulties."[98] But the more revealing comments came from critics. Louis Bell, then a physics fellow in Baltimore and later a prominent electrical engineer, reviewed the instrument in the *American Journal of Science* and argued that the rainband required the "eye of faith to comprehend it." It depended on eye estimations of intensity, which were dissatisfactory at best, becoming even more so when taking realistic conditions of observation into account. The movement of clouds, for instance, affected the light, and thus the reading. "It is no easy matter," Bell noted, "to compare [the rainband] with such a vague and variable thing as a mental scale of blackness."[99] In 1882, a prominent member of the Royal Meteorological Society, Ralph Abercromby, referred to the rainband dismissively as "simply a new sky prognostic." The occasion for the comment was significant—it came as a reaction to Smyth's attempt to turn his observations

95. Smyth, "Meteorological Spectroscopy in the Small and Rough," 28.

96. Smyth, "Rainband Spectroscopy," 552. Cf. Smyth, "Three Years' Experimentation," *Nature*, July 1, 1880, 195, and "Meteorological Spectroscopy in the Small and Rough," 29.

97. Smyth, "Rainband Spectroscopy," 87, original emphasis.

98. Cory, *How to Foretell the Weather*, 27–28.

99. Bell, "Rainband Spectroscopy," 347, 353.

to practical account with a letter to the newspaper offering harvest predictions based on rainband observations.[100]

Irritated at the way Abercromby brought "his mode of employing the instrument into the same category as the herdboy's confident advice," Smyth responded angrily. Prognostic, he claimed, was not necessarily a derogatory word. "Even a chart of isobars collected instantaneously from the whole extent of Europe by telegraph, and mapped down in a central office in London, is only another kind of weather prognostic—of a very grand and expensive kind truly."[101] Modern science, he argued, was not a matter of black or white, yes or no, but of degrees of grey. Rainband spectroscopy represented just such a "grey" case.[102] If we recall here Smyth's interest in chromatic observations, we can read this, properly, as an insistence on visual subtlety rather than uncertainty.

Indeed, criticisms like those of Abercromby were more wounding precisely because Abercromby and Smyth shared the same objective. Both sought to make meteorology more "scientific." Abercromby thus categorically rejected the very accomplishment Smyth claimed. Smyth concerned himself as much with the precision of his instrument as he did with the abilities of "its merest glance." In his article for *Nature* in 1880, he began with a detailed description of his equipment, which attempted to justify the spectroscope as a numerical device. His experiments with the meteorological spectroscope in an observatory setting concentrated on proving that it was "an essentially safe instrument for spectroscoping numerically anything within its power to spectroscope at all."[103] In a later description of the instrument, he compared the rainband spectroscope, managed with "a little extra nicety and tact," to recent "most accurate quantitative determinations" of relative proportions of silver and gold in alloys, or similar quantitative analysis of solutions of metallic salts. Furthermore, Smyth gave significant attention to the problems of calibrating a "mental scale of blackness," to use the words of his critic.[104] The key, he felt, was to compare the main rainband with a low sun band on the yellow side of the D line. As this band would vary with the angle of view and amount of light, it would provide the basis for a calibration.

100. The exchanges in the newspapers in September 1882 are reprinted in Cory, *How to Foretell the Weather*, 50–81; see 58–59 for Abercromby's letter in the *Times* (September 19, 1882).

101. Smyth, "Spectroscopic Weather Discussions," *Nature*, October 5, 1882, 552.

102. Smyth, "Rainband Spectroscopy," 87.

103. Smyth, "Three Years' Experimentation," *Nature*, July 1, 1880, 193.

104. Smyth, "Spectroscopic Weather Discussions," *Nature*, October 5, 1882, 551.

Smyth's spectroscope, at least as he understood it, transformed this awkward, descriptive observation into a systematic and scientific one. Like other ordinary meteorological instruments, the barometer, the thermometer, or the rain gauge, it was a suitable device for research observations. Yet it was also possible to use it to great effect as a *popular* instrument. It made the traveler or the observer at the periphery independent of centralized observations, like those circulating in London newspapers, which arrived on Smyth's Edinburgh doorstep too late to be of much interest. Smyth's instrument offered to bridge different interests in forecasting: local and practical needs for accurate weather prediction and the desire of scientific men for more systematic methods of observing the appearance of the sky. The instrument itself failed to gain more than a small following, probably because of the observing difficulties pointed out by Bell and influential British meteorologists like Abercromby.[105] Although he and his wife continued to use the instrument in their own observations, Smyth, himself, now concentrated on the solar spectrum and laboratory spectroscopy. But meteorology was not promptly dismissed, since the rainband spectroscope was not the only innovative meteorological instrument he investigated in the mid- to late 1870s. Smyth was also experimenting with a cloud camera. Turning the camera to the clouds must have seemed a natural accompaniment to his case for the rainband spectroscope, not least because the camera, just like the spectroscope, was confirming its promise as a scientific tool in astronomy.

As we have seen, clouds epitomized the challenge of scientific observation and communication. In addition, more emphatically than other kinds of meteorological data, clouds tied scientific observation to aesthetic experience and visual culture in general. For both these reasons, clouds were difficult to define as strictly scientific objects. The study of clouds offers the clearest examples of how ideas about weather wisdom could develop into a form of resistance to the claims of conventional scientific numeracy and precision. Smyth was an experienced artist, alive to both the aesthetic and scientific interest of the sky. By enlisting first the spectroscope and then the camera, Smyth argued that visual methods were the solution to the problems facing meteorological observations. But he also expressed his sense of the limits of that solution. To some extent, nature could not be captured by scientific facts, language, and instruments. For Smyth, a conservative Christian who believed in the imminent return of Christ, such a realization held profound religious significance.

105. Peterson, "The Zealous Marketing," suggests that the fact that instrument makers like Browning provided documentation with a purchase shows that rainband observing was technically difficult, but if so the remark must apply to all spectroscopes since the observing manual did not single out the rainband spectroscope.

Both the appeal of instruments and an acknowledgment of their limits can be traced in the unusual form of Smyth's cloud photography in the 1870s and 1890s. Because that project has been seen as idiosyncratic and aesthetic rather than religious, it is necessary first to consider the relationship between science and art in the techniques of and debates over cloud observation.

The Science and Art of the Clouds

It would be hard to overestimate the familiarity of the sky as a visual subject in Victorian culture. The middle class parallel to the seaman's glance at the sky was founded on aesthetic experience. Not all aesthetic traditions, of course, were equally concerned with mimesis or accurate observation of natural forms in ways that suited the purposes of natural philosophy—we can note for instance Alexander Cozens's method of "blot" landscapes, the distorted scale of John Martin's historical paintings, or some of J. M. Turner's work, notoriously condemned as unnatural. Yet the development of naturalistic painting in the late eighteenth and early nineteenth century had inspired close relationships between scientific and aesthetic traditions. In this sense, the reaction to Turner noted above indicated the strength of the expectations about the close observation of nature in art. And there are many well-documented links between the observational concerns of men of science and of artists such as John Constable, Cornelius Varley, John Turner, and John Ruskin.[106] Beyond such direct examples of intermeshed scientific and aesthetic interests, popular science texts give a vivid sense of conventional images of the sky. We need only look at the eighty-six woodcuts and ten colored lithographs in the English translation of Camille Flammarion's *The Atmosphere* (1873, edited by Glaisher) or at the numerous woodcuts of George Hartwig's *The Aerial World* (1886). Sunsets, snowfall, the auroras, rainbows, or storm and shipwreck appear in picturesque landscapes complete with fragments of shipwrecks, a tumbledown mill, or mountain peaks (fig. 5.14). Such dramatized scenes of atmospheric phenomena, often full-page size, far outnumber the occasional climatological map or the common, smaller engravings of instruments, which by contrast are stark, isolated objects embedded between paragraphs of the text. In short,

106. Axton, "Victorian Landscape Painting"; Badt, *John Constable's Clouds*; Bermingham, *Learning to Draw*; Dale, *In Pursuit of a Scientific Culture*; Hartley, "The Living Academies of Nature"; Klonk, *Science and the Perception of Nature*; Landow, "The Rainbow." Topper, "Natural Science and Visual Art," uses Constable and meteorology as one of his examples. Examples of contemporary discussion are Ansted, "The Influence of Certain Physical Conditions"; and Hulme et al., *Art-Studies from Nature*.

FIGURE 5.14 The conventional Victorian aesthetics of the sky shaped meteorological science. Here, in a typically dramatic image, a woman with two children rushes for shelter from a whirlwind. (Camille Flammarion, *The Atmosphere*, edited by J. Glaisher [New York: Harper, 1873], 346.)

popular accounts of meteorology were framed within conventional renderings of the natural world, which in style and subject, if not necessarily in quality, would be as much at home on the walls of art exhibitions. Victorians were well versed in the atmosphere as a subject of art as well as science.

Other examples of the close relationships of art and science in meteorology can be found in debates over painting and photography in the cloud atlas projects of the late nineteenth century. Capturing instantaneity in landscape was a goal of several English photographers in the 1850s, and involved a shift for this kind of subject from the long-exposure calotype photography to collodion photography, which required shorter exposure times and potentially gave sharper definition. John Dillwyn Llewelyn (a fellow of the Royal and Botanical Societies) took a series of photographs of waves in 1853–54, for instance, but even fast collodion photographs of sky in the late 1850s often merely caught a blue haze, and clouds were sometimes painted in after the photograph was developed.[107] One of the earliest English examples of photographs of clouds

107. Seiberling and Bloore, *Amateurs, Photography, and the Mid-Victorian Imagination*, 23–24.

was a volume published by Elijah Walton, a fellow of the Geographical Society, which actually consisted of photographs of *sketches* made in Egypt and Italy. It concluded with a critique of cloud forms in the old masters and in moderns like Turner. In a typical and revealing blend of audiences, Walton addressed his work to both the would-be artist and the "student of Nature."[108]

Nevertheless, fifteen years of photographic innovation and cloud study later, there was no consensus over a single proper method of scientifically illustrating the forms in the sky. Certainly there was no automatic assumption that photography was the preferable method.[109] Even the best modern photographic techniques and the authority of the international body of meteorology could still not resolve concerns about the kinds of images that should serve as standards. One of the foremost continental authorities on clouds, Hugo Hildebrandsson, director of the observatory in Uppsala, Sweden, collaborated in an expensive edition of ten quarto chromolithographs, reproducing oil paintings by Swedish and German artists that had been selected for their typical pictures of clouds. The artists, Hildebrandsson claimed, had "repeatedly retouched" their work in consultation with several eminent European meteorologists, and the lithography was paired with twelve photographic plates—the collection as a whole designed to explore the advantages and disadvantages of photography.[110] Despite the insistence of the editors of this collection that pictures should be considered as "intermediate between a picture and a diagram," and artistic beauty ignored, it was clearly impossible to ignore the artistic history of cloud portraiture, or to sever it from scientific illustrations.[111] Three years after Hildebrandsson's trial collection, in 1892, one of the first discussions of an international committee charged with development of a cloud atlas

108. Walton, *Clouds*, iii. Cf. a popular British cloud atlas consisting of reproductions of landscape paintings with the basic cloud nomenclature stamped onto the relevant portions of the sky: Kennedy, *The Hourly Weather Guide or Cloud Atlas*.

109. Marriot, "The Application of Photography to Meteorological Phenomena," 146–47; Ley, "Cloud Photography."

110. See Symons, "International Cloud Atlas"; "Current Notes: Atlas of Clouds"; and "Cloud Atlas." For the international cloud atlas, see Hildebrandsson, Riggenbach, and Teisserenc de Bort, eds., *International Cloud Atlas*. For a summary of the work done in the United States, see Bigelow, "Some of the Results of the International Cloud Work."

111. One obvious objection concerned color. Abercromby, complaining about an earlier English classification, argued that it had abandoned nuances of color: "In speaking of clouds, it might be said, that the form is their anatomy, and the color or chromatics their physiology . . . so in cloudland should form and color be jointly studied" (Abercromby, "Suggestions for an International Nomenclature of Clouds," 129–30). For relevant discussions of color and aesthetics in an earlier period, see Dolan, "Pedagogy through Print"; and Secord, "Botany on a Plate."

again protested against "any artistic conception or criticism of the atlas."[112] Yet it remained ordinary to include both photographic and oil or watercolor reproductions in cloud books, and several reproductions of paintings were eventually included in the International Atlas of 1896.

Smyth takes his place in this history of the representation of clouds both as a water-colorist and an eager experimenter with scientific photography. Before he developed his first cloud images, Smyth had an interest in photography dating back more than half a century.[113] At the Cape of Good Hope in the late 1830s, he was shown the daguerreotypes of another, more famous scientific resident in the British colony, Sir John Herschel. Later, at the advice of Herschel and others, Smyth took along a camera in his important Tenerife expedition in 1856, and used his experiences to mount a case for its essential role in natural history. Under some conditions, "the hand of man . . . fails entirely, not so the photographic hand of nature," Smyth declared in a lecture to the Royal Scottish Society of Arts. The camera, he argued, had overthrown the error-strewn observations of the past.[114] In the following decade, the camera steadily assumed more and more importance as a scientific tool. It was particularly successful in astronomy: it was used extensively in eclipse expeditions, and regular programs of solar photography were established at Kew in the 1860s. Thus, while Smyth's experiments with cloud photography in 1876 appear to be some of the earliest in Britain, they were not entirely a novelty. Clouds were a technically demanding subject for the camera, but consistent with the instrument's other applications to scientific observation.[115]

Smyth's first photographic apparatus for "cloud-taking" was stereoscopic, with two duplicate cameras separated by a variable distance of between five inches to five feet, with a single drop shutter that acted on both cameras at once.[116]

112. Abercromby, "The Classification of Clouds and the Cloud Atlas," 18.

113. Here I rely upon Schaaf, "Piazzi Smyth at Teneriffe."

114. Quoted in Schaaf, "Piazzi Smyth at Teneriffe," 32.

115. Brück and Brück note that Smyth believed he was the first to work in this field. I have not been able to locate his address to the Edinburgh Photographic Society in 1876, which they cite (Brück and Brück, *Peripatetic Astronomer*, 187). Hugo Hildebrandsson in Uppsala published some cloud photography in 1879, taken by Henri Osti, who may have been taking photographs as early or earlier than Smyth (Hildebrandsson, *Sur la classification des nuages*).

116. As Hankins and Silverman have shown (*Instruments and the Imagination*, 140–47), the authority of the stereoscopic camera in the first decades of photography arose from its imitation of interocular distance. Because the pair of cameras operated together like the human eyes, it was argued, the double photographs they produced gave an ideally faithful representation.

Smyth's early cloud pictures, therefore, were designed for use in stereoscope a viewer or lantern projection, to give the impression of three dimensions. But there were many technical difficulties. The wet collodion process, in which plates were coated with a photosensitive solution, inserted into the camera, exposed, and then removed and developed before the solution could dry, did not suit Smyth's observing conditions or his subjects. In "the smoky climate of Edinburgh," he noted briefly, wet collodion photography of clouds required too long an exposure time, leading to trouble developing a print before the solution dried. In addition, the three-inch portrait lenses he used in his cameras—lenses that maximized the focused area in the view—had a tendency to collect moisture, which distorted the photograph. Perhaps because of these technical difficulties, and perhaps because his interest turned to other projects, the cloud camera "fell out of use," as he recorded.[117]

Smyth returned to his cloud camera some fifteen years later, when his retirement offered both time for the experimentation and a new location away from a "smoky climate." The first of his new pictures, which he called "Meteorological Documents," was taken on February 18, 1892.[118] He had split up his earlier apparatus, leaving one camera at the Edinburgh observatory and taking one with him to his new home in Ripon, Yorkshire, where the attic became his photographic observatory. By taking only one of the cameras, of course, Smyth abandoned the stereoscopic effect, simply noting that it "was not really wanted by science at this latter date,"[119] and he concentrated on exploiting the technical improvements that had occurred in photography since 1876. These were considerable: dry collodion plates had supplanted the wet collodion plates, and dry plates were more durable, easier to use, and mass produced (therefore less expensive). He adopted a new lens; improved his drop shutter; adding steel springs to increase its speed, and a mechanism to vary exposure time; and continued to experiment with the plate suppliers, developing recipes and the print paper.[120] Smyth was delighted with the results, cheering the "spirited,

117. Smyth's early technical notes on his cameras are in a notebook labeled *Cloud Taking Camera, Edinburgh 1876; Clova 1892*; his latter notes are in the *Museum Copying Camera Notebook: Notes on Photography at Clova 1892–96* (Smyth Archive). The quotation is from *Cloud Taking Camera Notebook*, 3.

118. Smyth, *Cloud Taking Camera Notebook*, 9, Smyth Archive.

119. Ibid., 3.

120. To replace the unsatisfactory portrait lenses of the earlier cameras, he used first a Steinheil lens and then an expensive magnifying lens donated by Joseph Sidebotham, a wealthy Manchester manufacturer with interests in botany, astronomy, and photography. On Sidebotham, see Boase, *Modern English Biography*.

plucky pictures" that he now obtained with his camera.[121] He recorded in capitals his "GREAT AWAKENING" as he found a method to prevent the deterioration of his negatives; and he triumphed over a two-hour-exposure photograph of "FOG," which proved to him that God's fogs were "not, after all, so prejudicial to photographic science" as he had feared.[122]

Smyth's photographic work was clearly connected to the concerns present in his earlier development of the spectroscope as a meteorological instrument. In 1893, he turned his attention to "the literary work" of their presentation, producing documents that laid out those concerns explicitly.[123] Two collections of about 140 enlargements were produced from the dry collodion plates: one Smyth retained, and the other, more elaborate, he sent to the Royal Society in London labeled *Cloud Forms That Have Been*. He began *Cloud Forms* by discussing the reputation of meteorology, and its "utter incapacity . . . to make exact calculations and finite predictions."[124] A record of clouds would address some of meteorology's deficiencies. The documentation for each photograph included the rainband and the low sun band (his calibration measurement) alongside the technical descriptions of focus, aperture, and exposure time, and record of barometer, thermometer, and wind direction. A full meteorological diary accompanied the photographs as well, in order to show the "regular and unbroken course of the Natural Weather." The remarks that accompanied each photograph in his own copy often drew attention to the predictive nature of the cloud form, with labels such as "preparation for night rain" or "rain and ruin for poor farmers coming."[125] As we cannot know when these remarks were made (at the time of the exposure, or retrospectively when the events indicated by the appearance of the sky had unfolded), we cannot say whether the remarks were *literally* predictions, or merely exemplified for Smyth the *possibility* of prediction. In either case, the cloud studies assumed the scientific

121. Smyth, *Cloud Taking Camera Notebook*, 73, Smyth Archive.

122. Smyth, *Museum Copying Camera Notebook*, 21, Smyth Archive.

123. Ibid., 95.

124. Smyth's two-volume copy, titled *Clova Cloud Photographs*, is now held at the Royal Observatory Edinburgh; the Royal Society of London copy, titled *Cloud Forms That Have Been; to the Glory of God Their Creator and the Wonderment of Learned Men as Recorded by Instant Photographs Taken at Clova, Ripon in* 1894 *and* 1895, 3 volumes (privately printed), has more extensive introductory remarks and is the copy from which the subsequent quotations have been drawn. The quotation here is from *Cloud Forms That Have Been*, 1:3.

125. Smyth, *Clova Cloud Photographs*, plate V, no. 29; plate X, no. 57, Smyth Archive.

value of a moment's glance at the sky, as captured accurately by a modern instrument. Like the spectroscope, then, the camera could convert the canvas of the sky into a scientific record.

Though his 1876 photography was certainly pioneering, by 1890 Smyth was one of perhaps a dozen cloud photographers in Britain. A BAAS project from 1891 to 1896 had collected "many thousands" of photographs, and its reports illuminated some of the technical obstacles. Although successful photographs of the lower and more defined clouds—the cumuli—were common, clouds of higher altitudes, haze, and fog were difficult to photograph. These required developers that could intensify the image, a tricky process of judgment in the development stage; or a yellow filter to reduce the glare; or a even a black-coated glass mounted in front of the lens of the camera at an angle, which again could diminish the glare and allow a photographer to time the exposure more easily.[126] The BAAS cloud photography project was itself overtaken by international efforts to coordinate cloud observation. The International Meteorological Committee, a body that met regularly to coordinate regulations and standards for observations starting in 1873, had adopted in their meetings at Munich in 1891 and Uppsala in 1894 the nomenclature of Ralph Abercromby and Hugo Hildebrandsson over the objections of British delegates who thought it "entirely empirical."[127] The international classification was based on that of Howard but further subdivided the types in terms of their height: upper clouds at 9,000 m; intermediate cloud, between 3,000 and 7,000 m; lower clouds, 2,000 m; high fogs, 1,000 m; and the "diurnal ascending" clouds. By 1894, the Cloud Committee had produced the official description of the nomenclature and turned to the project of an official atlas, which eventually appeared in 1896. And in 1896–97, observatories in Europe and North America concentrated on the systematic study of clouds using the official atlas.[128]

Smyth was therefore obviously not alone in trying to forge techniques for cloud photography. He attended a BAAS meeting as late as 1892, and it seems reasonable to conclude that he would have known of the numerous discussions of cloud nomenclature and cloud photography appearing in the pages of the

126. For a description of this device, see Clayden, *Cloud Studies*, 165–81. There is a picture of this instrument in use in "Weather Watchers and Their Work," 188.

127. [Clayden], "Second Report of the Committee [on Cloud Photography]," *Report of the BAAS*, 1892, 77; Hildebrandsson and Hellmann, *Codex of Resolutions*, 17–18.

128. For a summary of the work done in the United States, see Bigelow, "Some of the Results of the International Cloud Work."

Quarterly Journal of the Royal Society of Meteorology, in the annual reports of the BAAS and in the *Journal of the Scottish Meteorological Society*.[129] Nevertheless, the only meteorological works to which Smyth referred directly were those of Herschel; and there is no evidence to suggest that Smyth's efforts were known to other meteorologists, nor that Smyth became involved with either the BAAS project or the related international efforts to develop a consensus on classification techniques—given his earlier altercation with Abercromby over the rainband spectroscope, perhaps this is not surprising. Moreover, Smyth's numerological researches, tracing prophetic significance in the dimensions of the pyramids of Giza, had isolated him from the scientific community, and he had resigned from the Royal Society in protest in February 1874.[130] Other than forwarding a copy of his cloud studies to the Royal Society, Smyth does not appear to have made further attempts to call attention to the work, in strong contrast with his active promotion of the spectroscope in meteorology. Although it fits contemporary research programs in some ways, then, it can be hard to assess Smyth's cloud work. Brück and Brück treat it as a failure, suggesting that "in the preparation of these beautiful photographs the artist had taken over from the scientist."[131]

We need not see Smyth's cloud photography, however, as a rejection of science. Instead, *Cloud Forms* addressed the question of intelligibility in an unusually self-conscious manner. The photographic record of particulars emphasized the action of selection, such as the interruption of a continuous sequence, or the reach of the instrument beyond the human senses. It emphasized a partial as much as an exact scientific view. Its visual methods were both typically ambitious—exact glimpses that proposed immediate insight—and admittedly tentative, acknowledging the constraints of any single moment or instance. In this doubled sense, Smyth's cloud photography expressed the tensions within meteorology, where visual knowledge was only partially incorporated into the body of science.

The introductory notes to the copy of *Cloud Forms* Smyth sent to the Royal Society traced these tensions. There Smyth announced that "religious feeling, and love of art and beauty, and hopes of scientific discovery, with practical results in the end, are all claimant in these last days." (The "last days" referred not only to Smyth's own age, but to his millenarian belief in the imminent Second

129. Abercromby spoke to the Royal Society on cloud classification in Edinburgh in 1888, when Smyth was still resident (Abercromby, "Suggestions for an International Nomenclature of Clouds").

130. Schaffer, "Metrology, Metrication, and Victorian Values," 449–59.

131. Brück and Brück, *Peripatetic Astronomer*, 252.

Coming of Christ.) Smyth then made a series of familiar comments on the current profligacy and inadequacies of the "grandest Government buildings" and observatory network, incessantly recording data with their "mechanical processes." He affirmed his commitment to "strict science" and to the goal of accurate weather prediction. Nevertheless, he went on to question scientific work in a more fundamental way, by suggesting that it was impossible to classify phenomena at all. Quoting from the Book of Job ("Yet who even now, all the world over, does fully understand the ultimate laws of existence and the duties of the clouds in the higher air?"),[132] he spoke of human ignorance and the complexity of the divine plan. The oddly phrased title of the work had the same message. *Cloud Forms That Have Been* emphasized the individual, unique moment or phenomenon, and the temporally limited view of a regular, uninterrupted course. For Smyth, cloud forms showed the power of God and displayed the parallel between the ephemerality of natural phenomena and of mortal life. Many years earlier, Smyth had commented in his diaries on the power of some astronomical illustrations. Colored prints of the universe impressed him with a humbling sense of "the strangely accidental scene of man's birth and life . . . to be tried for a time and no more."[133] Similarly, cloud photographs forced our "neglectful eyes" to God and to his imminent return, when God would restore his rule over "that more irrepressible and windy of all things, the mind and soul of educated man." The beauty of the forms and the difficulty of capturing them were equally significant to Smyth, giving his "mightily grown and very learned age" a much needed lesson in humility and reverence.[134]

As Smyth characterized it himself, his cloud project was "a labour of Love and Meteorologic Research, in days of old age and failing faculties."[135] We might easily dismiss Smyth's cloud photography, despite some idiosyncrasies in its presentation, as simultaneously conventional and obscure. Yet, if it was a scientific failure, it was a revealing one. Smyth showed how readily meteorology bound itself to motivations beyond those of the strict and regulated labor of science, to epistemological limits, to aesthetic interests, to religious proofs, to notions of knowledge as intuition—"meaning at a glance."

132. Job 36:27–28 (King James Version).

133. Smyth, *Journal*, May 15, 1876, Smyth Archive. He was referring to plate no. 39 of Amédée Guillemin, *Le ciel*. The work was well known in England and went into many editions, translated by Lockyer, as *The Heavens: An Illustrated Handbook of Popular Astronomy*.

134. These remarks are taken from Smyth, *Cloud Forms*, 7, 1, 5–6.

135. A comment Smyth marked on the title page to each volume of *Cloud Forms*.

The Storm Cloud of the Nineteenth Century: Clouds, Precision, and Language

Smyth's several meteorological projects from the 1870s to the 1890s merged cloud study, weather wisdom, and a critique of the ambitious reach of modern scientific culture. They show the grounds on which meteorological science played a critical role in the well-known polemics, institution-building, and educational and research agendas of the 1870s and 1880s.[136] Smyth's experience with weather instruments and visual culture argue against the picture of a professional scientific culture that was coherent, secular, and naturalistic. A more nuanced account of particular religious values, observations, and technological facility better represents the era. Smyth's critique of precision observation in meteorological science therefore joins George Airy's skeptical reaction to the Kew Observatory measurements described in chapter 4, whereas his religious response to clouds suggests some sympathy for the arguments advanced by Alexander MacLeod, who treated meteorology as the key demonstration of the limits of a deterministic view of the natural world. These relationships, between figures otherwise hardly sympathetic, suggests the need to explore any such critique carefully, avoiding any simple line dividing laymen and professionals, or religion and science. Accordingly, this chapter closes with an interpretation of a famously hostile critic of scientific culture, John Ruskin, in order to emphasize the common ground of Victorian intellectual life. Despite the cultivated extravagance of his denunciations of modern science, Ruskin in fact produced a typically Victorian claim. The definition of standards of natural knowledge would establish the basis of social and moral order and so must not support the exclusive activity of a few. And, like Smyth, he turned to the study of weather to make this point. In February 1884 Ruskin delivered a two-part lecture entitled "The Storm-Cloud of the Nineteenth Century" before a sold-out audience in the London Institution. Again like Smyth, his account of clouds emphasized the power of visual knowledge and used that to explore ideas about precision and scientific authority in modern culture.

Ruskin himself had impeccable credentials as an observer, having passed decades and thousands of pages relentlessly training his contemporaries in appreciation of art as a mirror of the natural world and as a foundation of

136. Turner, "Public Science in Britain" and *Between Science and Religion*; Barton, "'An Influential Set of Chaps.'"

moral experience. In many senses, Ruskin was the formal voice of Victorian aesthetic traditions.[137] In his lecture, Ruskin portrayed himself as an ordinary observer pitted against the arrogance of experts and the errors of instrumental science. Ostensibly, the lecture was a guileless description of meteorological phenomena as they had changed since Ruskin's boyhood. (Ruskin had been a member of the London Meteorological Society and published one paper in their proceedings in 1838—his first published work.) He introduced his subject by discussing sunsets and the definition and types of clouds to demonstrate that his close observation stretched over many years. Then he turned to what he maintained was a new southeasterly wind, with its accompanying characteristic cloud, which Ruskin claimed to have first noticed in 1871. Ruskin described the phenomenon in language as far from scientific as possible. The "plague-wind" was "a wind of darkness," which "looks partly as if it were made of poisonous smoke" and partly "of dead men's souls"; it withered roses and rotted strawberries. Although his language implied the cause was pollution—he called the cloud "a Manchester devil's darkness," and "sulphurous chimney-pot vomit"—Ruskin explicitly denied its similarity to smoky London fogs.[138] Instead, he swept to a denunciation of narrow modern science and its culpability in shaping a desolate industrial culture. "Blanched Sun—blighted grass—blinded man," he concluded somberly.[139]

By the time of this lecture in 1884, Ruskin was slipping from celebrity into notoriety, and many of the reviews of the lecture were mocking. As one newspaper described, the audience had been treated to "an address on weather wisdom, tempered with a few strictures on men who are not weather-wise, . . . by a chartered Libertine in Science and in Art."[140] Another declared that a more plausible hypothesis than climate change was that "Mr Ruskin, as he gets on in years, is more sensitive to disagreeable weather, and takes a more gloomy view of things in general."[141] Nevertheless, the lecture was widely reported in the London press and was soon sold in pamphlet form. *Storm-Cloud* was written up in scientific journals; scientific men sat in the lecture audience, including

137. On Ruskin, see Abse, *John Ruskin*; Hilton, *John Ruskin*; Hunt, *The Wider Sea*; and Landow, *The Aesthetic and Critical Theories of John Ruskin*. For contemporary responses to his work, see Bradley, ed., *Ruskin*.

138. Ruskin, *Storm-Cloud*, 33–35, 37–78.

139. Ibid., 40.

140. *Standard*, February 6, 1884, 5.

141. *Daily News*, February 6, 1884, 5.

Norman Lockyer, the editor of *Nature*; and Robert Scott, Fitzroy's successor as director of the Meteorological Office, preserved press clippings about the lecture in the Meteorological Office scrapbook.[142]

Yet none of the broader arguments of the lecture were new to Ruskin. In a set of essays on the "interpretation of myths relating to natural phenomena" published in 1869 as *Queen of the Air*, Ruskin pursued the same themes of defilement and the contrast of truths of modern science with other truths—in this case those of ancient philosophy and mythology.[143] His criticism of scientific culture, and its prominent spokesmen, Thomas Huxley and John Tyndall, had mounted steadily in the 1860s and 1870s. Tackling the current debates over materialism in an essay on miracles, prayer, and natural law in 1873, for example, Ruskin spoke of haunting doubts "of the security of our best knowledge, and discontent in the range of it," taking the position that neither the evidence for or against miracles could support dogmatic authority.[144] With works like *Eagle's Nest* (1872) on art and science and "A Caution to Snakes" (1880) on Huxley's natural history, it was clear that Ruskin had decided the chief dogmatists were the men of science. In all these texts, he trained his attention upon scientific observation. In *Eagle's Nest*, for instance, Ruskin compared the "mechanical" definition of sight with the willed act of seeing that was characteristic of the artistic gaze. Sight comes through the "animation" and "soul" of the eye, he pronounced, rather than its "lens." There is more than one way of seeing, Ruskin argued—"comparative sight is far more important than comparative anatomy"—and he urged physical science to investigate such subjects as the relation of the sense of color to health and mood, or the different modifications of sight possessed by different animals by study of the different physiology of their eyes.[145]

Storm-Cloud as an item in the Ruskin oeuvre has been interpreted in three principal ways—first, as a somewhat incoherent, idiosyncratic work that reveals Ruskin's mental strain in the early 1880s; second, as an expression of Ruskin's environmentalism (to give it an anachronistic label) and a prescient account of pollution; and third, as an allegory, with little or nothing to do

142. Reviews appeared in the *Times*, *Pall Mall Gazette*, *St James Gazette*, *Daily News*, *The Standard*, *Whitehall Review*, *Knowledge*, and *World*. See Ruskin, *Storm-Cloud*, xxiii–xxvii.

143. Ruskin, *Queen of the Air*, 300.

144. Ruskin, "The Nature and Authority of Miracle," 633.

145. Ruskin, "A Caution to Snakes" and *Eagle's Nest*, 199, 202. See Kirchoff, "A Science against Sciences"; and Landow, *The Aesthetic and Critical Theories of John Ruskin*.

with the ostensible account of meteorology.[146] None of these seems really satisfactory. Both the first and last of these interpretation slight the subject matter: even if mad, or nearly so, Ruskin surely turned to meteorology and clouds with a purpose. An allegorical explanation, furthermore, does not take the history of meteorology sufficiently into account (either its personal history for Ruskin, or history of its controversies in the previous two decades). The second interpretation is more interesting. On the one hand, it must be admitted that the essay is too freighted with emotional and symbolic expression to work well as a factual account of air pollution. Connecting *Storm-Cloud* with other contemporary arguments about pollution can be misleading: the consciousness of "new skies" in the same period has more to do with the aftereffects of the volcanic eruption of Krakatoa in 1883 than with any particular changes in the perception of the causes or effects of industrial pollution.[147] On the other hand, Ruskin's reading of the sky can be taken literally in another sense if we emphasize that the theme of the essay is in fact accurate observation. His most wildly figurative accounts of the plague-wind cloud are not then allegorical but intended as arguments about different kinds of precision. Ruskin's work suggests how the visual basis of scientific knowledge could persistently evoke the differences between the expert scientist and the ordinary observer. Without a tradition of shared experience, Ruskin's comparisons between the scientific meteorologist and other observers would have lacked force; once that tradition is acknowledged, Ruskin's lectures seem far less idiosyncratic.

Ruskin's life work had been to analyze visual art, to describe vision with prose. To use vision and language as his cudgels in an attack on modern society was a commonplace, not a novel, undertaking for him. But in a critique of science, they challenged the purpose of exactness, erecting a barrier against the authority of numbers and instruments. With its emphasis on language and vision, Ruskin's *Storm-Cloud* argued that poetic descriptions and a subtle visual sensibility were needed to convey a true account of natural phenomena. Moreover, only such attention to qualities not captured by measurement could lead to a proper judgment about the role of science in modern society. As Ruskin put it, "Even in matters of science, although every added mechanical power has its proper use and sphere, yet the things which are vital to our happiness and prosperity can only be known by the rational use and subtle skill

146. Danahay, "Matter Out of Place"; Fitch, *The Poison Sky*; Graham, *The Destruction of Daylight*; Wheeler, ed., *Ruskin and Environment*.

147. Zaniello, "The Spectacular English Sunsets of the 1880s."

of our natural powers."[148] The potentially misleading precision of instruments paralleled concerns about misleading language. For Ruskin, to discuss vision and language was simultaneously to denounce the arrogance and exclusivity of scientific methods, to support the common observer, to evoke the *practical* necessity for communication, and, finally, to mark the boundaries of a too-expansive, polluting force. His statement of the problem therefore expressed the place of weather wisdom in meteorology.

Conclusion

By the time Ruskin took issue with meteorological science at the podium at the London Institution, British experiments with scientific forecasts of the weather reached back over a quarter century. Throughout the whole period, weather forecasting remained an insecure enterprise. Any analysis of its difficulties must acknowledge the place of popular knowledge—weather wisdom—in shaping the ideals and the practices of meteorological science. Weather wisdom seemed to be irredeemably local and difficult to express in words or numbers. It was hard to move such knowledge around from place to place or person to person. In that sense, the Shepherd of Banbury, or Hardy's shepherd Gabriel Oak was the antithesis of the government meteorologist, and opposite to the ideal of universal science. Yet the reputed success of weather wisdom meant that meteorologists could never easily dismiss it. It represented a model of knowledge that scientific meteorology hoped to achieve: transparent meaning, rapidly grasped. The knowledge referred to as weather wisdom seemed to be a missing link in the epistemological chain between instrumental observations and accurate forecasting.

Meteorologists responded to the appeal of weather wisdom by emphasizing the instruments and techniques that replicated its qualities. This response was founded in contemporary interpretations of sensibility. The explanation for weather wisdom centered on sensation, a subtle but brute ability to decode natural signs that was physiological rather than intellectual. Since that physiological reaction was automatic or mechanical, it could be incorporated into machines, just as a leech's nervous system could blend into the wiring of Merryweather's Atmospheric Electro-Magnetic Telegraph. In the face of the apparent authority of weather wisdom, then, it was clear that scientific instruments promised a technological version of sensibility. Telegraphy, the machine that made weather forecasting possible in the first place by moving

148. Ruskin, *Storm-Cloud*, 66.

information more quickly than the wind brought the weather, was an exemplary "sensible" technology because of its significance as a physiological model of mental processes. Technologies like the telegraph and its offspring the synoptic map mediated between scientific meteorology and popular knowledge. Yet like any frontier, they faced in both directions. Were they cementing the exclusive authority of science, or demonstrating its accessibility to all? Although instruments offered to incorporate weather wisdom into the body of scientific meteorology, their success was never assured. Scientific meteorology was often ambivalent, pursuing precise observations but aware of what lay beyond their instrumental record. Critics inside and outside the world of professional science thus found in the subject an opportunity to demonstrate the limits of scientific knowledge. They seized on concerns with vision, language, and place, as Abercromby did when he suggested that the clouds of Borneo were not the clouds of Scotland. This odd subversion of the universality of scientific knowledge reflected the weakness of official meteorology, which was never sure that it could answer the challenge of local knowledge. In the 1870s, however, a different solution for centralized meteorology emerged. In that solution meteorologists and their public audiences turned their attention from local, popular knowledge and moved it far away—to the imperial setting of India. In India, colonial power and barriers of language and culture made it simpler to dismiss local and popular knowledge of the weather. In both social and geographic terms, the vast territory offered an ideal physical observatory that promised to allow meteorologists to capture the atmosphere at a safe distance from the vagaries of the British weather.

CHAPTER SIX

Science, State, and Empire

DESPITE the isobaric and isothermal curves in the atlases, weather did not really fit on any map. Its unbounded phenomena and global scale intensified questions about the political context for natural knowledge and the responsibilities of scientists and governments. Meteorology embodied the promise of state science in two different ways: as an example of government funding for research, but also as a model of control, both natural and social. "Climate is monarchy, weather is anarchy," *Blackwood's Edinburgh Magazine*, announced in 1875, developing an extended analogy between weather, climate, and government that played with notions of stability and instability in political and natural systems.

> Climate is a constitutional government, whose organization we see and understand . . . but weather is a red-hot radical republic, all excitements and uncertainties, a despiser of old rules, a hater of propriety and order. Climate is a great stately sovereign, whose will determines the whole character of the lives and habits of his retainers, and is therefore so little felt that it seems like liberty; but weather is a cruel, capricious tyrant, who changes his decrees each day and who forces us, by his ever varying whim, to remember that we are slaves. Climate is dignity, weather is impudence.[1]

The author of this essay, a regular contributor on continental politics named Frederic Marshall,[2] endowed "climate" with all the virtues of British consti-

1. [Marshall], "Weather," 611–12.

2. *Wellesley Index to Victorian Periodicals* 1824–1900, 4:342.

tutional order, while knowing, as did his readers, that the British Isles were notorious for "republican" weather. The irony drew attention to a distinction in British meteorology that became increasingly important in the 1870s, between the exciting, popular but uncertain prospects of forecasting the weather, and the orderly pursuit of climatological laws. More importantly, however, it shows how these distinctions were connected with allusions to government, responsibility, and tyranny. In part this connection grew from the assumed symmetry of natural and social order, in which the laws of nature extended to human society and the former would shape knowledge and control of the latter. But in addition, the pursuit of science seemed a litmus test of modernity. The state's role in the systematic discovery of natural order reflected its own sense of progress and order.

It seems appropriate to find this play with politics and the weather developed in *Blackwood's*, which was associated from its origins in 1817 with Tory and conservative politics. The second administration of Benjamin Disraeli (1874–80) promised to be more supportive of a close relationship between science and government than had been the previous Liberal governments of William Gladstone. As a mark of this relationship, Disraeli was elected a Fellow of the Royal Society in 1876.[3] Throughout several years of hard scientific lobbying in the 1860s and 1870s, meteorology was at the heart of debates about the relationship of science and the modern state. As we have seen, especially after 1867, the Meteorological Office reflected the ambitions of a scientific elite concentrated in London and in the Royal Society. Considered less directly, meteorology lent itself to questions of power and politics because its phenomena were so ambiguous in physical terms. Meteorology required a global reach because it involved a scale of natural events in which the continent of Europe, let alone the British Isles, looked almost insignificant. As one meteorologist insisted in the 1880s, the science "recognizes neither political nor superficially physical divisions of the land"; it is as "indifferent to political boundaries as to political opinion, and requires to be studied on the largest possible scale."[4]

This writer, Douglas Archibald, tried to suggest that scale erased political questions. Yet such a claim was manifestly false. In an era of global scientific

3. The closest account of these relations are given in MacLeod, *Public Science and Public Policy*; cf. Hall, *All Scientists Now*. Extensive documentation of the changing relationship and expectations is preserved in the evidence of witnesses to the Royal Commission on Scientific Instruction and the Advancement of Science, 1872 to 1875, as well as in debates over the endowment of research.

4. Archibald, "Indian Meteorology," *Nature*, August 23, 1893, 405. See also the continuation of the article August 30, 1893, 428–30, and September 13, 1893, 477–79.

researches, like the expeditions to study eclipses, or the pioneering exploration of the ocean depths in the ship significantly named *Challenger*, meteorology forced potentially sensitive comparisons upon a public used to associating science with national prestige.[5] Since meteorology was beyond borders, its study required more attention to different national styles and methods, in order to build the international cooperation necessary to share data. By the end of the 1870s, Americans were suggesting that they could predict the weather in Europe, and European meteorologists were arguing about whether "American" storms could translate themselves across the Atlantic. At the same time, research on cyclical patterns of rainfall entered into heated debates about the management of the British Empire—which, just like meteorology, involved a conception of governance on "the largest possible scale." Turning to rainfall, sunspots, and famine prediction, meteorological science shifted away from the anarchy of British weather to India's intense but regulated tropical climates. As Victorians argued over the future of the empire in the 1870s, meteorology provided a natural foundation for British imperialism, while India provided a new field in which to trace atmospheric laws and prove the relevance of science to the modern state.

Cooperation and Control in the Science of Meteorology

The place of meteorology in Victorian concerns with state science has already been examined in earlier chapters. After 1867, the Royal Society governing committee pressed for research meteorology and precision observation, detached from any obligation to provide public notifications of the weather. This approach met considerable resistance from critics who opposed the endowment of research and from the public audiences interested in continuing Fitzroy's experiments with weather forecasts and storm warnings. The controversy pitted London, with its scientific elite, central office, and Kew Observatory, against the periphery. The scale and reach of atmospheric phenomena wove into questions about the scale and reach of central government authority. To explore the geography of these issues more closely, we can turn first to the relations of the Scottish Meteorological Society with the Meteorological Office.

The Scottish Meteorological Society, the largest scientific society in Scotland after the Royal Society of Edinburgh, had been founded in 1854. In its first decades, the Society's interests were very much directed to practical meteorology, including forecasting. Its first president sponsored several generous

5. Deacon, *Scientists and the Sea*; Rozwadowski, *Fathoming the Ocean*.

prizes on weather patterns and herring fishery, for instance, or averages in soil temperatures, rainfall, and sunshine as they affected various crops throughout Scotland.[6] It was by far the largest and best-organized meteorological society in Europe, with about 570 members and 80 instrument stations in 1865, operated by observers who underwent regular inspections.[7] By the 1870s, the permanent secretary of the SMS, Alexander Buchan, had acquired an international reputation, and he often served as Britain's second official representative to European congresses alongside the delegate from the government Meteorological Office. The significance of Scottish interests emerge most clearly in a "storm history" produced by the Astronomer Royal, Charles Piazzi Smyth.

On October 3, 1860, a year after the *Royal Charter* storm had inspired Fitzroy with a dedication to weather forecasting, another ugly gale hit British coasts, this time concentrating on Scotland. The storm was sudden and devastating, completing its damage in the early morning hours, when few witnesses could record its beginnings. Those in its path on land woke to find "roofs destroyed, harvested crops blown away, trees large and small levelled to the ground."[8] At sea, fishing boats, small coasters, and six large steamers were caught in the storm and wrecked, several without a trace. One of the six, the *Edinburgh*, was owned by R. M. Smith, a good friend of Smyth; indeed, Smyth and his wife had traveled to St. Petersburg on board the vessel the previous year. The couple could vividly imagine the effects of a storm in that part of the North Sea and of course were well acquainted with the *Edinburgh*'s captain and crew. In 1860, on the morning before the storm, Smyth and his wife, Jessie, had been down to the harbor to see the *Edinburgh*, set sail and had dropped off a box of photographs and stereoscopes for transport to the St. Petersburg observatory. With these memories, Smyth felt the storm come very close to home.[9]

Despite his personal and professional interest in the storm, Smyth only published his analysis a decade later in 1871. This delay added to the significance of the report: spanning the first experiments with forecasting under Fitzroy and the Royal Society reforms after his death, Smyth's analysis was a comment on both. But it was the latter reforms that forced Smyth to publish

6. See the *Journal of the Scottish Meteorological Society* (*JSMS*) starting in 1864; Watt, "Early Days of the Society."

7. This efficiency contrasted with the British Meteorological Society based in London, whose membership only reached 358 in 1877 and which did not maintain strict control over its data. See Symons, "History of the English Meteorological Societies."

8. Smyth, "Hyperborean Storm," T83 [pagination irregular].

9. Brück and Brück, *Peripatetic Astronomer*, 70, 77–78.

his work. Smyth's report indicated how the Scottish response to meteorology came to center on the relations of science with the state. What defined a truly national science? Under the circumstances of widespread debate about forecasting, Smyth wrote in 1871, the "interesting storm" had an "after-glow" and "duty called me . . . to take up the long-delayed investigation."[10] Gathering observation from lighthouses, ships' logs, local and national observatories, and colleagues in Holland, France, and Russia, Smyth mapped out the passage of the storm. The storm reached about 700 miles across, with its destructive effects concentrated over about half that size, and it traveled from west to east, reaching speeds of forty miles per hour. Smyth showed its passage over twenty-four hours in a dramatic chart whose clustered isobars showed the steep barometric gradations of the storm (fig. 6.1). In the accompanying text he emphasized its unusual suddenness. The storm "gave too little Barometric notice and was too local, as well as too quick in its movements" for any transmission of warnings. Perhaps, he speculated, a cable between Scotland and Iceland might have been of service; perhaps an observatory ship anchored offshore with a cable attached might have been similarly useful. But, Smyth recognized, these were expensive and impractical solutions: "it is doubtful whether all the revenue of the nation could maintain ship and cable in such a position."[11] Skilled seamanship and a scientific appreciation of the cyclonic changes of the wind could help ships ride out the storm, but telegraphic networks would not be the savior.

These conclusions meant that Smyth's storm history was not intended to stand as a straightforward defense of weather forecasting. Smyth endorsed that work, and paid tribute to Fitzroy's humanitarian and scientific goals, but he was honest about the possibilities of forecasting. The unusual speed of this storm, as noted, did not allow it to stand as a good example of what could be done with warnings of the type Fitzroy had developed. No telegraphic web could master every storm, Smyth pointed out, and "if another such storm takes place, [men] must bear it as their fathers did."[12] Instead, there were other significant lessons in the tempest.

In Smyth's eyes, the storm history was a demonstration of ideal scientific exchange. He sought to show how this should occur, freely and cooperatively among different interests—commercial and scientific, local and international. Above all, Smyth used the storm to underscore the tense relations between

10. Smyth, "Hyperborean Storm," 86'" [pagination irregular].

11. Ibid., 134.

12. Ibid.

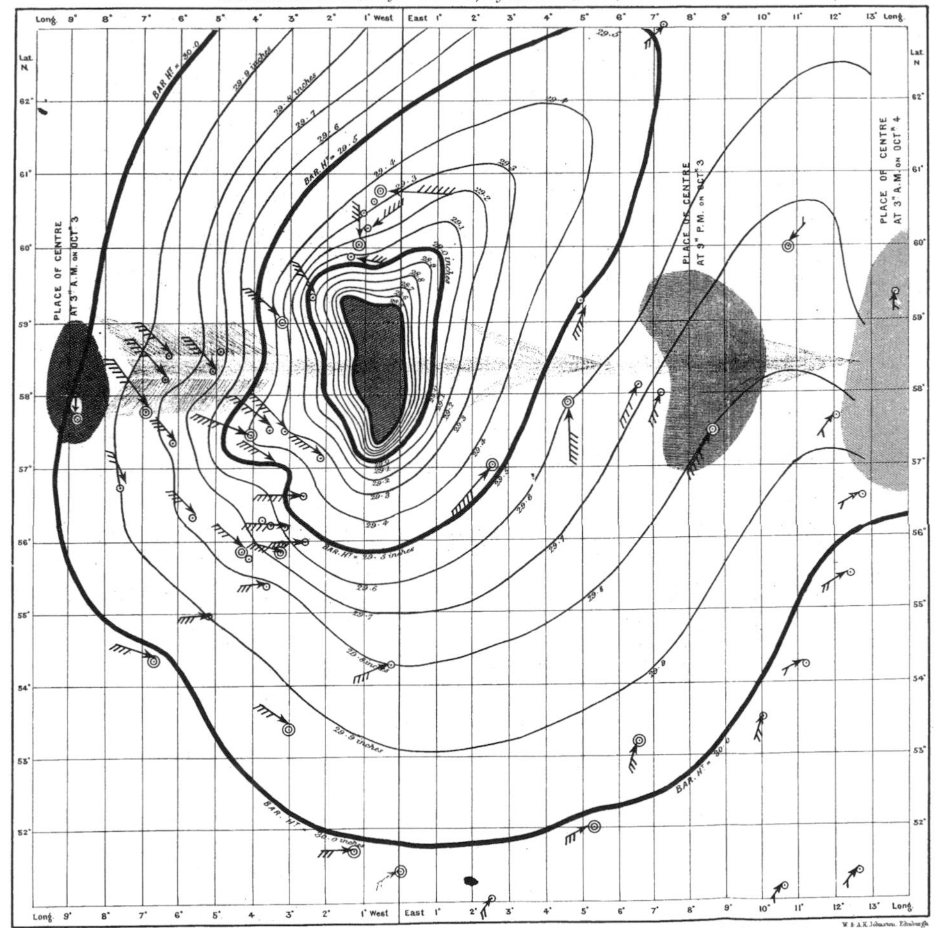

FIGURE 6.1 The Edinburgh astronomer Charles Piazzi Smyth published this synoptic map as a critique of the centralization of scientific work in London. The free exchange of weather data that made his Hyperborean storm analysis possible contrasted with the strained relations of the Meteorological Office and the Scottish Meteorological Society. (Charles Piazzi Smyth, "Hyperborean Storm," *Edinburgh Astronomical Observations* 13 [1874]: plate 56, opp. T83.)

Edinburgh and London. The cancellation of storm warnings epitomized the dangerous influence of narrow, metropolitan interests on government actions. The analysis offered Smyth an occasion to support forecasting "in the Provinces, so-called, though really sitting in the metropolis of the ancient kingdom of Scotland."[13] Smyth had strongly supported the protests in 1867. He was one of the few prominent scientific men who publicly denounced the Royal Society decision, and he did so in terms that made his views of scientific work very clear. Endorsing the local objections to the Royal Society's decision, Smyth agreed that the systems failed to meet "hypercritical" standards of "exact science of the highest order," but he maintained that "high science is one thing, and storm warnings so completely another, that it is not fair to measure its use and right to existence by a test derived from anything else of so entirely different a nature." Violent storms, Smyth went on, could be predicted and "the means, too, for this end are as abundantly in our hands now as they are likely to be for years, and even ages to come."[14] A too-narrow view of science had blinded scientific spokesmen in London to the real practical value of imperfect knowledge for the larger community. As the Hyperborean storm blew, Smyth commented acerbically, London, with its "pleasant Westerly breeze" and "mere microscopic fall of the barometer," failed to notice the devastation elsewhere in the nation.[15]

Smyth (and other Edinburgh meteorologists) were particularly incensed that in 1869 the reformed Meteorological Office, now operating on a vastly increased budget, had begun to charge the Scottish society for copies of their data. To the Scots, this added injury to insult. They had argued in recent years for a share of the funds available to meteorology, in recompense, for instance, for their observations to the registrar general, as well as general restitution for the underfunding of Scottish universities and scientific institutions compared to their English counterparts. The imbalance and insult was the more marked considering that the exchange value, as it might be called, of the periphery was naturally high in meteorology. When Smyth was writing his report in 1871, this particular tempest was just beginning to brew—its subsequent history extended to the establishment and funding of Ben Nevis, an important mountain-top observatory for the rest of the century.[16] Nevertheless, the

13. Ibid., 86' [pagination irregular].

14. Smyth's letter was reprinted in *Communications to the Board of Trade* (Parliamentary Papers).

15. Smyth, "Hyperborean Storm," 85–86.

16. Burton, "History of the British Meteorological Office to 1905," 121–33, 196–207; *Royal Commission on Scientific Instruction and the Advancement of Science: Fourth Report*, 547, and

relationship between Scottish and southern interests was already sufficiently strained. Smyth, for instance, complained about the petty accountancy that bullied him over "a certain little remnant of Barometers and Thermometers employed in the small and most economical expedition to the Peak of Teneriffe in 1856."[17] This treatment seemed symbolic of general grievances, and he called for "entirely Scottish Institutions under Home Rule" to carry out a true "Meteorology of Scotland."[18]

Smyth's account of meteorological affairs painted a picture of a distant government, resisting the legitimate concerns of its citizens and badly advised by scientific officials. His own analysis modeled a different approach to scientific meteorology. It respected utilitarian concerns, without being seduced by the promise of weather telegraphy. It responded to local, particular pressure—the gentlemen pressing Smyth to produce an explanation and description of the storm—but it did so by integrating evidence from a wide geographical area and many different communities in meteorology. All of these observers, from domestic and foreign merchant navy captains to British and foreign observatories, to the network of the Scottish Meteorological Society, contributed freely to help build the record. This was science as a universal enterprise, rather than science as a clique of powerful individuals controlling affairs for their own interests in London. Smyth's references to home rule thus expressed the underlying issues of governance that were involved in meteorology. The Meteorological Office embodied a major commitment to science on the part of the government, a commitment that scientific men were anxious to see sustained and developed. When trying to formulate the direction of research, those concerned with meteorology could not escape the political and administrative questions surrounding government science.

The preoccupation with reorganizing meteorology in the 1860s and 1870s in Britain reflected a disorder that was both administrative and theoretical. These forms of uncertainty fed each other. Smyth had emphasized international cooperation in meteorology as a solution that was dictated by the scale of the phenomena. Another solution was geographic in a different sense. This was the division of the science itself into different spheres of interest, a carving up of the territory of the discipline that would lead to new ways to conceptualize, contain, and control the diffuse phenomena of the atmosphere. Both

Royal Commission on Scientific Instruction and the Advancement of Science: Eighth Report, 482–83 (Parliamentary Papers).

17. Smyth, "Hyperborean Storm," 85–86.

18. Smyth, "Hyperborean Storm," T136 [pagination irregular].

meteorology and its administration would be rationalized. In the 1870s, some meteorologists in Britain began to turn to "natural" distinctions, promoting a new definition of the science that separated "physical meteorology" from "climatic" or applied meteorology. Dividing up meteorology in this way seemed to advance research and facilitate demarcations between different kinds of workers, clearly separating high and low science. Yet this intellectual distinction in meteorological work was confusing and made little sense to contemporaries outside of the history of administrative disputes in which it was embedded—over what forms of institutions should be responsible for what kind of work, and who should fund them. The reshaping of meteorology introduces the sense of territory and borders that permeated science in Victorian Britain. It presented claims about geographical control that connected science firmly with the enterprises of an imperial state.

Balfour Stewart, the former director of Kew Observatory, was a principal figure in the attempts to demarcate branches of the science of meteorology in the 1870s. Stewart proposed a distinction between physical and climatic studies in April 1874 to the Devonshire Commission. Climatic work (defined as "the branches of meteorology which relate to human interests") was best done by individuals and independent societies, Stewart argued, whereas physical meteorology ("the physics of the earth's atmosphere") was a prime candidate for centralized, endowed research.[19] In an earlier article in *Nature*, Stewart's view that strict boundaries were needed to control meteorology emerged in a vivid, half-evolutionary, half-anatomical image. Meteorology was like an "organism of very low development, that had just begun to exhibit the slightest possible tendency to split into two." What was now needed was "the application of the knife." Following this vivisection, Stewart asserted (in a disconcerting shift of metaphor), meteorology could rush separately into the "arms of physiology" and "the embrace of physical research."[20] The reference to physiology showed how closely Stewart associated climatic research with medical studies of healthy environments or mortality, which was certainly a facet but by no means the whole of meteorology relating "to human interests." Nevertheless, some meteorologists welcomed this distinction between climatic and physical meteorology, since it offered a clearly defined role for private societies. Alexander Buchan, secretary of the Scottish Meteorological Society, appeared

19. Evidence of Balfour Stewart, April 23, 1874, *Royal Commission on Scientific Instruction and the Advancement of Science: Eighth Report*, 517–20 (Parliamentary Papers).

20. Stewart, "Physical Meteorology I: Its Present Position," *Nature*, November 25, 1869, 103. The article was continued December 2, 1869, 128–29.

before the Devonshire Commission shortly after Balfour Stewart and agreed that "prosecuting the practical inquiries . . . can be best done by societies [and amateurs]."[21] For Buchan, the distinction was a way to escape the interference of the central office with their local observers and projects. Unfortunately for both Stewart's and Buchan's interests, weather forecasting—the central puzzle for a commission investigating government meteorology—sat uneasily in this picture. It was clearly practical work but, equally clearly, could not be relegated to private efforts because it demanded centralization and substantial funding. Primarily for this reason, Stewart's proposed divisions merely confused the Devonshire Commission.[22] A swift surgical split was no solution to the complexities of either government administration or scientific research.

Another of Stewart's reforming initiatives was equally revealing. Like his physiology and physical science split, this proposal sought to give a hard outline to a diffuse subject. In November 1869, Stewart—then still director at Kew Observatory—published a two-part article in *Nature*, revealing his views of meteorology and making suggestions for its development. In the first historical note on meteorology's "present position," Stewart measured it by an "astronomical standard" of development, arguing that science progresses through definite stages of observation, generalization, and prediction. (By clear implication, forecasting was a doomed effort to leapfrog to the ultimate "prophetical" stage of knowledge.) One way forward, Stewart suggested, was to enclose atmospheric phenomena in a barrier of instruments that could record the loss or gain of moisture per cubic foot of air. This would allow meteorologists to measure vertical movements of air, one of the most inaccessible of observations. Make a certain station, Stewart proposed, "the imaginary centre of a circle" and stud the circumference with other observing points. We could then tell "hour by hour" how much dry air passes in and passes out. If the "imports into the area of the circle" are greater than "exports," then the air in the central station should assume the quality of the imported air (dry or moist); if it did not, Stewart argued, we may estimate from the difference something about the quantity and quality of the ascending current of air.[23] In

21. Evidence of Alexander Buchan, *Royal Commission on Scientific Instruction and the Advancement of Science: Eighth Report*, 517–20 (Parliamentary Papers).

22. See the questions put to Stewart, *Royal Commission on Scientific Instruction and the Advancement of Science: Eighth Report*, 492–95 (Parliamentary Papers). It should be noted that these reforms were not confined to meteorology. At the same time, many of the same players (e.g., Francis Galton and Richard Strachey) were promoting a stricter geography as a physical science in the Royal Geographical Society. See, for example, Strachey, "Introductory Lecture on Scientific Geography."

23. Stewart, "Physical Meteorology II: Suggestions," *Nature*, December 2, 1869, 129.

an approving response, the Victorian doyen of physical science, Lord Kelvin, called the proposal "a meteorological blockade." (Kelvin further proposed that the ebb and flow of atmospheric electricity could be measured as well as water vapor.)[24] These suggestions, comparing observations to statistics of imports and exports, and vertical movements of the atmosphere to smuggling, evoked both the military and economic characterization of a blockade. According to Stewart, meteorology would progress by dividing the atmosphere into areas and strictly monitoring the traffic of air over its borders.

If Balfour Stewart's proposal to cordon off the air seemed both absurd and promising, it was because meteorologists clearly recognized the scale of atmospheric change as a key problem. All these debates about the proper scale of meteorological networks, or different spheres of interest, reinforced the fact that international cooperation in the science was as necessary as it was difficult to achieve. The meeting of ten nations in Brussels in 1853 to set standards for observations at sea by navies and merchant shipping was envisioned by its chief promoter, Maury, as a preamble to more extended and official cooperation in meteorology generally. However, almost every other participant was markedly pessimistic about the chances of achieving uniformity in instruments and measures. The European nations used a variety of temperature scales, for instance, and the most that was accomplished in 1853 was an agreement that thermometers should all have centigrade markings in addition, if necessary, to those of another scale. As the Council of the Royal Society noted, given the individual authority of national establishments, "to call upon countries already so advanced in systematically conducted meteorological observations to remodel their instructions and instruments" was fruitless. Instead, it could only suggest "cheap and rapid intercommunication of the results of the researches."[25]

It was nearly two decades before another international meeting was convened, in Leipzig in 1872. There a gathering of meteorologists drew up a list of inquiries, to be addressed the following year in Vienna. Yet in many ways the Vienna Congress simply resurrected the difficulties encountered at Brussels, underlining the constraints of state science. Only official delegates could attend the congress, which meant that active meteorologists not part of the national meteorological services could not participate (the Scottish Meteorological Society's secretary, Alexander Buchan, for instance, was excluded by these terms but had participated at Leipzig). France, by contrast, was in

24. Thomson, "Balfour Stewart's Meteorological Blockade," *Nature*, January 20, 1870, 306.

25. Christie, "The Reply of the President," 189–90.

disgrace after its failure to send *any* representatives—unresolved struggles between the Montsouris Observatory (under Charles Deville) and the Paris Observatory (under his foe Urbain LeVerrier) over which institution should control meteorology meant that France was not represented until the antagonists' deaths in 1876 and 1877 and the formation of a new Bureau Central Météorologique in 1878 resolved the issue.[26] In acknowledgment of the tensions of state-controlled science, some delegates floated plans for an independent international institute, which would be supported by contributions from all the nations. E. Plantamour, director of the Geneva Observatory, for instance, insisted that only such a centralized institution could emancipate meteorology from the monopolizing, "official" character of state science. His interventions received little serious attention, however, and instead a five to seven member "permanent committee" of meteorologists responsible for guiding international cooperation and planning subsequent meetings was established. The congress settled on advocating an international research fund, to Plantamour's disappointment. On questions of instruments and uniformity, the congress still floundered, echoing the earlier difficulties at Brussels. The British delegate Scott, naturally, promoted Kew Observatory standards, but he met opposition from the Russian and Belgian representatives (Wild and Quetelet), and the question was quietly postponed. Similarly, Scott prevented a unanimous vote in favor of "the uniform Metric system" since for him the question led straight into heated disputes over adoption of the "French" meter in Britain.[27] The debate over forecasting was equally tentative. Indeed, the work of Sub Committee V on "Weather Telegraphy and Marine Meteorology" appeared to have been designed *not* to reach conclusions: the Leipzig formulation of their inquiry declared that, in order to "be in unison with most authorities of science" the question of forecasting's practicability and utility should be left open.[28] Although international gatherings convened regularly thereafter, it is fair to say that the Vienna Congress was as much a demonstration of international obstacles to cooperation as it was a founding structure for that cooperation.[29]

26. Davis, "Weather Forecasting," 380.

27. Schaffer, "Metrology, Metrication, and Victorian Values"; Porter, *Trust in Numbers*; Kula, *Measures and Men*.

28. *Report on Weather Telegraphy and Storm Warnings*, 7.

29. *Abstract of the Report of a Conference Held at Brussels* (Parliamentary Papers); *Report of the Proceedings of the Meteorological Congress at Vienna*. After the Leipzig organizational meeting (1872) and the first congress at Vienna (1873), subsequent congresses were held every few years: e.g., Paris (1878), Rome (1879), Berne (1880), Copenhagen (1882), and Paris (1885). For a summary of

Yet at the same time, meteorology was knowledge that could be carried into the affairs of other nations. The international scale of meteorology evidently gave any successes in the science a supranational prestige. American meteorologists, in particular, to the chagrin and envy of their European counterparts, busily demonstrated the latter in the years immediately after the Civil War. Americans had never been slow to give their attention to meteorology, one of the earth sciences favored by the large spaces of their nation. For a decade in the 1830s and 1840s, meteorologists in Philadelphia and New England—James Espy, William Redfield, and Robert Hare—made American meteorology synonymous with controversy, trading acrimonious papers about the origins of storms, their rotary character, and the role of electricity in the atmosphere. By the late 1840s, Elias Loomis of New York University was making a name for himself with weather maps and papers on convection and cyclones. Maury spearheaded the 1853 Brussels conference on maritime observations from his base in Washington at the Naval Observatory there, and energetically lobbied for a national organization of observations. At mid-century, however, it was a rival of Maury's, Joseph Henry at the Smithsonian Institution, who seized the initiative and put in place a network for collecting data. Over the next decade, Henry's system grew to enormous size, coordinating hundreds of volunteer observers from the Atlantic to the Mississippi and cooperating with other government bodies like the Patent Office, the Department of Agriculture, and the Navy. By 1857, a Washington paper was publishing forecasts based on the Smithsonian analysis of telegraphed data from nineteen stations; three years later, the number of telegraph stations was more than forty and extended to the Midwest. The American response was generally favorable. The west-to-east movement of the atmosphere across the continent, as well as the northerly progress of storms up the Atlantic coast, considerably aided the U.S. system. After the Civil War, and a disastrous fire of 1865 at the Smithsonian, however, Henry's system was in disarray. The U.S. Signal Corps, a new unit of telegraphic engineers whose position was endangered by postwar cuts, quickly latched on to storm warnings as a means of justifying their existence. The U.S. Congress passed a weather service bill in 1870, giving the new system a $25,000 appropriation, trained and equipped military observers to take synchronous observations three times a day, and, most impressively, the authority to construct new telegraph lines to suitable observation points. Henry transferred manage-

---

the work of the congresses, see Hildebrandsson and Hellmann, *Codex of Resolutions*. Cannegieter, "The History of the International Meteorological Organization 1872–1951," gives an overview but deals only briefly with the early period.

ment of the Smithsonian observers to the Signal Corps in 1873. In 1874, the U.S. weather service budget was increased to a remarkable $400,000 (about £72,000)—not including the salaries of the military observers. Six years later, the American network consisted of 290 telegraph stations, submitting observations synchronously to the main signal office, which analyzed the data and produced deductions in an average of ninety minutes. It claimed 90 percent accuracy for its storm warnings, as opposed, one English journalist pointed out, to just over 50 percent in the British system of 1862. European scientists looked on in envy.[30]

The real blow to British pride came in 1877, when the *New York Herald*, under proprietor James Gordon Bennett, Jr. (the man who sent Stanley to find Livingstone, a journalistic coup of several years earlier), began to transmit storm warnings to England via the Atlantic cable.[31] The first warning was issued on February 14—the *Herald* weather bureau considered the storm to have arrived five days later. The *Herald* kept its methods vague, but apparently gathered information on storms from ships arriving in New York after a North Atlantic crossing; in addition to its usual data from the Signal Corps, from British Columbia, and from Central America and Mexico. Then it issued warnings to British and French coasts.[32] It was a well-publicized, and for British meteorologists, somewhat humiliating proceeding. *Gentleman's Magazine* in 1879 published a table of twenty-seven warnings and the resulting weather: seventeen were reckoned "a perfect success," and eight "partial," with only

30. For an example of European reactions, see "The Weather of the United States," 721–23. On the history of American meteorology, see Fleming, *Meteorology in America*; Moyer, *Joseph Henry*; and Bates and Fuller, *America's Weather Warriors*.

31. On Bennett, see *Dictionary of American Biography*. Bennett sent Stanley to find Livingstone in 1869; their famous meeting took place in 1871. Bennett was a keen yachtsman and promoted shipping intelligence in the paper. On the *Herald*'s rise as the premier example of sensationalist reporting, see Crouthamel, *Bennett's New York Herald*. The elder Bennett, who made the newspaper's reputation and fortune, retired in 1866; his son and successor was a reputed playboy, but the *Herald* had the "largest circulation in Europe of any American paper" (Crouthamel, *Bennett's New York Herald*, 158). It claimed circulation domestically of 131,000 daily in 1865 and started a European edition in 1887 (ibid., 151–53).

32. The *Herald*'s typical coverage of the weather in February 1877 was a brief paragraph on national weather and one summary sentence on "from our report this morning, the 'probabilities' for today" (*New York Herald*, February 3, 1877, 2). It also printed separately the chief signal office midnight report from Washington, in common with most U.S. papers. The official *Herald* yacht, which traveled the harbor collecting information from recently arrived ships, began printing weather observations on February 17, 1877; and the paper also occasionally printed weather reports from Plymouth, England, "made for the Herald" (e.g., February 18, 1877).

two misses.[33] In vain, European meteorologists insisted there was no evidence that "American" storms traveled intact, let alone at a predictable pace and direction, across the Atlantic. Robert Scott at the Meteorological Office argued that storms "change their character *en route* . . . so that it is all but impossible to predict which storm, out of several starting from the States, will reach us."[34] In a series of articles in *Nature*, however, Jerome J. Collins of the *Herald* took for granted the American mastery of meteorology, pointing out its economic and military significance. "I have watched with the greatest interest," wrote Collins complacently, "the progress of the recent campaign in Bulgaria, and have frequently announced in New York many days in advance the changes of weather that impeded the Russian progress, endangered the Danube bridges, and filled the Balkan passes with snow." He advised Europeans to build a military weather service immediately, a simple task because their work would consist of forwarding the American information his newspaper and countrymen could supply.[35]

Whatever the merits of the *Herald* system, it was clear that American meteorology was a challenge. As late as 1879, the International Meteorology Committee was unable to persuade European observatories to submit their instruments to a standardization procedure; in contrast, the American delegate to the international body had organized a global synchronous daily observation in 1874 and began producing international weather charts, based on combined European and American coastal observation, supplemented by any data that could be obtained from ships. The London meteorologist Symons commented that the American maps were published three times a day (compared to the single British map) and were superior on many grounds, including aesthetics. "For good looks, the Americans leave our English maps hopelessly in the rear."[36] Amid the work of establishing paths for international cooperation in meteorology, the problem of weather prediction gave a naturalistic framework to political rivalries and conflicts. Famously, Vilhelm Bjerknes's development

33. Thompson, "American Storm-Warnings," 613.

34. Scott, "Forecasting the Weather"; [Scott], "The Weather and Its Prediction." Scott was most irritated that the *Herald* supplied only warnings, not the data, so meteorologists in Europe had little chance of judging the reasoning for themselves. Cf. "The American Storm Warnings," 28; and the letter from a loyal supporter in Birmingham, Thomas L. Plant, who found British warnings "so reliable that we need not trouble ourselves about warnings from America" (*Times*, April 17, 1878).

35. Collins, "The American Storm Warnings," *Nature*, May 2, 1878, 4. The article was continued May 9, 1878, 31, and May 16, 1878, 61–61.

36. Symons, "Daily Atlantic Weather Maps," 48.

of the concept of the weather front in the years immediately after World War I, drew on an analogy with trench warfare, but it is clear that military models of the atmosphere were well established in weather prediction much earlier.[37] By the 1870s, an awareness of the grand scale of atmospheric events combined with the controversies over weather prediction; both led to an acute sense that control of knowledge and observers across a wide territory was the key to reining in the disorder of meteorology.

India: Epitome of Meteorology

By far the most illuminating site of attempts in the 1870s to define the proper political, fiscal, and intellectual territory of science in the modern state took meteorology a long way from London, Paris, or Vienna into the terrain of new theories of "cosmical meteorology" based on the relationship of solar energy and the terrestrial atmosphere. Meteorology developed into one leg of a speculative triangle of cosmical forces, the others being solar activity, tracked by the appearance of sunspots, and terrestrial magnetism. Such speculations tied meteorological research to the proposal of a state observatory for physical astronomy, a key ambition among some circles in London science for a decade. But it also projected meteorology onto the jewel of the empire. India offered an ideal natural laboratory for the science, and an ideal space in which to demonstrate the political importance of science in a global age. At the same time, the model of Indian meteorology vindicated the policy of political centralization and active engagement with the empire that characterized Conservative policy regarding India in particular and the British empire in general in the years after the Indian Mutiny of 1857. If India gave geographical expression to hopes for a coherent meteorology, then meteorology promised India a natural exemplar of rational, centralized management. In the political climate of the 1870s, when the word "imperialism" was newly minted, meteorology linked Calcutta and London as tightly as it linked sun and earth.[38]

37. Friedman, *Appropriating the Weather.* Also see Stewart, "Physical Meteorology I: Its Present Position," *Nature,* November 25, 1869, 101–3; and "Physical Meteorology II: Suggestions," *Nature,* December 2, 1869, 128–29. Cf. a note of 1882 on observation centers as "outposts" and "pickets" and weather as "a series of attacks" that pour in on the country (Ley, "Weather Forecasts," *Nature,* November 9, 1882, 29).

38. This perception among Victorians that the bonds between different geographies, economies, and human societies had a foundation in global physics offers an interesting ingredient for the current debates about orientalism. Said, *Orientalism,* 329–52; Cannadine, *Ornamentalism.* It will also be apparent in what follows that I do not think that the case of Indian meteorology bears

With Indian affairs hovering in the background, any discussion about the boundaries of meteorological science in Britain went beyond metaphor to literal concerns with geography, territory, and political control. India was Britain's most important imperial possession, and, especially after the political changes in 1857, much in the public eye. Around the mid-nineteenth century, the East India Company's steady accumulation of territory and power had collapsed. Their old authority developed first from trading posts around Madras and Bombay and, subsequently, from direct control over the Bengal region in 1765 as the center for a system of alliances with other local rulers. In the first two decades of the nineteenth century, a series of war and annexations placed more land under direct rule of the East India Company. In the 1840s, the company pressed its control north and west to Sind and the Punjab; in the 1850s, it concentrated on the provinces south of Nepal. The 1857 rebellion against the company therefore ignited generations of resentment and took two years and great brutality to suppress. At its conclusion, the administration of India passed from the company to the crown, governed through a secretary of state and council in Britain and a governor general, or viceroy, in India, who oversaw several provincial governments. After 1857, Britain maintained about 60,000 troops in India, and British administrators, civil servants, and educators were all part of the regular human traffic to centers like Madras, Bombay, and Calcutta. Economically, too, India's role steadily intensified: a growing web of banking, shipping, and insurance ties invisibly crisscrossed the distance between British and India. By the late 1860s India absorbed a fifth of all British exports, principally cotton, iron, steel, and machinery. In addition, huge flows of capital left Britain for India, as British investors, protected by guaranteed interest rates, financed the railways, irrigation, and roads that preserved India's economic and military stability. In 1877, Queen Victoria and Disraeli sealed the special position of these possessions with a flourish, making Victoria henceforth the Empress of India.[39]

India's significance to Victorian Britain is evident. The place of the empire in scientific life, however, was more obscure but equally important—a reality

out Lewis Pyenson's account (Pyenson, *Civilizing Mission*) of imperial physical science as a straightforward reflection of the practices and theories of the scientific metropolises in Europe, though the investigations into colonial science that he pioneered deserve much further exploration. Cf. Grove, "The East India Company, the Raj and the El Niño."

39. Moore, "India and the British Empire"; Cain and Hopkin, *British Imperialism*; Juergensmeyer, ed., *Imagining India, Essays on Indian History*. For an examples of ideas about India immediately after the Mutiny of 1857, see Chesney, *Indian Polity*; and compare Martineau, *Suggestions towards the Future Government of India*.

so omnipresent that contemporaries sometimes accorded it little conscious notice. Yet accounts of nature in the nineteenth century were produced in an imperial context. Many of the great figures of the scientific world had worked in the colonies or traveled on voyages of exploration, and these experiences colored their subsequent, influential careers. Two examples showing the particular influence of India will suffice. Joseph Hooker was one of the first Europeans in Tibet. A botanist, he began as a naturalist on Ross's Antarctic expedition of 1839–43, then traveled through Nepal, Tibet, and Bengal botanizing in the late 1840s and early 1850s. He returned to succeed his father as director of the Kew botanical gardens in London and was president of the Royal Society from 1873 to 1878. For Hooker and his fellow official naturalists, Darwin and Huxley, their imperial explorations made their reputations, leading them to the first rank of British scientific men. Hooker was a gentleman naturalist, but equally typical were active scientific men from military backgrounds and colonial postings. Edward Sabine, president of the Royal Society in the decade before Hooker (1861–71), was an artillery officer who had seen action in the 1812–14 war between Canada and the United States. Like Hooker, he had experienced the extremes of polar and tropical environments. He traveled as scientific observer in Arctic expedition of Captain Ross in 1818 and Captain Parry in 1819. In the 1820s he sailed to the West Indies to undertake magnetic measurements, research that culminated in 1840 in the establishment of a chain of government observatories in Toronto, St. Helena, the Cape of Good Hope, and India. From about 1837, like Hooker, his environment was metropolitan—London, the Royal Society, and Woolwich—but it is easy to understand his ready interest in global phenomena.[40]

These examples are typical of the way that the world of British science cascaded outward to the bounds of the empire. As exploration, colonization, and observatory networks spread in the nineteenth century, the base of comparisons for a local or European experience of nature grew more and more extensive. As the historical picture of the interconnections between the British and imperial settings grows more clear, it seems less and less accurate to think in terms of older models of colonial traffic: amateur observers channeling a flow of statistics and specimens to experts at home. Instead, historians have turned to notions of reciprocal exchange, to interactions with indigenous traditions and local circumstances, and perhaps most significantly, to considerations of proximity rather than distance. The presence of the empire

40. Allan, *The Hookers of Kew*; Huxley, *Life and Letters of Sir Joseph Dalton Hooker*. For Sabine, see *Dictionary of Scientific Biography*.

provided a constant pressure to assert the universality of science, to *integrate* distant, exotic, heterogeneous phenomena with familiar ones. The fascination of the wettest spot on earth, Chirra Punji, in Assam, for example, was that its torrential rainfalls of thirty inches in twenty-four hours (details relayed by Joseph Hooker in his popular travel narrative, *Himalayan Journal*, in 1855),[41] could be imaginatively placed alongside the rain of Manchester or London. Meteorology, as we will see, gave an exceptional illustration of the mental proximity of many geographically distant possessions.[42]

India was prominent in meteorological literature. A brief list of important publications would include travelers' observations, official reports, and works on meteorology and navigation. In 1850 Colonel William Henry Sykes, a member of the Board of Directors of the East India Company, and a former officer in the Bombay Army, published a long paper in the *Philosophical Transactions of the Royal Society* on hourly observations he took at Deccan from 1825 to 1830.[43] Richard Strachey, as a young member of an influential family of Anglo-India administrators, produced meteorological observations in his travel through the Himalayas and Tibet in 1847–48. Newly back in Calcutta in 1857 after several years in England, he pressed the Indian authorities to develop a system of meteorological observations. James Glaisher of the Greenwich Meteorological Department produced a survey of Indian meteorology for a government report on the sanitary conditions of the Indian Army in 1862. In 1863, the Schlagintweit brothers, a trio of explorers and surveyors employed by the East India Company, collected as many meteorological records as they could find on their travels (from more than two hundred stations), and published a comprehensive summary of meteorological results based on their own observations made in 1854–58.[44] On the important subject of storms, the accounts of India included

41. Hooker, *Himalayan Journal*, 2:283–84. Hooker later produced the standard reference on Indian botany: *Flora of British India*.

42. Martin, *The Indian Empire*, 1:489, where the comparisons between tropical and English rainfall refer to London, Edinburgh, and Glasgow. For recent accounts of the historiography of colonial science, see MacLeod, "Nature and Empire"; Grove, Damodaran, and Sangwan, eds., *Nature and the Orient*; and MacKenzie, ed., *Imperialism and the Natural World*. I have also found the following helpful: Baber, *The Science of Empire*; Sangwan, "Reordering the Earth" and "From Gentlemen Amateurs to Professionals"; Kumar, "Calcutta"; and Prakash, *Another Reason*. There is also a survey of events and themes in Bose, Sen, and Subbarayappa, eds., *A Concise History of Science in India*.

43. Sykes, "Discussion of Meteorological Observations Taken in India." The philosopher and naturalist Alexander von Humboldt's comments were published in his *Cosmos*, 1:313–14.

44. Hermann de Schlagintweit, "Numerical Elements of Indian Meteorology [Series 1]," "2: Insolation and Its Connexion with Atmospheric Moisture," 111–12, and "3: Temperatures of the Atmosphere and Isothermal Profiles of High Asia." The data on which the memoir was based

Henry Piddington's *Horn-Book of Storms for the India and China Seas* (1844) and Alexander Thom's *An Inquiry into the Nature and Course of Storms in the Indian Ocean South of the Equator* (1845).

As this list suggests, the significance of India for meteorology could be explained as a consequence of Victorian scientific archives on the one hand, and of life in the tropics, on the other. Meteorological tables were another exotic sample to bring home, or another instance of the practical knowledge required for the extension of British interests over the globe. None of this exposes the particular interest of the science, however, and so another approach must be considered. Meteorology did play a small role in the grandest scientific project in India, the Great Trigonometrical Survey. That enormous enterprise, initiated at the beginning of the century, recorded the geodesic details of the Indian landscape in an almost never-ending fashion as British territories accumulated. Meteorology had particular bearing on one of the trickiest technical challenges of the survey, the triangulation of the wide plains of the Indus region. Triangulation depended on sighting distant points; where no natural landmarks were found, as in the plains, high towers had to be constructed. For economy, these towers had to be placed as far apart as feasible. Accordingly, taking an observation from a distance with the precision required for geodesy was a challenge; it was considerably affected by atmospheric conditions, as water vapor in the air modified the refraction of light and apparent position of the object. Objects appeared to change their height depending on cloud cover, humidity, and type of soil across the sight line.[45] This scientific reason for noticing the weather in India, however, must have paled beside the more immediate concerns of survey officers, or any other transplanted European: the heat and rain of the tropics.

Henry Francis Blanford, who later became India's chief meteorologist, had a typically daunting initiation to the Indian climate. Graduating from the London School of Mines in 1855, he arrived in Calcutta in September, the height of the rainy season. He had been appointed to the Indian Geological Survey, and his first assignment took him to a region near Talchir, in Orissa, on the east coast of India. There he struggled with mountainous terrain and

---

seemed to be considered almost useless by later meteorologists, but they were, however, still anxious to extract the original data (Henry Blanford to Richard Strachey, November 8, 1877, Strachey Papers, India Office, MSS Eur F 127/186/55–58). The Schlagintweit brothers achieved more recognition for their ethnographic collections than any other aspect of their work. See Armitage, "The Schlagintweit Collections."

45. Edney, *Mapping an Empire*, 40–43.

dense tropical forests as well as the October cyclone season.[46] After field work had taken its toll on his English constitution, Blanford left the survey in 1862, but he remained in Calcutta, appointed to the Presidency College, where he was encouraged to develop a professional interest in meteorology. In 1875, he became the chief meteorological reporter to the government of India, directing the coordinated system of observations through the country and editing an impressive set of research publications. He could not, however, escape the penalty of the Calcutta climate. In broken health, he retired to the south of England in 1887 and declined slowly to his death five years later.[47]

As Blanford would have known before leaving London, the rigors of the Indian climate were a large part of India's exotic difference from the West. Even though many regions of India were not tropical at all, the three cities where British administration and hence British residents were concentrated—Calcutta, Madras, and Bombay—all experienced tropical weather. The year was divided into three seasons, a cold season (November to mid-February), a hot season (until mid-June), and a rainy season (mid-June to October). English residents described the monsoon for those at home with grim relish (fig. 6.2). Then the ground lay "like a soaked sponge," the air was "a vapour bath," collars wilted as they were placed on the neck, books disintegrated, and boots grew green with mold overnight.[48] Beyond the danger of exposure to infectious diseases such as malaria, cholera, and dysentery, the medical hazards of this climate were treated seriously. Sartorial precautions flourished. The pith or cork helmet, protecting the head and spine from the direct rays of the sun, was considered indispensable covering by mid-century; and medical advisors similarly urged Europeans to wear a garment of flannel—known as the cholera belt—even in the hottest weather as a protective layer between their skin and air.[49] In 1864, the Indian government officially moved its summer headquarters from steamy Calcutta to the "hill station" of Simla, effectively following rather than leading an annual migration of climate relief. Even allowing for such measured behavior, the climate was considered a constant strain on a European constitution, draining the pores, relaxing the nerves, and making

46. On the region, see Hunter, "Orissa and Orissa Tributary States."

47. On Blanford, see *Dictionary of National Biography*; and obituaries in *Quarterly Journal of the Royal Meteorological Society* and *Symons's Monthly Meteorological Magazine*.

48. Wilkins, *Daily Life and Work in India*, 64–66. See also the description of the drier heat of the Punjab and the northwest provinces by Rev. Merk, quoted in Blanford, *The Climates and Weather of India*, 127–29.

49. Kennedy, "The Perils of the Midday Sun"; Cohn, "Cloth, Clothes, and Colonialism."

FIGURE 6.2 This engraving, c. 1861, of Bombay during the monsoon gave a characteristically dramatic account of the tropical climate as weather taken to the extreme. (Robert M. Martin, *The Indian Empire* [London: London Printing and Publishing Co., 1858–61], 3:371.)

true assimilation impossible. "It is doubtful whether there is any part of the country where a European colony would permanently thrive, so as to preserve for successive generations the stamina and energy of the northern races."[50] Not surprisingly, colonial conceptions of disease resisted the insights of germ theory as it was then developing in Europe. James L. Bryden, for instance, a surgeon in the Bengal Army and statistical officer to the sanitary committee of the Indian government, argued in 1869 that disease could not be separated from the weather. Mapping the disease on the land, and comparing that to distribution of rain, Bryden showed that the regularity of the outbreaks were connected to meteorological patterns. The parallels between the epidemics were as "fixed and stable" as Indian's meteorological phenomena. For Bryden, the statistics of disease and weather made clear "the extreme subordination of

50. Martin, *The Indian Empire*, 1:491; Arnold, *Colonizing the Body*; Curtin, *Death by Migration*. For an example of Indian medico-climate writings in the 1830s, see Ainslie, "Observations of Atmospheric Influence." Standard guides during this period were Johnson, *Influences of Tropical Climates on European Constitutions*, first published in 1813 but much reprinted; and King, *Madras Manual of Hygiene*.

epidemic cholera to meteorological influence." He insisted that "the highways by which epidemic cholera travels are, in this country, aerial highways, and not routes of human communications."[51] To a Briton in India, climate was an authoritative explanation for disease.

These topographical and medical reasons for interest in the weather meant that meteorological data was collected wherever British official life established itself, in civil and military hospitals, on board government ships, in prisons and police stations as well as in the government observatories at Madras, Bombay, and Calcutta. Its quality, however, was haphazard. In 1857, Richard Strachey, a scientific army engineer, proposed the reform and rationalization of these heterogeneous observations. Strachey had recently returned from London a new fellow of the Royal Society (1854), and was undoubtedly aware of the formation of the Meteorological Department there. His proposals, however, made through the auspices of the Royal Asiatic Society, came to nothing for several years, put aside amid the upheaval of the rebellion and its aftermath. But in October 1864, an immensely destructive cyclone swept through Calcutta, taking 50,000 lives and destroying much property. A clamor for storm warnings, along the lines of Fitzroy's establishment in Britain, arose in Bengal. Spurred to act, the governments of the several jurisdictions in India—the Madras, Bombay, and Bengal presidencies, the northwest provinces, the central provinces or Deccan, and the Punjab—appointed meteorological reporters in the mid-1860s. As a critical account of these developments in the *Calcutta Review* argued some years later, however, this first effort at an Indian system of meteorological observations continued the fragmentary, ill-equipped, and untrained efforts of earlier periods. The reporters usually fell under the direction of a Sanitary Commission, "a body almost necessarily devoid of any special qualifications" for meteorology, according to the *Calcutta Review*. There was no uniform pattern of observation, let alone standardization of instruments, so that data were essentially useless for comparisons or aggregation. In some cases, unsupervised natives recorded observations irregularly; overburdened medical officers kept their instruments indoors; or picked the "most likely" observation from a number of unreliable records. When systematic quality data was collected, it moldered in archives in India or London, like the twenty-one volumes containing the data of the Bombay magnetic and meteorological observatory from 1841 to 1864. "What may be their final destination," the *Calcutta Review* noted sarcastically, "is not very clear, but it does not appear

51. Bryden, *Epidemic Cholera in the Bengal Presidency*, 49, 90, 92.

that they have ever been utilized for obtaining a knowledge of the meteorology of Western India."[52]

I will return to this critique and the proposed solution to India's mismanaged meteorological enterprises later in this chapter. Yet whatever the shortcomings of the system, its existence testified amply to the colonial rationales for collecting meteorological observations. Beyond these archive-producing, health-preserving pressures for a meteorological record, however, lay another fundamental motivation for studying meteorology in India. This motivation was paramount to those at home, who after all had limited interest in the medical anxieties of Englishmen abroad. In brief, India offered a special situation for the study of meteorology. As Blanford outlined it, the chief problem facing meteorologists was the inaccessibility of their subject, due to the vast extent of the atmosphere and the fluctuating and interconnecting forces that shaped it, "We are in the position of a commander who can find no eminence from which he may gain a bird's eye view of the combat," he claimed, alluding to Stewart's military blockade metaphor. Yet

> [c]ould we but find some isolated tract of mountain, plain and ocean, under a wide range of latitude, girdled round by a giant mountain chain that should completely shut in and isolate some millions of square miles of the atmosphere, resting on a surface vast and varied enough to exhibit within itself all those contrasts of desert and forest, of plain, plateau and mountain ridge, of continent and sea, that we meet with on the earth's surface,

then the progress of meteorology would be assured. India provided just such a model (fig. 6.3). "As England is the epitome of stratigraphic geology, so is India an epitome of atmospheric physics," Blanford declared.[53] Moreover, as part of the region's appeal, the frequent cyclones of the Indian ocean were suitable for the study of the origin of storms, still a great preoccupation of meteorologists.

52. "Meteorology in India," 287, 289. Strachey's intervention is noted in this review, 274. Official documents relating to meteorological observations in India 1866–71 were printed as a parliamentary paper in 1874: *Organization of a Meteorological Department in India* (see *Copies of the Despatch* . . . in the bibliography). The inaccuracy of Indian records before 1875 was notorious in the literature: Symons recalled, for instance, "the days when Indian rain gauges were taken indoors at night and locked up for safe-keeping." Symons, "On the Climates of the Various British Colonies," 59–60. For an account from an observer's point of view, see Fayrer, *Rainfall and Climate in India*, 26.

53. Blanford, *The Indian Meteorologist's Vade Mecum*, 99. Blanford also wrote a popular account of Indian meteorology, *The Climates and Weather of India, Ceylon, and Burmah and the Storms of Indian Seas* (1889). Two letters to Richard Strachey in 1877 gave an account of his work: June 5, 1877, and November 8, 1877, Strachey Papers, India Office, MSS Eur F 127/186/49–58.

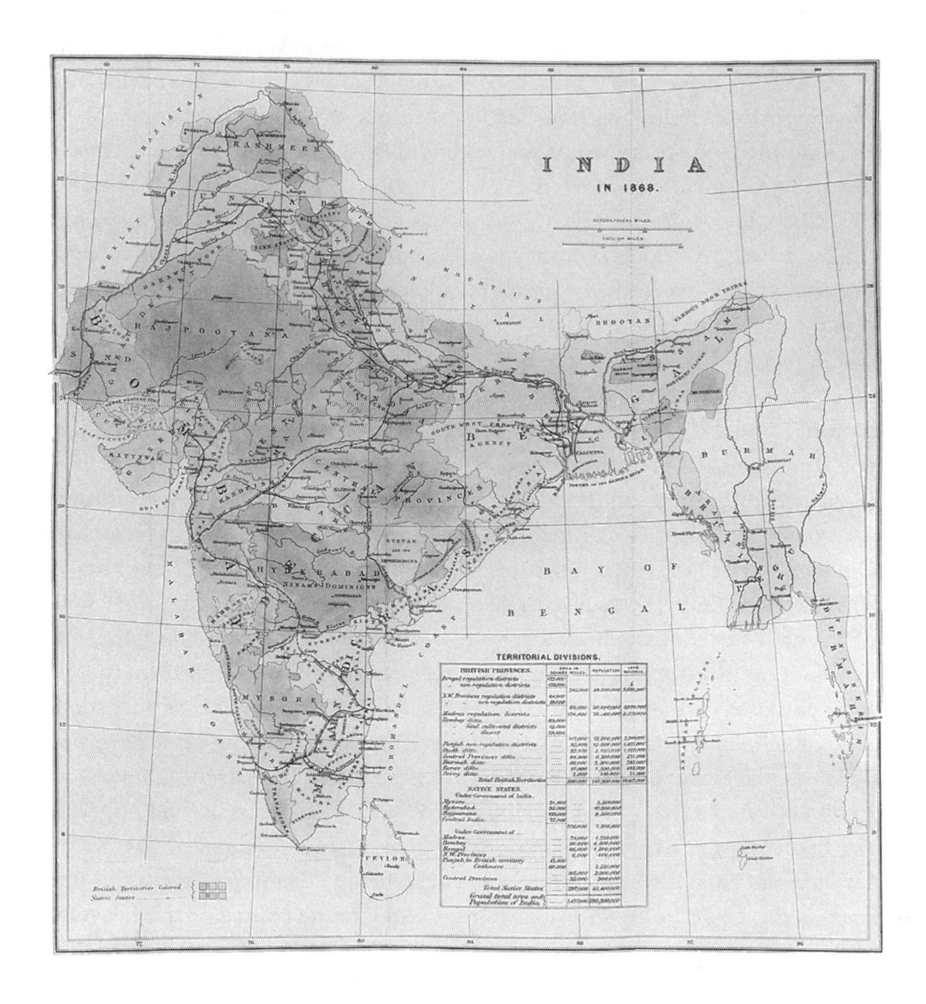

FIGURE 6.3 The Indian empire offered a solution to the interpretation of the atmosphere. This geographical space, enclosed by oceans and mountains, and marked by the regularity of its seasons, represented a natural laboratory of extreme but simplified weather. (George Chesney, *Indian Polity: A View of the System of Administration in India* [London: Longman, Green and Co., 1868], frontispiece.)

The assiduous meteorologist Charles Meldrum, at the observatory on the island of Mauritius, claimed that "no gale of any magnitude could occur at a distance of 1500 to 2000 miles" without his instruments recording the disturbance.[54] He, Blanford, and other meteorologists based in India became active figures in the theory of storms. In short, the empire made it possible to resume national scientific achievement. India was Britain's continent—the place where, like their successful American counterparts, English meteorologists could analyze the atmosphere over a large enough, coherent enough territory to produce significant results.

Coupled with this striking geographical logic, India appeared to be a natural laboratory for meteorology for a further important reason. A tropical climate seemed to hold the key to unraveling the laws of the atmosphere. Alexander von Humboldt, the great student of global forces, had argued this point earlier in the century. Humboldt had a naturalist's fascination with the fecundity of the tropics, and his writings had inspired a generation of scientific travelers, including Charles Darwin. But the scientific attraction of the tropics was not just their exotic variety; it was also their regularity. Nearer the equator, the length of the days varied less, eliminating great differences in daily solar exposure, or what Humboldt referred to as "the excessive complication" and "perpetual local variations" of temperate regions. By contrast, the tropical setting made it possible to trace natural laws on the earth's surface with the same precision as in the planetary sphere. "The regions of the torrid zone," Humboldt wrote in his *Cosmos: A Sketch of the Physical Description of the Universe*, "not only give rise to the most powerful impressions by their organic richness and abundant fertility, but they likewise afford the inestimable advantage of revealing to man . . . the invariability of the laws that regulate the course of the heavenly bodies, reflected, as it were, in terrestrial phenomena." Humboldt referred here not only to "the contrasts of climate and vegetation exhibited at different elevations," the vertical law of natural phenomena for which he was famous, but also the regular variations of the atmosphere.[55] Humboldt's characterization exerted a powerful influence. "We must dismiss from our minds much that constitutes the mental stock-in-trade of the European meteorologist," pointed out Blanford. "Order and regularity are as prominent characteristics of our atmospheric phenomena, as are apparent caprice and uncertainty those of

54. Meldrum, "On Synoptic Charts of the Indian Ocean," 145. On Meldrum, see *Dictionary of National Biography*.

55. Humboldt, *Cosmos*, 1:34, 36. On Humboldt's biogeography, see Dettelbach, "Humboldtian Science"; and Rupke, "Introduction."

their European counterparts."[56] Writ small, this order appeared in the confined daily fluctuations of the barometer, which Humboldt thought exhibited a regularity so mechanical it could be used to tell the time, a kind of natural clock, accurate to fifteen minutes.[57] Writ large, the regularity was represented by the monsoons, "the primary contrast of land and water."[58]

The monsoon, India's most notable meteorological feature, was atmospheric change reduced to its essentials. Twice a year, the winds reversed direction, blowing from the northeast in the cold season, after October or November, and switching to southwest in April or May. The transition depended on the surface heat of the land, which in turned depended on the seasonal position of the earth with respect to the sun. During the cold season, air flowed northeasterly from the colder, higher-pressure region on land toward the warmer air over the ocean; conversely, as the land baked hotter, the temperature differential grew smaller and then shifted, bringing relatively cooler air from the southwest off the ocean onto the land. Filled with moisture from the ocean, this wind, the stronger monsoon, brought the heavy rain associated with the word, and the shift in wind abruptly inaugurated the rainy season of June to September or October. The simple binary picture was summarized in a cartoon commemorating the foundation of the Meteorological Department in 1875, "Indian Meteorology; or, the Weather Classically Treated" (fig. 6.4). There, sufferers with their pith helmets and umbrellas were positioned on a grilling iron held by the sun god Apollo and sprinkled liberally by Neptune. In the background, the British danced in the only place they were comfortable, in the hill stations of Simla. The relatively stable pattern of the atmosphere in India, dominated by the monsoon, presented the meteorologist with fewer fluctuations to unravel. It also allowed him to study "vertical" rather than "horizontal" effects, as one meteorologist put it. That is, the stability of the systems across large homogenous regions meant that observations recorded "expansive and contractive" movements of the atmosphere more than they did in Europe.[59]

Perhaps more important than any of these scientific opportunities, the appeal of India as a site for science had another element that deserves emphasis here—an appeal connected to political as well as geographical realities.

56. Blanford, *The Indian Meteorologist's Vade Mecum*, 144.

57. Humboldt, *Cosmos*, 1:34–35, 313–14. On Humboldt and the tropics, see Dettelbach, "Global Physics and Aesthetic Empire."

58. Blanford, *The Indian Meteorologist's Vade Mecum*, 144.

59. Archibald, "Indian Meteorology," *Nature*, August 23, 1893, 406.

FIGURE 6.4 *The Indian Charivari* greeted the formation of the Meteorological Department of India under Blanford with a cartoon: "Indian Meteorology; or, the Weather Classically Treated." It showed the residents of India alternately roasted on a grill by Apollo and sprayed by Neptune. The oblivious transfer of western mythology contrasted with William Hunter's discussion of Hindu rain and wind gods in his article on sunspots and weather prediction in India. (*The Indian Charivari*, July 23, 1875, 21; British Library.)

It was the fundamental fascination of the imperial situation: the imposition of Western conceptions of order upon a vast, confusing subject. The whole, long fascination of the West with India reflected this hopeful sense of the progress of reason and order. Under the Raj, then, India itself as a political entity offered a model of reason literally wrestled from a jumble of disorderly territories. The emphasis on Indian's geographical coherence as a reason for meteorological studies that we saw in Blanford's writings was not a novelty. It was lifted entire from a commonplace of British conceptions of India—that the Raj created a nation by recognizing a set of "natural" boundaries. With this natural, reasonable conquest, British law and order overcame the chaos that was an inevitable consequence of dissolute petty rulers, the errors of the Hindu and Moslem religions, and the unreason of castes. (The latter, as a misguided form of classification, could be equally portrayed in terms of natural boundaries. "[L]ike other attempts to cramp the human intellect, and forcibly to restrain men within bounds which nature scorns to keep, this system, however specious in theory, has operated like the Chinese national shoe, it has rendered the whole nation cripples.")[60] Blanford wrote a geography school textbook as well as his volumes on Indian meteorology, and the echoes between the works are clear. India was "really a collection of many countries . . . languages, religions, and civilization[s]," he began. "Yet no country is more distinctly marked off by natural boundaries"—the mountains of the Himalayas and Tibet, the lower but inaccessible mountainous regions in the northwest, and the ocean surrounding the southern triangular peninsula. These natural borders made it, as another geographical account noted, "a compact dominion."[61] For Blanford and his readers, the geography and the geographers of India literally gave the nation its identity.[62] In parallel fashion, then, meteorologists could use the natural boundaries of India as an enlarged version of Balfour Stewart's cordon, to impose coherence on the messy phenomena of the atmosphere.

Meteorology—or, to be exact, the hopes for meteorology—thus paralleled British rule—or the hopes for British rule—in India. Analogies ran every way. In reciprocity, for instance, statesmen conducted the debate about the management of the empire in environmental terms. The great political question of the 1870s was whether to enlarge or confine conceptions of the empire—to disentangle Britain economically and politically from the colonies, or to develop a permanent, global vision of British power. Writing in 1877 on "Greater or Lesser Britain," the former governor of New Zealand, Julius Vogel, compared the urgency of the question to deforestation and climate change. A difficulty can be compounded by delay, Vogel insisted: "the time will come when . . . regularly flowing rivers will become fitful torrents, when the earth, deprived of its moisture and its soil washed into the ocean, will cease to produce as it

60. Thomas Ward (1811), quoted in Cohn, "Notes on the History of the Study of Indian Society and Culture," 144.

61. Blanford, *An Elementary Geography of India*, 1; [Hunter], "India," 731. On the idea of natural boundaries and frontiers in general, see Kristof, "The Nature of Frontiers and Boundaries"; Pounds, "The Origin of the Idea of Natural Frontiers in France"; Prescott, *The Geography of Frontiers and Boundaries*; Curzon, *Frontiers*; and Dutta, *Imperial Mappings*.

62. Cf. a modern account: "As a political entity . . . India was a creation of the same power that drew the frontiers" (Embree, "Frontiers into Boundaries," 84). On the history and influence of Western ideas about India, see Embree, "The Idea of India in Classical Western Political Thought," 28–40.

did before." The political relationship between Britain and her colonies was a force of nature; its control required calm reason, not "angry passions."[63] The picture of managing natural forces had direct bearing in India, in which many regions were dependent on irrigation. But such comparisons were obvious in a period when Indian land, peoples, and economy were seen as one huge natural resource. Furthermore, these arguments about coherence and management opposed the other naturalistic account of government in India—an evolutionary model of gradual independence. This model was particularly associated with Liberal ideology. Lord Blachford, the permanent undersecretary of the Colonial Office from 1860 to 1871, and one of the founders of the *Guardian* newspaper, was one such Liberal voice. In an October 1877 essay on the empire, he explained that the idea of colonial self-government "took possession of men's mind like a scientific discovery . . . [it was] one of those irresistible pressures which arise out of the order of nature."[64] In the teeth of these claims, Conservatives welcomed the counterweight of a compelling analogy of scientific order. The symmetry of social, political, and natural order had become a commonplace of Victorian thought. Science conducted in an imperial space thus necessarily fostered consideration of the relations between government and knowledge.

Famines, Imperial Economy, and Imperial Government

By the 1870s, the promise of India's model space began to center on a particular set of speculations concerning sunspots. The nature of these dark areas of varying size on the surface of the sun was a matter of controversy. At the end of the eighteenth century William Herschel had proposed that the spots represented holes in the sun's atmosphere, allowing a glimpse of the darker and cooler core underneath. Several decades later, Gustav Kirchhoff, who pioneered the chemical evaluation of the sun's spectrum in 1859, argued for the opposite view: sunspots were the condensation of gases on the surface of the sun after they emerged from the bright molten core. By mid-century, inquiry shifted from their origins to their periodicity. The discovery of a cyclical pattern in sunspots, approximately eleven years between years of maximum solar activity, emerged from the exemplary observation over four decades of an otherwise obscure German, Hofrath Schwabe. Schwabe's work

63. Vogel, "Greater or Lesser Britain," 82. The essay was first published in *Nineteenth Century* 1 (1877): 809–31.

64. Rodgers, "The Integrity of the British Empire," 124.

was publicized by Humboldt in the third volume of *Cosmos*, in 1851, and solar observations became part of the physical astronomy research programs then developing in Europe. According to some, it promised to become the basis for "physical meteorology," a truly scientific study of the atmosphere.[65] The key element of these speculations was evidence of a shared cycle of intensity in sunspots, terrestrial magnetism, and rainfall—what Balfour Stewart called "three corners of a triangle" of cosmic relations.[66] By 1870, researchers in the earth's magnetism had found a "close and intimate relationship" between the sunspot cycle and magnetic phenomena, like auroral displays, and the small fluctuations of magnetic needles (although the peak of magnetic phenomena lagged several months behind the peak of sunspots).[67] A host of supporting atmospheric evidence emerged, as meteorologists eagerly seized the opportunity to apply a theory to their mounds of laboriously recorded data. Cyclones in the Indian Ocean, river levels, barometric pressure, mean annual temperature, famines in India, all formed patterns that correlated with the sunspot cycle. These analyses bound physical astronomy and meteorology tightly together as models of modern research science—the former drawing on the instrumental advances of photography and spectroscopy, the latter moving away from "a collection of mere statistics" into a physical science or the "Meteorology of the Future," in the grandiose language of *Nature*'s editor, Norman Lockyer, in 1872.[68]

Why the enthusiasm for "sunspottery," as one critic called it? Inside the pages of scientific journals, the reasons for the heat generated by sunspot research often went unstated. Overtly, the focus was that given in Lockyer's article: the conversion of an empirical, troubled science into a promising, exact physics of the atmosphere, appropriately aided by modern developments in the queen of sciences, astronomy. Even those most concerned with the practical implications of the pattern of sunspots and rainfall could convey this impression. In 1875 William Stanley Jevons turned to sunspot and meteorological correlations to supply an explanation for roughly decennial economic crises. Jevons was a well-respected economist and was then a professor at Owens College in Manchester, moving in 1876 to the University of London. In his sunspot articles, he sought to make the case that "the periodic variation of tropical harvest is connected with the solar period, and that this harvest

65. Hufbauer, *Exploring the Sun*.

66. Stewart, "Suspected Relations between the Sun and the Earth," *Nature*, May 17, 1877, 45.

67. Stewart, "Suspected Relations between the Sun and the Earth," *Nature*, May 10, 1877, 26.

68. Lockyer, "Meteorology of the Future," *Nature*, December 12, 1872, 100.

variation operates so as to stimulate and determine the naturally rhythmic fluctuations of European trade."[69] Many contemporaries greeted Jevons's sunspot theory as a jest, and it has remained something of a blot on his reputation as a serious economist.[70] Nevertheless, Jevons himself placed great weight on the theory, defending it with vigor and collecting his sunspot articles for publication with other writings on economic fluctuations.[71] Like Lockyer, Stewart, and others, Jevons treated the question as an exemplary case. The sunspot theory gave Jevons a basis from which to attack his compatriots for their inadequate attention to scientific reasoning, the importance of which he had outlined in the recent *Principles of Science* (1874), a treatment of logic and scientific method. Jevons had characterized meteorology in an 1857 letter to his sister from Australia where he was engaged in meteorological observations as "a sort of difficult *scientific exercise* rather than a science itself."[72] This didactic approach was at the center of his interest: he used sunspots, as he put it in his first essay on the subject, "to sketch out the course of inductive argument and enquiry" following "the general principles of inductive logic."[73]

Yet, as Jevons was well aware, the immediate interest in sunspot research linked natural knowledge to far less abstract considerations. Just as the statistical projects of the nineteenth century made a more thorough analysis possible, so its economy made the discovery more urgent, he admitted. "An empire on which the sun never sets, and whose commerce pervades every port and creek of the sunny south cannot wisely neglect to keep a watch on the great fountain

69. Jevons, *Investigations in Currency and Finance*, xxxiii. For Jevons, see Schabas, *A World Ruled by Number*; and Black and Konecamp, eds., *Papers and Correspondence of William Stanley Jevons*. Peart, "Sunspots and Expectations," gives an account of all Jevons's contributions to sunspot debates.

70. It has since attracted some interest: see Mirowksi, "Macroeconomic Instability and 'Natural' Processes in Early Neoclassical Economics"; Peart, "Sunspots and Expectations"; and Klein, *Statistical Visions in Time*. Jevons's speculations were not unprecedented: he linked his work to that of William Herschel at the close of the eighteenth century on the correlation between corn price series and sunspots (Hufbauer, *Exploring the Sun*, 38–40).

71. Jevons, *Investigations in Currency and Finance*. His interest in meteorology in fact had deep roots, dating back to time spent in Australia in the 1850s. He wrote an enthusiastic article on Fitzroy for the *Spectator* in 1864, and two years later he advocated using telegraphed warnings of a drop in barometric pressure to prevent coal mining explosions. See the letter to the *Times*, December 22, 1866, published in Black and Konecamp, eds., *Papers and Correspondence of William Stanley Jevons*, 3:144.

72. From a June 17, 1857, letter to his sister Henrietta advising her against taking up meteorology, "a most troublesome extensive, and to most uninteresting subject." Black and Konecamp, eds., *Papers and Correspondence of William Stanley Jevons*, 2:291–92.

73. Ibid., 5:92, original emphasis.

of energy."[74] Indeed, the beginnings of Jevons's own studies into solar cycles overlapped with his work on an analysis of the limits of British coal supplies. With its pessimistic implications for the future national wealth, Jevons's work on energy supply and economies was attracting great attention. The press leapt immediately on this second, equally newsworthy speculation by Jevons at the BAAS meeting in 1875.[75] Outside the pages of scientific publications, then, the future promise of physical science was far from a question of logical philosophy: it was rooted in the economics and politics of Victorian imperial affairs. In particular, sunspots found their way into one of the great dramas of Victorian imperialism, the Indian famine of 1877 and 1878. Famine was the test of empire. Stories of emaciated natives who were eating grass or selling their children vied in the daily papers with lengthy discussions of the proper response of government to the crisis. If sunspot research made long-term climate prediction possible, the periodic famines that devastated India could be anticipated and controlled. In a sense, the Indian monsoon and the Indian economy mirrored each other in the concentration of their effect. The Indian economy was single-mindedly focused on its grain harvests. Indian taxes to administer the country and pay dividends to British creditors came from the satisfactory arrival of the rains.[76] Control of famine meant India could be governed more effectively. It could be freed both from terrible human misery and the political spectacle of famine relief measures.

Both the misery and the politics of famine were communicated to the British public above all by William Wilson Hunter, a man rightly called "a splendid example of the ambitious Scot turned loose on the empire" (fig. 6.5).[77] Hunter was the son of a Glasgow manufacturer, educated at the university there. Casting about for a career in 1860, Hunter attempted the new Indian civil service examinations, becoming one of the examples of post-Mutiny "competition-wallahs," who won their posts and promotions in examination rather than through patronage networks of the East India Company. After he placed high in the exam, Hunter chose a Bengal posting and became an assistant magistrate and tax collector in Birbhum in 1862, hearing disputes on land tenure and petty

74. Jevons, *Investigations in Currency and Finance*, 235.

75. Jevons, *The Coal Question*. "I found the reporters had got my abstract and telegraphed it everywhere" (Harriet Jevons, *Letters and Journal of W. Stanley Jevons*, 340).

76. As one imperial historian summarizes, "the *Raj* depended upon wringing a surplus from an overwhelmingly agricultural economy" (Moore, "India and the British Empire," 81). On the Indian economy, see Cain and Hopkin, *British Imperialism*; and Crouzet, *The Victorian Economy*.

77. Greenough, "Hunter's Drowned Land," 238.

crime in the mornings and reading and writing in the afternoon. He became a noted contributor to Calcutta newspapers, with regular columns on Indian life and customs. Within the decade, his publications on Bengali history, language, and politics had attracted the notice of the governor-general and viceroy, Lord Mayo, who became a patron of the earnest young man. His pleas for a systematic account of the land and peoples of India were taken up. In 1871, at age thirty-one, he became the director-general of statistics, a post developed expressly for him, in which his chief task was the production of a comprehensive gazetteer of India. He himself amassed twenty volumes on Bengal and Assam in the mid-1870s and supervised the production of hundreds of other local volumes, as well as overseeing the authoritative *Imperial Gazetteer*.[78] The first edition, nine volumes of dense information on geography, climate, population, agricultural practices, trade, legal systems, and everything else indispensable to a modern administrator, appeared in 1881; a second, fourteen-volume edition, incorporating the results of the 1881 census, came out in 1885–87. Hunter's work on the *Gazetteer*, much of which he carried out from England, gave him a position of authority on Indian affairs. He wrote the 1890 *Encyclopaedia Britannica* article on India, giving an influential account of "the struggle between the blind forces of nature and those of organised society." There his political ideals of stewardship were clear. "The British Empire is seen as the latest stage in a vast gigantic organic growth, . . . its fabric raised slowly, with many mistakes, much blind groping at higher ideals, until it shelters a seventh of the population of the globe."[79] From early in Hunter's career, meteorology wove itself into those ideals.

Hunter's first work, *Annals of Rural Bengal* (1868), was a history of the introduction of East India Company authority in the region at the end of the previous century, including the famine of 1770, when at least third of the population died—"an appalling spectre on the threshold of British rule in India," as Hunter noted.[80] The modern lens through which Hunter traced this particular history was obvious. His subject was the conditions and responsibilities of English rule just as that rule was being imposed anew, this time under the aegis not of a trading company but a reformed imperial civil service. As he wrote, Hunter was himself in the thick of the 1866 famine in Orissa, the first real test of the post-Mutiny government. His *Annals* deliberately established the contrast between the humane intervention of modern rule with

78. On the project, see Hancher, "An Imagined World."

79. Skrine, *Life of Sir William Wilson Hunter*, 303.

80. Hunter, *Annals of Rural Bengal*, 29.

FIGURE 6.5 William Wilson Hunter was director-general of statistics in India, a post that allowed him to spend most of his time writing in Britain. He became a highly influential voice for Indian affairs and in 1877 wrote a paper energetically espousing sunspot theory as a means of predicting and managing famine. (Francis Henry Skrine, *Life of Sir William Wilson Hunter* [London: Longman, 1901], frontispiece.)

the wholly inadequate official response in the eighteenth century. At the same time, however, it emphasized that the demand of humanity had only been partially met. In 1866, magistrates and officials like Hunter tried to determine how much poverty or how many deaths made a famine, whether to supply rations or provide employment as relief measures, and whether to prohibit exports from regions with shortages, or whether to simply take measures to ensure free circulation of food supply. But ignorance of local conditions and disputes over policy often made their efforts ineffectual. Visiting Midnapur in

July, Hunter saw peasants too weak to cook their government ration of rice, barely chewing the hard grains as they lay on the ground. Appointed school inspector later that year, he toured the affected region to assess the impact of the famine and cholera on the education system and sent graphic accounts of starvation and ruin to the Calcutta papers.

Leaving history for contemporary analysis, Hunter's next work addressed the conditions of modern rule directly. By 1869, he was collecting statistical information in the Bengal region on prices, area of cultivation, relative weight of each of the four Indian harvests, means of transport, and percentage of population directly dependent on agriculture. He presented this study, *Famine Aspects of Bengal Districts*, to the Indian government in the autumn of 1873 and published it in London in 1874. Out of this scheme of data, Hunter proposed, could emerge a prediction of the chances of famine, which would prevent the desperate situations of previous shortages. In other words, as Hunter put it, administrators would gain "a formulated system of Famine Warnings."[81] An embossed cover of the published work presented the title differently as *System of Famine Warnings*, perhaps deliberately echoing the storm warnings then being developed in Calcutta, and discontinued in London.[82] Hunter's earlier *Annals* had established him as the rising voice on India, and his new work also gained attention. By this time, the response of British opinion to famines had developed into a policy whose primary consideration was the absolute prevention of death by starvation as a condition of civilized rule.[83] The *Times* was unequivocal, reviewing Hunter's *Famine Aspects* in January 1874: "We must decline to reason at all on the assumption that the weak must perish. England will require that unless by accident not a life shall be lost for want of food with such ample warning beforehand, and when so small an amount of food will suffice to preserve life." As famine continued, the *Times* returned to its theme, arguing against a repetition of conditions in 1866 and referring to a "new spirit" of empire. "We entreat the government of India to draw with a bolder hand upon the resources of the Empire."[84] Although a month later, Hunter was still predicting a "solemn and terrible ordeal" for Bengal, late rains in January and February provided relief, and the danger of widespread famine

81. Hunter, *Famine Aspects*, 2.

82. This heading is embossed on the copy held in Princeton University, Firestone Library.

83. Critics, with some justice, argued in later relief campaigns that this policy merely made it impossible to count deaths by starvation, which were instead recorded as a consequence of various diseases.

84. *Times*, January 19, 1874, 12; January 20, 1874, 6; March 17, 1874, 10.

had passed by October. Hunter nevertheless sought to fix British attention on the recurrent nature of the problem. Famines, he argued, "formed the natural penalty for inadequate means of internal transit and for an unhusbanded and uncontrolled water supply."[85] The solution must be to remove conditions of local isolation with irrigation, road and railway projects, allowing supply and demand to operate unrestricted.[86]

The next famine loomed at a particularly troubling political moment. Victoria had been declared empress of India on January 1, 1877, amid some skepticism at home and abroad. The pageant of the Imperial Assembly marking the occasion in Delhi unfolded simultaneously with dire news of suffering in Madras.[87] Eventually, the death toll neared six million. The government of India and the affected provinces responded with rapid, even practiced, measures. Yet, as thousands crowded relief works, public sympathy for the victims of famine grew. Famine politics pitted free trade against an ideology of stewardship of the colonies.[88] A specially appointed famine commissioner, with orders to reduce expenditure, toured the famine districts with such dispatch and apparent arrogance that opinion was outraged. At the high point of the famine, in August 1877, the viceroy issued a policy statement affirming his government's commitment to local responsibility, labor rather than handouts, and noninterference with trade (no enforced imports of grain). The response of the *Times* was typical. Present famine policy, it criticized, assumed that all Indians were "only too ready to 'sponge' on the benevolence of the State." Fellow subjects were "hustled and bullied as if they were a race of hereditary paupers, instead of the victims of a great national calamity."[89]

Arriving in Madras on *Imperial Gazetteer* business in January 1877, Hunter found himself again in the midst of famine. By this point, his own conven-

85. *Times*, March 17, 1874, 10.

86. For the details of the 1873–74 famine, see *Final Reports of the Measures Taken in Consequence of the Scarcity of* 1873–74 (Parliamentary Papers).

87. Digby, *The Famine Campaign in Southern India*, 1:45–47.

88. On famines and famine policy, see Srivastara, *The History of Indian Famines*; McAlpin, *Subject to Famine*; and Digby, *The Famine Campaign in Southern India*. For examples of the political discussion, see the collection of speeches on the subject by Ashley Eden, Lord Lytton, and Sir John Strachey, *The Indian Famine of* 1877; and Cotton, *England and India* and "Has India Food for Its People?" *The Colonies*, a London weekly paper associated with the Colonial Institute, carried lively discussions, including comment on a paper by Robert H. Elliot, "Indian Famines and How to Modify the Causes that Lead to Them," on November 24 and December 1, 1877.

89. *Times*, May 24, 1877, 12. The viceroy, Lord Lytton, felt that Madras officials were overreacting and expended too much on relief. "It is a struggle with exasperated lunatics which must be conducted without breaking any of the furniture" (quoted in Gopal, *British Policy in India*, 116).

tional views had shifted. In part, this change reflected his sense that the old commitment to preventing starvation had failed, mired in a political split between the "Manchester men"[90] and those who saw government responsibility in much more extensive terms. But in part it reflected a deeper study of "the natural penalty," as Hunter had called it.[91] Famine was a Malthusian scenario, and especially after Darwin had rendered Malthus as the natural law of struggle for existence, famine was conceived as an inevitable feature of the environment. Richard Strachey's analysis of the causes of the Indian famine proceeded in this way, announcing in its opening words that "[l]ife . . . is passed in a continued struggle." Strachey suggested that social conditions in tropical lands, like India, were more subject to the pressures of nature. Just as the "gifts of nature are most profuse," so too does India "encounter in their extreme form the devastating forces of tempest, drought, flood and disease." Both disaster and fecundity were natural conditions of the tropical environment that no human authority could reverse. The management of famine, then, in Strachey's Malthusian account, hinged on training the population to understand and accommodate nature. India must draw from her own resources—developing its own labor, irrigation, and taxation systems—rather than relying on external help. The charitable funds flooding in from England were misguided; they prevented Indian from building up the moral qualities of "self-reliance" and "self-sacrifice" that would alone provide a lasting solution for "periodical suffering."[92] When these conventional assessments focused on natural law and the Indian environment as the cause of famine, it seemed logical that Hunter's more activist temperament should turn to nature for solutions. In February 1877, he decided he had found one: the cyclical pattern of droughts and sunspots.

Hunter came to these views after meeting the astronomer of Madras, Norman Pogson, a strong advocate for the connection between solar and atmospheric research.[93] Analyzing the solar and meteorological records with Pogson, Hunter found that five out of the six famines in Madras since 1810 had occurred during years of minimum solar activity. He immediately used his

90. A reference to the influence of Richard Cobden and John Bright on William Gladstone's Liberal Party. For an example, see Skrine, *Life of Sir William Wilson Hunter*, 294.

91. *Times*, March 17, 1874, 10.

92. Strachey, "Physical Causes of Indian Famines," in *Report of the Royal Commission on Indian Famines*, 1881, appendix 1:1–7 (Parliamentary Papers). This commentary was a version of the talk Strachey gave to the Royal Institution on May 18, 1877, which was widely reported in the press.

93. See the obituary "Norman Robert Pogson."

journalistic experience to great effect, sending his first memorandum on the subject not only to the Indian authorities but also to the local *Times* correspondent and the Reuters agent. Both went to their telegraphs. By the time Hunter, who had been on a short visit to India, returned to England in April, the connection between sunspot research and the Indian famine was the talk of the nation. With typical energy, Hunter threw himself into a veritable campaign, enlisting influential men of science to his support and encouraging their research. Balfour Stewart published a summary of sunspot cycle evidence in *Nature* in early May, proposing that sunspots might be viewed as solar meteorology, "a species of celestial rainfall," whose presence indicated solar activity just as rainfall indicated atmospheric activity.[94] Just after the final installment of Stewart's article appeared, the theory was denounced as bold and premature speculation in a packed lecture on the causes of the famine by Richard Strachey to the Royal Institution. Strachey argued for greater local responsibilities and initiatives; "a system of public relief in times of distress, not guarded by the sense of specific local responsibility, is a source of grievous abuse, misery and demoralization."[95] Hunter, responding to this challenge, called on Norman Lockyer as soon as he arrived in London a few days later, and as he recorded in his diary, "interested him" by showing that Strachey intended to "discourage solar researches in India."[96] Lockyer cabled Balfour Stewart in Manchester immediately, and the same evening all three attended what the newspapers called a "long and animated" session of the Royal Society, where Strachey's objections were trounced.[97] By that autumn, Hunter had become the unofficial source of famine news in England, receiving weekly reports from the viceroy by telegram at his home in Lanarckshire and editing their contents for distribution to the British press.[98] As the distress mounted in

94. Stewart, "Suspected Relations between the Sun and the Earth," *Nature*, May 10, 1877, 26. Cf. Gooday, "Sunspots, Weather and the Unseen Universe."

95. This remark was quoted in the *Illustrated London News* report on the lecture, May 26, 1877, 502.

96. Skrine, *Life of Sir William Wilson Hunter*, 269.

97. *Times*, June 7, 1877, 5. For a discussion of these exchanges as an early example of correlation, see Porter, *The Rise of Statistical Thinking*, 274–78. Hunter (who may well have been biased) thought that Strachey's calculations were summarily dismissed by his audience, with Strachey retreating in grumbling disarray (see W. Hunter to Richard Strachey, May 28, 1877, Strachey Papers, India Office, MSS Eur F 127/186/39–40). Blanford was unimpressed by Hunter, calling the article in *Nineteenth Century* "just what might have been expected—pretentious and shallow—but very readable, and it will take as a nine-day wonder" (Henry Francis Blanford to Richard Strachey, November 8, 1877, Strachey Papers, India Office, MSS Eur F 127/186/55–58).

98. Cf. the account of the press and imperial news in Nalbach, "'Software of Empire.'"

Madras, readers of the *Times* could see reports on the Indian famine in one column balanced by exchanges on the rainfall cycle controversy and with their own daily weather forecasts in another column. The juxtaposition underlined the universal nature of weather concerns.[99]

The high point of this publicity was a pair of articles for the influential monthly *Nineteenth Century* in November 1877. In "Sunspots and Famines," the editor of *Nature* and the editor of the *Imperial Gazetteer*, Lockyer and Hunter, teamed up to assess the significance of the sunspot cycle and of Indian meteorological research in general. In "Indian Famines," Major George Chesney, a hero of the Mutiny and a military reformer, roused his readers to a "distinct perception of our obligations" in India—that is, more intervention, more public works, and more public charity rather than less.[100] As a pair, the articles showed that support of science was like support of irrigation—a necessary, practical requirement of good government. Lockyer and Hunter listed all the correlations established by researchers between sunspots and temperature, magnetism, rainfall, and storms in the preceding decade, but they paid special attention to cyclones and shipping, one of the most easily grasped economic consequences of the natural cycle. At Mauritius, the government observer, Charles Meldrum, a dedicated meteorologist, had been one of the first to call attention to the pattern of meteorological activity, showing the "strict relation" between cyclones and sunspots.[101] Such research necessarily carried observation into the mainstream of practical concerns and commercial life. "While we have been quietly mapping out the sunspot maxima," Lockyer and Hunter noted, "the harbours were filled with wrecks and vessels coming in disabled from every part of the great Indian ocean." Similarly, they drew attention to an analysis carried out at Lloyd's showing that posted shipping losses had been 15 percent greater during the previous two years of peak sunspots than the losses in the years of minimal sunspots. Both the practical importance of

99. Skrine, *Life of Sir William Wilson Hunter*, 260, 268–69, 273–74. On the British response to the famine see Digby, *The Famine Campaign in Southern India*; Philindus, "Famines and Floods in India"; or the representative comment in the *Times*, May 18, 1877, 5; June 7, 1877, 5; and August 3, 1877, 10. The *Illustrated London News* also published weekly counts of those receiving relief and of the charitable relief fund in England. Upward of £26,000 was sent. See *ILN*, vols. 77–80 1877–78 passim, and October 20, 1877, 381, for a picture of victims in Madras.

100. Chesney was author of *Battle of Dorking* (1871), a fictional account of an invasion of England, designed to alert Britain to the weaknesses of its defense, and several works on Indian administration and military. His *Indian Polity* called for centralization. See *Dictionary of National Biography*.

101. Hunter and Lockyer, "Sunspots and Famines," 592.

the famine question and the geographic situation of India combined to ensure that there was "no place better suited for research."[102] The conclusion was unequivocal: "a study of the rainfall is one of the first duties of a civilized government in India."[103]

Almost inevitably, the outcome was a government inquiry on famines—a royal commission was appointed in 1878, and it reported in 1880. A Famine Relief Code stressed the need for vigilance in noting early signs of distress, articulated procedures for relief works and food distribution, and confirmed the need for famine insurance taxes (the latter had begun after the 1874 famine). Despite the development of this code, in many ways the commission fulfilled Hunter's doleful prediction of 1874, that famines would be buried in "mere masses of detail" once the immediate crisis had passed. Public attention to Indian affairs shifted from the starving hordes in the south to hostile Afghans in the north, as the British worried about the growing Russian influence at the outskirts of their Asian empire.[104] Lytton, the Indian viceroy, invaded Afghanistan in November 1878, and the Tory defeat in the general elections of 1880 owed more immediately to that Indian crisis than to the equally partisan quarrels over famine policy. The famine commission decided that the scientific evidence on the relationship between sunspots and rainfall was promising but inconclusive and contented itself with printing a selection of different points of view and urging the Indian government to support further investigation. Although this conclusion was an anticlimax, it did not represent a collapse of either the principle of famine prediction, or the support of scientific research. After all, if the arguments about the eleven-year sunspots cycle were well founded, it seemed the government had a decade to prepare for the next crisis (in fact, there were seven major famines in the next thirty years, with particularly widespread and horrifying outbreaks in 1896 and 1899). Meanwhile, the solar physics laboratory in London had received an annual grant of £500 in November 1879, a notable triumph for Lockyer (fig. 6.6).[105] A photographer named Meins, a royal engineer trained in Lockyer's facilities, was dispatched to India to supervise solar observations. However inconclusive, the debate over sunspots

102. Hunter and Lockyer, "Sunspots and Famines," 601.

103. Chesney, "Indian Famines," 605.

104. See Gopal, *British Policy in India*.

105. Lockyer had managed to obtain some space at the South Kensington School of Science in 1873, but 1879 marked the first direct funding for a solar physics laboratory. His work, and the solar physics observatory, was not secure until a reorganization in 1886 and his appointment as a professor at South Kensington in 1887 (Meadows, *Science and Controversy*, esp. 113–34).

and rainfall gave incentive to the efforts of Indian meteorology. As Blanford, who was not a blind enthusiast for the solar-meteorology connection, noted in 1880, even an erroneous hypothesis was of value: it gave "system and definite purpose to research."[106] Most importantly, the famine debates made it clear that science—even the prosaic measurement of rainfall—could contribute to public affairs. With its usual satirical gift, *Punch* published a cartoon in 1879 of Disraeli, holding his umbrella and tapping the barometer thoughtfully, trying to gather a political forecast from the fluctuations of the instrument (fig. 6.7).

The Indian Observer: Discipline and Command in Science

The activities of Lockyer, Stewart, and Hunter doubtless showed that they recognized the tide of their opportunities to promote their interests and seized it. Yet Indian meteorology represented more than opportunism. It tied together the debates about the role of science in the modern state. First, it presented utilitarian justification—the prevention of famine—that vindicated abstract research. Both meteorology and solar physics were planks of the endowment of research campaign of the 1870s, and both attracted in equal measures opprobrium from critics of government support for research and vocal support from its advocates. Hunter's widely publicized link to famines gave authoritative evidence that such apparently tedious, apparently remote scientific measurements would result in knowledge of direct benefit to those who administered the vast territories of India. Some of the flavor of these beliefs can be conveyed in a mythic and romantic description of the Indian monsoon, which was almost certainly from the pen of Hunter rather than Lockyer in their joint article of 1877. "It seemed as if the continent . . . had only to sit still and receive in her lap the treasures which the winds gathered from distant tropical seas." The article described for English readers the Sanskrit gods of Indra, the benevolent "personification of the Watery Atmosphere," and Vayu, "the Wind . . . or storm gods," whose struggles determined each autumn's monsoon season. With its monsoon and winds, its Indra and Vayu, India was transformed into a passive, waiting cornucopia of scientific wealth: observers had only to begin to collect data to penetrate to the prize of meteorological knowledge.[107] It was a powerfully appealing vision, and helps explain the rapid development of an Indian meteorological network. In 1875, the central government had eighty-

106. Evidence of H. F. Blanford, *Further Report by the Committee on Solar Physics*, 32 (Parliamentary Papers).

107. Lockyer and Hunter, "Sunspots and Famines," 594.

FIGURE 6.6 The successful researcher and editor Norman Lockyer in a series of caricatures of eminent men in *Punch*. Lockyer, known for his sunspot photography and his advocacy of the connection between sunspots and atmospheric conditions, is shown "Illuminating the Sun" with a candle labeled "Science." The wings that propel him to the clouds are the pages of *Nature*, the weekly science journal that he edited. ("Punch's Fancy Portraits, No. 167: Norman Lockyer, F.R.S., Illuminating the Sun," *Punch*, December 22, 1883, 299.)

four observatories, organized under seven main regions, and a decade later there were twenty more. The place of the observer in Indian meteorology, and the contrast with the networks of observers at home, cemented the power of its model. In conclusion, then, we should turn to story of the Indian observer, who became a lauded and indispensable element of the impressive Indian meteorological network in the 1880s and 1890s.

Initially, this seems a contradictory outcome, given the widespread reputation of Indian observers as classic instances of disorganization, negligence,

FIGURE 6.7 As sunspots and science became tied to monsoons and famine, the weather of India became a question that was evidently both natural and political, as seen in this picture of Disraeli, tapping the political barometer in the lobby of the House of Commons. ("Punch's Essence of Parliament," *Punch*, April 12, 1879, 159.)

and deliberate incompetence. In 1877, Symons in London recalled the stories of "the days when Indian rain gauges were taken indoors and locked up every night,"[108] while Blanford in Calcutta noted succinctly that Indian meteorology "afforded instructive examples of how untrustworthy a Meteorological Register is like to be" when left to "the unremunerated zeal of an untrained subordinate." Falsified entries, according to Blanford, were legion. He pointed to the "salutary" surprise of observers when they realized that fraudulent entries were detectable—if the series didn't display regular diurnal variations for instance, or were inconsistent with the records of other instruments at the same time. "The observers learned," Blanford noted, "that such practices were likely to be detected, and that it was necessary to gain a better knowledge with the empirical relations of the several parts of their registers before they could successfully forge such observations." Even so, of course, deception was possible. One economical observer simply turned back the pages of his records and repeated the observations of a previous year.

Blanford's view of native observers reflected British racial arrogance, and race was obviously a crucial element of the observing network. Among native observers, Blanford commented, a "sense of duty is probably not exceedingly strong "; they were "fully sensible to the charms of sensuous ease" and worked "in the slip-shod fashion that is habitual to persons of imperfect education."[109] At the same time, however, race was the solution to the problem of disciplined observation networks. Native employees, for instance, were the answer to difficult or unhealthy observation stations, as the story of the solar photographer Meins trained by Lockyer and sent to Leh in 1878 makes clear. When Meins died mere months after beginning his observations, a victim of the climate, there was some debate in England and India about a successor. Should another highly trained engineer be sent out? Or should a local solution be sought? The British joint commissioner at the spot, Ney Elias, opined that while "an European would certainly do better work than a Native," it was extremely difficult to find the right sort of European—one who would not "mind the solitude, the rough living, the cold &." Moreover, the wrong sort of man was detrimental to British authority generally: "it would be most awkward to have a troublesome or disreputable European." Yet the (unspecified) "temptations" of "the

108. Symons, "Indian Meteorology," 162.

109. Blanford in *Administrative Report of the Meteorological Reporter to the Government of Bengal for the Year* 1867–68, 4; *Administrative Report of the Meteorological Reporter to the Government of Bengal for the Year* 1870–71, 3; and *Administrative Report of the Meteorological Reporter to the Government of Bengal for the Year* 1872–73, 2–3.

solitude and the cold" were considerable. The solution, thought Elias, was a temporary appointment of an Englishman to train a native of Upper India, someone "well-selected and English-speaking." Of course this was also the economical solution—it was proposed to pay the native observer three-fifths the salary of a noncommissioned officer, or 150 rupees rather than 250 rupees a month.[110] Thus, if native observers represented the problems of observation—being racially characterized as lazy, sloppy, or fraudulent—nevertheless, their race was also the solution to observation. They could apparently cope with environmental and social conditions better than Europeans, and they were manifestly cheaper to employ.

Strict control and training were also simpler among a distinctly subordinate class of native observers than the traditional European networks with their mixtures of employed and voluntary labor. Much less delicate lines of control than those within observers' networks in Britain extended from supervising officials to observers in India. Some of the contrasts emerge from the records of Elizabeth Iris Pogson. As meteorological reporter to the government of Madras from 1881, she was one of seven such officials covering the Indian network of more than a hundred stations, and she reported directly to Blanford, as chief meteorologist to the government of India. (The other regions were Bengal and Assam, the northwest provinces, the Punjab, Bombay, and the central provinces, plus a scattering of single observation posts.) Pogson did her work efficiently and described her supervision visits in greater detail than some of her colleagues, perhaps because her gender made a specific display of her authority over the observers more critical. She supervised eighteen observation stations spread through the Madras-Mysore and Hyderabad area. At one station in Kurnool, the observer, whose work had been "perfectly useless and unreliable," was carefully reinstructed by her on the proper method of reading the barometer. "Anxious to learn," he improved rapidly, she reported, but then balked at the task of reading the anemometer as well. Thereupon Pogson advised him to resign and, "after a little reluctance," he agreed. She promptly appointed a less recalcitrant observer and subsequently reported no further problems with the data.[111] Clearly an important part of the Indian context was a ready supply of workers willing to work with British scientific men (or women) for a salary of a few rupees per month.

However, if race was a key ingredient of observation in India, this is not to

110. *Further Report by the Committee on Solar Physics*, 35 (Parliamentary Papers).

111. *Administrative Report of the Meteorological Reporter to the Government of Madras for the Year 1884–1885*, 9.

say that race in itself was the issue. Blanford was almost as severe about the casual observing practices of medical men as he was about natives. He insisted on the importance of *physicists*, noting that even motivated observers, when bereft of training in the physical sciences, become "an impediment [rather] than an assistance."[112] Similarly, Elizabeth Pogson's appointment hints at the way gender resolved meteorological work into a simpler hierarchy. As a woman, Pogson's appointment was clearly unusual, but she was the daughter of the Madras astronomer Norman Pogson, and a fellow of the Royal Meteorological Society. Her father, while interested in meteorology at least to the extent of drawing Hunter's attention to the sunspot-famine correlation in 1877, was one of the least effective of Blanford's contributors because he was unable to spare the staff for producing the requisite data summaries. The longstanding rivalry between the Madras government, which the elder Pogson represented, and Blanford's "Indian" or central government, ruling over the whole of India from a Bengal base, made their interactions tense. His daughter's appointment may then have reduced the friction between the men, because like racial relations, it presented an obvious hierarchy—woman to man as well as provincial reporter to the chief reporter of India—and replaced the much more equivocal exchanges of the astronomer of Madras (tackling meteorology as an extra burden) with the chief reporter based in Calcutta. In short, race was perhaps the chief feature that facilitated strict lines of command, but other factors also assisted in creating exemplary scientific order. To race and the labor economics of race, and occasionally to gender, can be added experience of a military mode of administration. It would be hard to avoid direct or indirect influence with the ruling structures of Indian society, and the history of many of India's scientific men, including Blanford, was tied to the survey projects administered through the army. It is worth remembering, too, that the successes of the American network appeared to be tied to the professional and military training of their Signal Corps observers as much as to the huge government monies that supported a continental observation network.

For Britain, then, the colonial framework provided an exposition of the kind of disciplined relationship indispensable to serious scientific enterprise. This was the heartening possibility that India offered to British observational science. The *Calcutta Review*, which complained so loudly in 1871 about the poor quality of observations in India, had been equally clear about the remedy. "The mere concentration of the management in the hands of one qualified

112. *Report on the Administration of the Meteorological Department of the Government of India for 1878–1879*, 39.

chief will at once remove most of the causes of that imperfection [in the meteorological observations]." Scientific researches in Germany, the *Review* thought, showed the way: in "the laboratories of great chemists," the leadership of a single authority allows "science [to be] advanced by a judicious division of labour." The article suggested that such an authority, like John Tyndall, the head of the Royal Institution, could successfully establish the direction of colonial science.[113] (Tyndall, though apparently tempted, had declined the offer to spend a few years on Indian meteorology in 1867.) In any event, Blanford, on the spot at Calcutta University, got the post, and from 1875 he directed a centralized, monitored system of observatories of the type that the *Review* had envisioned. In the end, it was the observation network, rather than sunspot correlations, that led to seasonal predictions. Pressure patterns over India were very persistent, and air pressure determined the course of the winds, which in turn determined the vapor supply and rains. Inductively, with an increasingly good picture of normal patterns against which to compare those of the current year, Blanford was able to offer monsoon forecasts a few months in advance, saying whether rains would be heavy or light, late or early. A first effort occurred in 1874, and by 1878, he was framing long-term forecasts regularly, even though he preferred not to publicize the conclusions unless "an emergency" made the forecast "of unusual importance."[114] This steady and cautious attitude was typical of Blanford and was part of the reason he succeeded so well in his post. According to one approving record of his appointment, he brought the qualities most necessary to an imperial meteorologist: "fixity of purpose [and] faculty of command."[115]

Conclusion

In the key decade of the 1870s, as leaders in Britain sought to reform relations between government and science, meteorology exposed all the possibilities and all the fault lines of state science. From its inception, the Meteorology Department provided an example of reasonably generous government funding. Because of the scale of its phenomena and the costs associated with telegraphic data collection, it was typical, almost to excess, of those sciences that could only

113. "Meteorology in India," 278, 284.

114. *Report on the Administration of the Meteorological Department of the Government of India for 1878–79*, 2.

115. "Meteorology in India," 289. See Tyndall's account of the offer in Add. MSS 60631/113–24, British Library. For the reaction to Blanford's appointment, see Symons, "Indian Meteorology," 612.

hope to progress through a unified effort bankrolled by the state. The creation of a government office to study meteorology offered, symbolically, only the first, promising step in the careful rearing of the science along modern principles. Accordingly, the two major institutions of the British scientific world, the Royal Society and the BAAS, were closely involved with the origins of the Meteorological Department and were active during key points in its evolution, moving to direct control after 1867. Yet if meteorology fitted the program of this decade's ambitions in many ways, it was perhaps even more effective as an example of the difficulties that vision of science faced in British culture. First and foremost, acrimony continued to surround a research-oriented meteorology, as conceived by the Royal Society, because it resisted the accepted principle of public funding—that the funded work should benefit the nation. In meteorology, it was almost impossible to avoid the question of utility that so annoyed Balfour Stewart, Norman Lockyer, and other advocates for endowment of research. From their point of view, the lesson of meteorology was that practical benefits could divert a scientific enterprise from "true" scientific goals, obscuring distinctions between good and bad science.

Shifting meteorology away from British shores, however, offered different possibilities. India provided a new sphere for the work of science. A science of the weather in that tropical setting provided opportunity for abstract research *and* practical benefit, both on a grand scale. India was Britain's continent, an ideal laboratory in which to develop a scientific command of the unruly forces of the atmosphere. And rather than the awkward short-term forecasts published in the *Times*, Indian meteorologists used a different temporal frame. They could give their attention to seasonal monsoon forecasts and solar cycles as well as a storm-warning network. Similarly, the knowledge gained and applied in India offered potential control of a different scale of natural events: not a single storm, however devastating, but the monumental and sustained crisis of famine—disorder of a magnitude that threatened to cripple a government, its trade, and its population. In their intensity, the consequences of ignorance seemed to match India's torrential rains.

Indian meteorology, therefore, defused the theoretical and administrative uncertainty that surrounded the subject at home. Even more importantly, however, it suggested how science and the state could become mutually reinforcing models of rational order. *Blackwood's Edinburgh Magazine* in 1875, with its model of weather and anarchy, climate and constitution, expressed this more fervently than most. In its politicized portrait, meteorology was the most unmanageable of sciences. Observing meteorology, "our admiration for nature's capacity of lawgiving ought to increase immensely; for the statutes which she has invented for the government of weather must be far more wonderful than those which she enforces elsewhere." But the turbulence of the atmosphere was only an illusion. Its real lesson was the commanding reach of science: "the existence of the strictest order amidst indescribable disorder, of a recognised predominant will where all wills appear to be contending for the mastery; of an accepted absolute commander, where all looks like flagrant disobedience; of ever present reason amidst what seems to be the wildest incoherence."[116] In the 1870s, as Victorians debated the natural foundations of empire, the atmosphere provided the context for explaining British rule in India. Whether India would evolve into self-government, or further centralize its unwieldy administration and enter into a tighter, permanent state of British rule was the question at the heart of imperial affairs in this decade.[117] Meteorology provided a timely example of underlying law and practical command. By the 1890s, these expectations had borne some fruit. The Meteorological Department of India was highly praised for its scientific output; Blanford's textbook was one the most up-to-date summaries of meteorological theory available in English; and no one could fault Blanford or his successors on the scientific standards of their analysis.[118] However, the department was also noted for its practical labors: its cyclone warning service and, more uniquely, the June long-term monsoon forecasts, which gave an estimate of the general character of rainfall for the following four months. These rains determined the quality of the principal Indian harvest, so the forecast of regions where rainfall was expected to be significantly lower than average marked out those that were vulnerable to drought.[119] Meteorological research, it seemed, could control the anarchy of the weather just as the Raj controlled its chaotic and immense possessions. In return, the prospects of meteorological science in India suggested the importance of command in science. The atmosphere, like empire, required discipline on a monumental scale.

116. [Marshall], "Weather," 627.

117. Chesney, *Indian Polity*. For a wide-ranging modern account, see Cohn, "Representing Authority in Victorian India."

118. Archibald, "Indian Meteorology," *Nature*, August 23, 1893, and "Twenty Years of Indian Meteorology," *Nature*, July 8, 1897, 226–28; August 30, 1883, 428–30; September 13, 1883, 477–79.

119. The forecast was based on such information as the pressure conditions on the land during the hot season and snowfall of the previous season in the Himalayas, which modified monsoon winds by affecting temperature and pressure in the northwest as it melted (Dallas, "The Prediction of Droughts," 15). Dallas was the assistant meteorological reporter to the Government of India in the 1890s.

CONCLUSION

METEOROLOGY in Victorian Britain was a science of expectations, not only in the literal sense that it dealt with statements about future weather, but also figuratively, in that it summarized what Victorians thought science could or should be. The question of weather prediction became a means for contemporaries to identify critical ways in which the authority of modern science shaped the prevailing culture. Through their experiences with the science of the weather, Victorians explored ideas about expertise, the growth of public institutions, and the effects of a burgeoning print culture within which intellectual leadership took shape. None of these developments was uniquely confined to meteorology, or indeed to science, yet weather forecasting was particularly effective in outlining the claims made for natural knowledge and its leaders in modern society. Meteorology revealed these claims so well precisely because it was *not* conspicuously successful—because the difficulties of authoritative weather prediction exposed the uncertainties involved in defining proper science for a wide range of people and interests in the mid-nineteenth century. The weather prophet Patrick Murphy's lucky hit in his almanac of 1839, Robert Fitzroy's suicide in 1865, or the analysis of sunspot cycles to predict famine in India in the 1870s all showed meteorology's distinctive qualities. In each of these developments, Victorians negotiated the character and effects of knowledge, discipline, participation, and public life as features of modern society.

The status of weather prediction as reliable knowledge grew from the difficulty of understanding atmospheric change. It was challenging to collect enough observations over wide areas of land and sea in forms that could be coordinated and compared. But on a more fundamental level, the work of coordinating many individual observations highlighted the effort of interpretation. Throughout the nineteenth century, meteorologists struggled not only with the process of data collection but with the relationship of particular observations to the general laws of atmospheric change. In philosophical terms, this was the problem of induction: how could one build a general understanding from the accumulation of facts? The observing programs of the 1840s through the 1860s were inspired by the assumption that more data was necessary, and the hope that such data was perhaps also sufficient, for the transformation of meteorology into a mature science. And yet by the 1870s, years of extensive and mechanized observations had not redeemed this promise. The pressure of these circumstances on meteorologists was a recurring theme. Reading the Meteorological Office publication *Weather Charts and Storm Warnings* in 1876, Francis Galton underlined in frustration the statement that atmospheric change was "mainly regulated by the distribution of barometrical pressure over the globe," jotting a sardonic note about cause and effect in the margin: "The author seems to regard the pressure as a sort of wild beast having volition of its own."[1] It was all too easy to transplant instruments and observations to the center of inquiry and veer away from more fundamental questions.

Coupled to this philosophical and practical difficulty with observations in meteorology was the definition of probable knowledge. The development of weather observation networks coincided with the introduction of a new approach to probabilities: the statistical management of large amounts of data to reveal underlying patterns. As much as meteorology lent itself to the new program of statistical study of natural phenomena, however, the challenge of actually predicting the weather was much more difficult. Calculating the chances that one could expect that a particular day would bring sunshine or rain made sense within an older philosophy of probabilities. It aligned meteorological science with ideas about reason and the role of individual judgment that many Victorians continued to find sympathetic. But it was flatly out of touch with statistical thinking, the reinterpretation of probabilities as a form of precise knowledge based on analysis of a cluster of observations. At the same time, statistical techniques to manage the multiple variables that characterized a question of weather prediction had not yet been developed. Changing ideas in this period about what probability meant and what problems it could manage made it particularly difficult to speak of weather prediction as a science.

1. See the copy of Scott's *Weather Charts and Storm Warnings*, 121, annotated in Galton's hand, Galton Papers.

Concerns with the nature of observations or probabilities cannot be separated from the history of institutions, technologies, or political culture in which they developed. Meteorological science thus gives the historian a way of analyzing the intellectual world of the Victorians and showing how scientific ideas and practices blended into broader debates. The most obvious of these contexts in meteorology was the call for discipline and organization. Observation networks were enterprises built around the ideals of voluntary participation, the control and subordination of individual parts, and the public value of natural knowledge. These elements, according to Victorian men of science, would transform fragmented individual accounts of natural phenomena into a coherent and productive science. Meteorology thus promised to demonstrate the value of collective effort, coordinated under the expert guidance of a centralized authority. That description fits the project of the marine observations that prompted the foundation of the Meteorological Department in the Board of Trade, and it fits too the private or semi-public networks like George James Symons's rainfall observers or James Glaisher's observers producing statistics for the General Register Office. A similar picture of science and community inspired Dr. Merryweather's instrument at the Great Exhibition in 1851. With his leeches arranged in a circle of counselors, linked by telegraphy to the nation as a whole, Merryweather envisioned knowledge as a vast spatial and social exchange.

The technology that had inspired Merryweather, electric telegraphy, made those visions of community and control more intense. Telegraphy spectacularly expanded the scale and speed of scientific exchange. But even more significantly, by offering the possibility of weather forecasting, it concentrated attention on the question of the immediate benefits of science, especially to the shipping industry. That in turn meant meteorology became part of long-standing discussions about how to characterize the relationship of natural knowledge to economic prosperity. The authority of Victorian men of science hinged on a careful definition of this relationship, close enough that there could be no doubt about the importance of scientific research, but not so close that it tainted the scientific gentleman with the aura of trade and business. There was no monolithic position on this relationship, within the scientific community or without, and the range of opinion on this important issue only increased the significance of meteorology as a test of how to evaluate scientific work. This was the background to the shift of meteorological science to the natural laboratory of India. Invested with these strong associations of discipline and public responsibility, the science of meteorology logically combined with questions about the management of the empire. Famine prediction in imperial India therefore can be seen as the natural heir to the standard Victorian accounts of meteorological science as a collective enterprise, renewed by a technological conquest of distance and disciplined leadership from the center.

The emphasis on collective endeavor in meteorology, however, contained a dilemma. Wide participation in the science was critical to its development as a coherent body of knowledge, but it also disrupted that coherence. The distinctive features of meteorological science were the almanacs and astrology, on the one hand, and the installation of meteorology in a government office on the other. Because of the almanacs, meteorologists struggled to define their relationship to popular knowledge and the illegal world of astrological prophets. The government office set the stage for a clash between a conception of science as practical knowledge and a conception of publicly funded research. These circumstances increased the range of interests in meteorological science and created particularly controversial juxtapositions of individuals and claims. Eminent Victorian philosophers often carried debates about meteorology to places that seem inappropriate to the modern picture of the professional scientist: in newspapers, almanacs, or lectures on art as well as scientific journals and monographs or official commissions of inquiry. Although certainly subtle distinctions about what sort of publication to use or which scientific meetings to attend were significant, all involved shared the conviction that public exchange was the way to evaluate science. This conviction connected figures as diverse as John Herschel, Zadkiel, and John Ruskin. Meteorology was typical of Victorian scientific culture, a fluid and heterogeneous world in which messy disputes arose precisely because of commitment to reason and natural knowledge as common property, in principle accessible to all.

As revealing as is this diversity of places and people, it represents differences of degree rather than kind for a history of Victorian science. Studying meteorology does offer a particular advantage, however, for an analysis of the relations between popular and elite knowledge. This advantage emerges not because the science brought such a diverse range of individuals and visions of science into contact. Nor is it simply a question of the presence of strong traditions of popular knowledge. These qualities, as just noted, could be considered typical rather than extraordinary. Rather, meteorology was a science in which the nature of popular knowledge and its relationship to the world of observatories and precision instruments was explicitly a subject of discussion. This takes us back to the importance of failure in the science. Because weather prediction was often inaccurate, the apparent accuracy of weather wisdom was an important consideration. It was critical for some meteorologists to establish the distinction between weather wisdom and their own approaches; others sought to incorporate the features of weather wisdom that seemed to be responsible for its authority. Investigating these relationships between organized meteorology and weather wisdom provides a richer picture of both meteorology and Victorian science in general. For instance, it suggests a new way to think about instruments and maps in the science. Rather than tools to construct modern forms of objectivity through quantitative and mechanical records, some instruments can be seen as far more tentative objects, used by meteorologists as a substitute for natural precision. They could represent connections between popular and elite knowledge, instead of a stark boundary. Similarly, weather maps had an appeal based partly on their representation of global and quantitative data, and partly on their endorsement of a weather-wise visual sensibility, meaning at a glance.

A history of weather prediction then leads directly to a picture of the complexities of Victorian intellectual life and the values it upheld. But that picture suggests a need to think further about the development of global sciences. Is there something paradoxical about studying a history of meteorology in Britain? Why insist on the importance of national framework for understanding a science like meteorology in this period, when contemporaries so clearly recognized that its dynamics were global and required new commitment to widespread, coordinated observation? One answer is that the organization of the science through government institutions, and the importance of public expectations for the science, reinforced the effect of specific national contexts on the science. In Britain between the 1850s and the 1880s, there was certainly increasing exchange of observations with other countries, but little real pressure to approach the study of the weather more internationally. The emphasis was rather on developing a central institution to control meteorological work on a national scale, an endeavor that meshed with the ambitions of the scientific metropolitan elite. Nearly two decades passed between the Brussels conference that proposed to coordinate marine observations, for example, and the next international meeting, in Leipzig in 1872, to set common goals and standards in meteorology. Regular international congresses followed, but coordinating meteorological work among different national observatories and standards continued to present significant obstacles.

On the particular question of weather prediction, for example, the difference of opinion within European nations precluded anything but an agreement to differ. In 1878, nearly two decades after the introduction of government forecasting, an international meeting in Vienna included a report that presented the results of a questionnaire about storms warnings and forecasts distributed to observatories and prominent individual meteorologists. There was no agreement about what information to distribute—the observations themselves, opinions, or predictions—nor what kind of forecasting was possible (in the case of storms, for example, whether only to signal strength of wind, or direction as well). To many, the system in place in the United States was a model; others found it too ambitious. Although there was general favor for storm warnings along the lines introduced by Fitzroy in nations with strong maritime interests, like Portugal, there was more skepticism in landlocked capitals like Paris, Berlin, or St. Petersburg.[2] Philosophically, the science of meteorology was global; in practice, global science developed in distinctively different political and geographical landscapes, and contemporaries insisted on the importance of the differences.

The question of global versus national contexts for the history of nineteenth-century meteorology could also take a reflexive turn. It can be argued that the idea of global science was itself a particular challenge on both a philosophical and practical level. Meteorologists were particularly aware that their science was based on connecting local experiences and global phenomena. At its most vivid level, these shifts in perspectives were the basic premise of nineteenth-century mapping techniques, the isobars and isotherms, linking up places where barometers and thermometers gave the same numbers. The arguments about sunspots hinged on similar questions about general knowledge: could the solar variation signaled by changes in sunspots explain weather, or did the particularity of terrestrial conditions make such literally over-arching explanation impossible? Such issues were related to critical narratives about the progress of meteorology, in which the relationship of particular fact to general law corresponded to the transition from a local knowledge of airs and places into a universal physical science. Victorian men of science thus wrestled hard with the problems of integrating local and general perspectives. Any irresolution in international cooperation reflected their experience with those questions as well as with the more obvious problems of agreeing on standard registration forms and instruments. For the historian, this suggests that the meaning of global science needs to be investigated through exactly these shifting contemporary characterizations of meteorology as global, national, or local science.

But there is another kind of irresolution in the history of Victorian weather prediction. How does the story end? One could point in conclusion to a partic-

2. *Report on Weather Telegraphy and Storm Warnings*.

ular historical momentum that governs the narrative. The particular circumstances of the mid-Victorian era—including the economic rationale for storm forecasting and the pressure for public funding of scientific institutions—gave way in the 1880s as steam power replaced sail, and science became a more secure profession. With the passage of time, it seemed simultaneously less important to resolve the philosophical status of forecasting and less likely that a decisive and concerted effort on the part of meteorologists would produce a definitive dynamics of the atmosphere. Meteorology and weather forecasting, however it was evaluated, simply became less prominent. But fading controversy did not represent resolution. Another narrative of meteorology would represent cycles rather than a dying momentum. The Meteorological Office ceased to publish warnings and forecasts after Fitzroy's death and the inquiry of 1866, but warnings resumed at the end of 1867. By 1879, Meteorological Office forecasts for eleven different districts of the British Isles began to appear regularly, provided not only to newspapers but also available as a direct service to subscribers and for individual inquiries. Three years later, a failure to predict an October storm sparked a correspondence in *Nature* about observations, instruments, and public funding that could have been lifted word for word from the time of Fitzroy's experimentation with storm warnings twenty years earlier.[3] The uneasiness about the scientific status of prediction remained, too. In 1926, more than seventy years after the founding of the Meteorological Department under Fitzroy, his successor Napier Shaw commented that "it is hardly an exaggeration to say that meteorologists have a natural aversion from the iteration of the duty of forecasting."[4] Many of the features of the Victorian experience with weather prediction, in other words, remain remarkably persistent.

Victorian meteorology thus continued to shape the science long after the energy of the controversies of the 1860s had dissipated. The resilience of Victorian patterns of thinking of leadership, visual knowledge, discipline, and collective knowledge emerges in Lewis Fry Richardson's renowned 1922 book *Weather Prediction by Numerical Process*. Richardson's calculations of air pressure, density, and humidity in this book were a mathematical performance of the hydrodynamical theories of the atmosphere developed by Vilhelm Bjerknes and others in Bergen, Norway, in the years immediately after World War I.

3. Carlisle, "Weather Forecasts," *Nature*, November 2, 1882, 4; Ley, "Weather Forecasts," *Nature*, November 9, 1882, 29; Carlisle, "Weather Forecasts," *Nature*, November 16, 1882, 51–52; Carlisle, "Weather Forecasts," *Nature*, November 26, 1882, 79.

4. Shaw, *Manual of Meteorology*, 1:9.

Directly contrary to his hopes, Richardson's work convinced most readers that such numerical analysis of the weather was completely impractical: his six-hour forecast for a single location had taken him six weeks to calculate. Serious application of numerical methods did not emerge until after another world war, when John von Neumann at Princeton chose weather prediction as a project to demonstrate the potential of the electric computer, and thereby turned meteorology, as Richardson had foreseen, into a science of calculations.[5]

Richardson's book is a recognized landmark in the history of meteorology and its transformation in the early decades of twentieth century. Less evidently but equally importantly, the book reveals how the framework for the development of modern meteorology was created in the nineteenth century. In anticipation of the critics who found his methods impossibly laborious and slow, alongside his own calculations Richardson described a model for a large calculating center. Working on a scale that any one individual could not match, this imagined system would make weather prediction an attainable science. In Richardson's scenario, a large number of human computers raced through differential equations for several layers of blocks of the atmosphere, altogether representing two hundred square kilometers. The remarkable details of his center summarize the concerns we have just finished tracing in this history of Victorian meteorology. Here then is another ending, one which looks both forward and backward.

"Imagine," Richardson wrote, "a large hall like a theatre" whose walls, ceiling and floor are all "painted to form a map of the globe." Each computer works upon equations for the regions corresponding to his or her place in the theater, and their work is fed to overhead displays so that adjacent computers can incorporate their results instantly. In the center, "a tall pillar" supports a "large pulpit" in which stands the central director and his assistants. From his pulpit, the director "maintain[s] a uniform speed of progress in all parts of the globe . . . like the conductor of an orchestra in which the instruments are slides and calculating machines." Instead of a baton, however, this scientific conductor has "a rosy beam of light" to wave on the walls representing a region whose computers are moving too quickly, and a "beam of blue light" for those regions whose computers need to improve their speed. Pneumatic tubes transfer the calculations to another room, from which a coded version is sent by telephone to a radio station. In an adjacent building, researchers toy

5. See Fleming, ed., *Historical Essays on Meteorology*; Friedman, *Appropriating the Weather*; Nebecker, *Calculating the Weather*.

with small hydrodynamic models, in the unlikely event that such laboratory experimentation can be channeled to help the work of the computers.[6]

Richardson's fantasy suggests the enduring effect of the features that were explored and established by Victorian meteorology. The need to work on a global scale, to coordinate and control individual activities, to create a hierarchical system that could decipher knowledge and then transmit it to waiting audiences—these were the challenges of modern weather prediction. The leader who is simultaneously a director, a preacher in his pulpit, and a conductor creating harmony from a miniature world of instruments and calculators was an evocative image. As an account of the scientist, it responded to interpretations of collective knowledge and scientific leadership that had developed in the mid-nineteenth century. The tension familiar to Victorians between researchers, working on their models, and the practical computers, working with their data, is there as well, as is the integration of the whole system with modern communication technologies. Finally, the visual and panoramic sensibilities associated with the study of the weather appear dramatically in the mapped walls, the beams of light, and the commanding panoptic pillar. In 1922, Richardson intended his calculating theater as a prophecy, marking out the future of meteorology as a science of computers. But it was also testimony to the past, and to the influence of Victorian experiences with meteorology.

6. Ashford, *Prophet or Professor?* 91–92. This famous fantasy of numerical methods deserves comparison with other, fictional fantasies of the control of the weather in the early twentieth century, like Jones, *The Great Weather Syndicate*. For a related discussion of the conductor and scientific knowledge, see Winter, *Mesmerized*, 309–20.

Bibliography

Note: Contemporary newspapers and annuals are cited in the notes only.

Manuscript Collections

Airy, George, Papers. Royal Greenwich Observatory Collection. Cambridge University Library.

Fitzroy, Robert, Correspondence. Additional MSS, British Library, London.

Galton, Francis, Papers. University of London, London.

Glaisher, James, Papers. Royal Astronomical Society, London.

Herschel, J. F. W., Papers. Royal Society of London, London.

Lloyds Committee Minutes. Lloyds of London, London.

Meteorological Office, Correspondence and Papers. Public Record Office, London.

Minutes of the Meteorological Committee of the Royal Society, 1867–1877. Meteorological Office Library, Bracknell.

Minutes of the Meteorological Council, 1877–1905. Meteorological Office Library, Bracknell.

Royal Meteorological Society Papers. British Meteorological Office Archives, Bracknell.

Smyth, Charles Piazzi, Archive. Royal Society of Edinburgh, Edinburgh Royal Observatory.

Strachey, Richard, Papers. Additional MSS and India Office, British Library, London.

Parliamentary Papers

Abstract of the Report of a Conference Held at Brussels Respecting Meteorological Observations. 1854 (4) XLII. 443.

Communications to the Board of Trade Respecting the Discontinuance of Storm Signals as Heretofore Practised by the Board. 1867 (206) LXIV. 185.

Copies of the Despatch from the Secretary of State on the Organization of a Meteorological Department in India. 1874 (183) XLVIII. 25.

Correspondence and Papers Relating to a Committee to Report on the Methods of Conducting Observations in Solar Physics. 1878–79 (179) LVII. 721.

Final Reports of the Measures Taken in Consequence of the Scarcity of 1873–74. 1875 (123) LIX. 547.

Further Report by the Committee on Solar Physics. 1882 (3411) XXVII. 495.

Letter from the Meteorological Committee of the Royal Society to the Board of Trade Concerning the Resumption of Storm Warnings. 1867 (388) LXIV. 209.
Letters (in Continuation of No. 388 of 1867). 1867–68 (10) LXIII. 389.
Preliminary Report by the Committee on Solar Physics. 1880 (2547) XXV. 805.
Report of a Committee Appointed to Consider Certain Questions Relating to the Meteorological Department of the Board of Trade [Galton Report]. 1866 (3646) LXV. 329.
Report of the Meteorological Committee of the Royal Society for 1867. 1867–68 (4045) LXIII. 295.
———— *for 1868*. 1868–69 (4180) XXIII. 57.
———— *for 1869*. 1870 (143) XXVII. 17.
———— *for 1870*. 1871 (382) XXIV. 17.
———— *for 1871*. 1872 (579) XXIV. 145.
———— *for 1872*. 1873 (808) XXVI. 383.
———— *for 1873*. 1874 (1103) XXI. 15.
———— *for 1874*. 1875 (1307) XXVII. 17.
———— *for 1875*. 1876 (1536) XXVI. 15.
———— *for Sixteen Months Ending 31st May 1877*. 1877 (1872) XXXIII. 631.
Report of the Meteorological Council to the Royal Society for the Year Ending 31st March 1879. 1880 (2458) XXV. 689.
———— *for 1879–80*. 1881 (2741) XXXVII. 677.
———— *for 1880–81*. 1882 (3100) XXVII. 71.
———— *for 1881–82*. 1882 (3415) XXVII. 209.
———— *for 1882–83*. 1884 (3915) XXVIII. 625.
———— *for 1883–84*. 1884–85 (4294) XXVIII. 467.
———— *for 1884–85*. 1886 (4603) XXIX. 507.
———— *for 1885–86*. 1887 (4914) XXXV. 425.
Report of the Meteorological Council 1904–5: Appendix II: An Account of the Work of the Meteorological Office during the Past Fifty Years. 1906 (2829) XXVII. 1.
Report of the Meteorological Department of the Board of Trade. 1857 (2234) XX. 283.
————. 1857–58 (2421) XXIV. 389.
————. 1862 (3004) LIV. 435–521.
————. 1863 (3132) LXIII. 27.
————. 1864 (3334) LV. 125.
Report of the Royal Commission on Indian Famines. 1880 (2591) LII. 387.
————. 1880 (2735) LII. 479.
————. 1881 (3806) LXXI. Parts 1–3.
Report of the Treasury Committee Appointed to Inquire into the Condition and Mode of Administration of the Annual Grant in Aid of Meteorological Observations. 1877 (1638) XXXIII. 731.
Returns of Wrecks, Casualties, and Collisions, 1861 to 1870. 1871 (139) LXI. 669.
Royal Commission on Scientific Instruction and the Advancement of Science [Devonshire Commission]. *First and Second Reports*. 1872 (536) XXV. 1.
————. *Fourth Report*. 1874 (884) XXII. 1.
————. *Eighth Report*. 1875 (1298) XXVIII. 417.

Other Published Sources

Abbe, Cleveland. *Mechanics of the Earth's Atmosphere: A Collection of Translations*. Washington: Smithsonian, 1910.
————. "Meteorology." *Encyclopaedia Britannica*. 11th ed. 1910–11. 18:264–91.
Abercromby, Ralph. "The Classification of Clouds and the Cloud Atlas." In *Report of the International Meteorological Conference at Munich 1891*, 18–20. London: HMSO, 1892.

———. "Modern Developments of Cloud Knowledge." *Journal of the Scottish Meteorological Society* 8 (1888): 3–18.

———. *Seas and Skies in Many Latitudes; or, Wanderings in Search of Weather.* London: E. Standford, 1888.

———. "Suggestions for an International Nomenclature of Clouds." *Quarterly Journal of the Royal Meteorological Society* 13 (1887): 154–66.

———. *Weather: A Popular Exposition of the Nature of Weather Changes from Day to Day.* London: Kegan Paul, Trench, 1887.

Abse, Joan. *John Ruskin, the Passionate Moralist.* New York: Knopf, 1981.

Administrative Report of the Meteorological Reporter for the North-West Provinces and Oudh [for the years 1877–78, 1878–79]. Allahabad, 1878–79.

Administrative Report of the Meteorological Reporter to the Government of Bengal [for the years 1867–68, 1870–71, 1872–73, 1874–75]. Calcutta: Bengal Printing Company, 1868–75.

Administrative Report of the Meteorological Reporter to the Government of Madras [for the years 1881–82 through 1885–86]. Madras: Government Press, 1881–86.

Agassiz, Louis. *Essay on Classification.* Edited by E. Lurie. Cambridge, MA: Belknap Press, 1962.

Ainslie, W. "Observations of Atmospheric Influence." *Journal of the Royal Asiatic Society* 2 (1835): 13–42.

Airy, George. *Autobiography of Sir George Biddell Airy.* Edited by W. Airy. Cambridge: Cambridge University Press, 1896.

———. *Essays on the Invasion of Britain by Julius Caesar, the Invasion of Britain by Plautius and by Claudius Caesar, the Early Military Policy of the Romans in Britain and the Battle of Hastings, with Correspondence.* London: Nichols and Sons, 1865.

Alberti, Samuel J. M. M. "Amateurs and Professionals in One County: Biology and Natural History in Late Victorian Yorkshire." *Journal of the History of Biology* 34 (2001): 115–47.

Alborn, Timothy. "A Calculating Profession: Victorian Actuaries among the Statisticians." In *Accounting and Science: Natural Inquiry and Commercial Reason*, edited by M. Power, 81–129. Cambridge: Cambridge University Press, 1994.

Allan, Mea. *The Hookers of Kew, 1785–1911.* London: Joseph, 1967.

Allen, David Elliston. *The Naturalist in Britain: A Social History.* London: A. Lane, 1976.

Allingham, William [Unus de multis, pseud.]. "Modern Prophets." *Fraser's Magazine* 16 (1877): 273–92.

Alter, Peter. *Reluctant Patron: Science and the State in Britain 1850–1920.* Oxford: Berg, 1987.

Altick, Richard. "Past and Present: Topicality as Technique." In *Carlyle and His Contemporaries*, edited by J. Clubbe, 112–28. Durham, NC: Duke University Press, 1976.

———. *Shows of London.* Cambridge, MA: Belknap Press, 1978.

"The American Storm Warnings." *Symons's Monthly Meteorological Magazine* 13 (1878): 27–28.

[Amos, Sheldon.] "Circumstantial Evidence." *Westminster Review* 83 (1865): 158–194.

Anderson, James. "On the Statistics of Telegraphy." *Journal of the Statistical Society* 35 (1872): 272–313.

Anderson, Katharine. "Instincts and Instruments." In *The Transformation of Psychology: Influences of Nineteenth Century Science Philosophy, Technology and Natural Science*, edited by Christopher Green, Marlene Shore, and Thomas Teo, 158–74. Washington, DC: American Psychological Association, 2001.

———. "Looking at the Sky: The Visual Context of Victorian Meteorology." *British Journal for the History of Science* 36 (2003): 301–32.

———. "The Weather Prophets: Science, Reputation and Forecasting in Victorian Britain." *History of Science* 37 (1999): 179–216.

Anderson, Olivia. *Suicide in Victorian and Edwardian England.* Oxford: Clarendon, 1987.

Anderson, Patricia. *The Printed Image and the Transformation of Popular Culture 1790–1860*. Oxford: Clarendon, 1991.
Anderson, R. G. W., James Bennet, and W. F. Ryan, eds. *Making Instruments Count: Essays on Historical Scientific Instruments Presented to G. L'Estrange Turner*. Brookfield, VT: Varorium, 1993.
Ansted, David. "The Influence of Certain Physical Conditions on the Origin and Development of Art." *Art Journal* 30 (1868): 172–73.
Appleton, Charles. "On the Endowment of Research as a Productive Form of Expenditure." In *Essays on the Endowment of Research*, edited by Mark Pattison, 86–123. London: H. S. King, 1876.
Armitage, G. "The Schlagintweit Collections." *Earth Sciences History* 11 (1992): 2–8.
Arnold, David. *Colonizing the Body: State, Medicine and Epidemic Disease in Nineteenth Century India*. Berkeley: University of California Press, 1993.
Arnold, Thomas. *Sermons*. 3 vols. London: Longman, 1874.
Ashford, Oliver. *Prophet or Professor? The Life and Work of Lewis Fry Richardson*. Bristol: Hilger, 1985.
Ashworth, William J. "The Calculating Eye: Baily, Herschel, Babbage and the Business of Astronomy." *British Journal for the History of Science* 27 (1994): 409–41.
———. "John Herschel, George Airy and the Roaming Eye of the State." *History of Science* 36 (1998): 151–78.
"Attempts to Foretell the Weather." *Intellectual Observer* 6 (1864): 103–8.
Austin, Jill. "A Forgotten Meteorological Instrument: The Rainband Spectroscope." *Weather* 36 (1981): 151–55.
Axton, W. F. "Victorian Landscape Painting: A Change in Outlook." In *Nature and the Victorian Imagination*, edited by U. C. Knoepflmacher and G. B. Tennyson, 281–308. Berkeley: University of California Press, 1977.
Baber, Zaheer. *The Science of Empire: Scientific Knowledge, Civilization and Colonial Rule in India*. Albany: SUNY Press, 1996.
Backhouse, T. W. "The Problem of Probable Error as Applied to Meteorology." *Quarterly Journal of the Royal Meteorological Society* 17 (1891): 87–92.
Badt, Kurt. *John Constable's Clouds*. London: Routledge and Kegan Paul, 1950.
Baigrie, Brian, ed. *Picturing Knowledge: Historical and Philosophical Problems Concerning the Use of Art in Science*. Toronto: University of Toronto Press, 1996.
Barlow, Derek. *Origins of Meteorology: An Analytical Catalogue of the Correspondence and Papers of the First Government Meteorological Office*. London: Public Record Office, 1996.
Barrell, H. "Kew Observatory and the National Physical Laboratory." *Meteorological Magazine* 98 (1969): 172–77.
Barton, Ruth. "'An Influential Set of Chaps': The X Club and Royal Society Politics 1864–85." *British Journal For the History of Science* 23 (1990): 53–81.
Bates, Charles C., and John Fuller. *America's Weather Warriors 1814–1985*. College Station: Texas AandM Press, 1986.
Baxendell, Joseph. *On the Recent Suspension by the Board of Trade of Cautionary Storm Warnings*. Privately printed, 1867.
Beer, Gillian. "The Reader's Wager: Lots, Sorts and Futures." In *Open Fields: Science in Cultural Encounter*, 273–95. Oxford: Clarendon Press, 1996.
———. "Travelling the Other Way." In *Cultures of Natural History*, edited by N. Jardine, J. Secord, and E. C. Spary. Cambridge: Cambridge University Press, 1996.
Bell, Louis. "Rainband Spectroscopy." *American Journal of Science* 30 (1885): 347–54.
Bentham, Jeremy. *Principles of the Civil Code*. In *The Works of Jeremy Bentham*, edited by John Bowring, 1:297–364. New York: Russel and Russel, 1962.

Berland, Jody. "On Reading 'The Weather.'" *Cultural Studies* 8 (1994): 99–114.
Bermingham, Ann. *Learning to Draw: Studies in the Cultural History of a Polite and Useful Art*. New Haven, CT: Paul Mellon Centre for Studies in British Art by Yale University Press, 2000.
Best, Geoffrey. "Evangelicism and the Victorians." In *The Victorian Crisis of Faith*, edited by A. Symondson, 37–56. London: Society for the Propagation of Christian Knowledge, 1970.
———. *Mid-Victorian Britain*. London: Fontana, 1979.
Bigelow, Frank. "Some of the Results of the International Cloud Work." *American Journal of Science* 8 (1899): 433–44.
Bilham, E. G. "George James Symons, F.R.S." *Quarterly Journal of the Royal Meteorological Society* 64 (1939): 593–99.
Binfield, Clyde. "Jews in Evangelical Dissent: The British Society, the Herschell Connection and the Pre-Millenarian Thread." In *Prophecy and Eschatology*, edited by M. Wilks, 225–70. Oxford: Blackwell for the Ecclesiastical History Association, 1994.
Birt, William. "On Atmospheric Waves and Barometric Curves." In *A Manual of Scientific Enquiry: Prepared for the Use of Her Majesty's Navy, and Adapted for Travelers in General*, edited by J. F. W. Herschel, 337–57. London: J. Murray, 1849.
Black, R. Collinson, and Rosamund Konecamp, eds. *Papers and Correspondence of William Stanley Jevons*. 7 vols. London: Macmillan for the Royal Economic Society, 1972–81.
Blanford, Henry F. *The Climates and Weather of India, Ceylon, and Burmah and the Storms of Indian Seas*. London: Macmillan, 1889.
———. *An Elementary Geography of India, Burma and Ceylon*. London: Macmillan, 1890.
———. *The Indian Meteorologist's Vade Mecum*. Calcutta: Thacker, Spink and Co., 1877.
———. *A Practical Guide to the Climates and Weather of India, Ceylon, and Burmah and the Storms of Indian Seas*. London: Macmillan, 1889.
Block, Edward, Jr. "T. H. Huxley's Rhetoric and the Popularization of Victorian Scientific Ideas." In *Energy and Entropy: Science and Culture in Victorian Britain; Essays from Victorian Studies*, edited by P. Brantlinger, 205–28. Bloomington: Indiana University Press, 1989.
Blomefield, Leonard. *Chapters in My Life*. Bath: privately printed, 1889.
Blum, Ann Shelby. *Picturing Nature: American Nineteenth-Century Zoological Illustration*. Princeton, NJ: Princeton University Press, 1993.
Boase, Frederic. *Modern English Biography*. London: Frank Cass, 1965.
Booth, Gordon Kempt. "William Robertson Smith: The Scientific, Literary and Cultural Context from 1866 to 1881." PhD diss., University of Aberdeen, 1999.
Bose, D. M., S. N. Sen, and B. V. Subbarayappa, eds. *A Concise History of Science in India*. New Delhi: Indian National Science Academy, 1971.
Bowler, Peter J. *The Fontana History of the Environmental Sciences*. London: Fontana, 1992.
Boyer, Paul. *When Time Shall Be No More: Prophecy Belief in Modern American Culture*. Cambridge, MA: Harvard Belknap, 1992.
Bradley, J. L., ed. *John Ruskin: The Critical Heritage*. London: Routledge, 1984.
Brennan, Theresa, and Martin Jay, eds. *Vision in Context: Historical and Contemporary Perspectives on Sight*. New York: Routledge, 1996.
[Brewster, David]. "The Weather and Its Prognostics." *North British Review* 25 (1856): 173–204.
Brief Sketch of the Meteorology of the Bombay Presidency [for the years 1876–81]. Bombay: Government Press, 1876–81.
Brock, W. H. "The Development of Commercial Science Journals in Victorian Britain." In *Development of Science Publishing in Europe*, edited by J. A. Meadows, 95–122. Amsterdam: Elsevier Science Publishers, 1980.
Broman, Thomas. "The Habermasian Public Sphere and 'Science in the Enlightenment.'" *History of Science* 36 (1998): 123–49.
Brooke, John H. *Science and Religion*. Cambridge: Cambridge University Press, 1991.

Brown, S. J. *Chalmers and the Godly Commonwealth*. Oxford: Blackwell, 1982.
Browne, H. B. *The Story of Whitby Museum*. Hull and London: A. Brown, 1949.
Browne, Walter. *The Moon and the Weather*. 2nd ed. London: Balliere, Tindall, and Cox, 1886.
Bruce, Robert V. *The Launching of Modern American Science 1846–1876*. Ithaca, NY: Cornell University Press, 1987.
Brück, H. A., and Mary T. Brück. *Peripatetic Astronomer: The Life of Charles Piazzi Smyth*. Bristol: A. Hilger, 1988.
Brück, Mary T. "The Piazzi Smyth Collection of Sketches, Photographs and Manuscripts at the Royal Observatory, Edinburgh." *Vistas in Astronomy* 32 (1989): 371–408.
Bryden, J. L. *Epidemic Cholera in the Bengal Presidency: A Report on the Cholera of 1868 and Its Relation to the Cholera of Previous Epidemics*. Calcutta: Superintendent of Government Printing, 1869.
Buckle, Thomas. *The History of Civilization in England*. 2 vols. London: J. W. Parker, 1857–61.
Bud, Robert, and Susan E. Cozzens, eds. *Invisible Connections: Instruments, Institutions, and Science*. Bellingham, WA: SPIE Optical Engineering Press, 1992.
Bud, Robert, and Deborah Jean Warner, eds. *Instruments of Science: An Historical Encyclopedia*. New York: Garland, 1998.
Burgess, A. H. *On Admiral FitzRoy's Storm Signals*. London: privately printed, 1864.
Burrows, J. W. *A Liberal Descent: Victorian Historians and the English Past*. Cambridge: Cambridge University Press, 1981.
Burton, James. "History of the British Meteorological Office to 1905." PhD diss., Open University, 1988.
————. "Meteorology and the Public Health Movement in London During the Late Nineteenth Century." *Association of Open University Graduates Journal* 1 (1990): 5–9.
————. "Robert FitzRoy and the Early History of the Meteorological Office." *British Journal for the History of Science* 19 (1986): 147–76.
Butler, T. B. *Philosophy of the Weather and a Guide to Its Changes*. New York: D. Appleton, 1856.
Cain, P. J., and A. G. Hopkin. *British Imperialism: Innovation and Expansion 1688–1914*. London: Longman, 1993.
Camerini, Jane R. "The Physical Atlas of Heinrich Berghaus." In *Non-Verbal Communication in Science Prior to 1900*, edited by R. G. Mazzolini, 479–512. Florence: Leo S. Olschki, 1993.
Cannadine, David. *Ornamentalism: How the British Saw Their Empire*. New York: Oxford University Press, 2001.
Cannegieter, Hendrik G. "The History of the International Meteorological Organization 1872–1951." *Annalen der Meteorologie* 1 (1963): 1–280.
Cannon, Susan Faye. *Science in Culture: The Early Victorian Period*. New York: Science History, 1978.
Cannon, Walter B. "The Problem of Miracles in the 1830s." *Victorian Studies* 4 (1960): 6–31.
Capp, Bernard. *English Almanacs 1500–1800: Astrology and the Popular Press*. Ithaca, NY: Cornell University Press, 1979.
Cardwell, D. S. L. *The Organisation of Science in England*. Rev. ed. London: Heinemann, 1972.
Carpenter, William. *Principles of Mental Physiology with Their Applications to the Training and Discipline of the Mind, and the Study of Its Morbid Conditions*. 4th ed. New York: Appleton, 1891.
Casella, Louis P. *An Illustrated and Descriptive Catalogue of the Philosophical, Meteorological, Mathematical, Surveying, Optical and Photographic Instruments*. London: privately printed, 1860.
Castle, Terry. "The Female Thermometer." *Representations* 17 (1987): 1–27.
Cawood, John. "The Magnetic Crusade: Science and Politics in Early Victorian Britain." *Isis* 70 (1979): 493–518.
Chadwick, Owen. *The Victorian Church*. 3rd ed. 2 vols. London: Adam and Charles Black, 1972.

Chalmers, Thomas. *The Efficacy of Prayer Consistent with the Uniformity of Nature*. London: Clap and Hull, 1832.

———. *The Evidence and Authority of the Christian Revelation: On the Miraculous and Internal Evidences of the Christian Revelation and the Authority of Its Records*. 2 vols. New York: Robert Carter, 1845.

———. "On Prophecy." *Church of England Preacher* 1 (1837): 86–89, 131–35.

———. *On the Power, Wisdom and Goodness of God, as Manifested in the Adaptation of External Nature to the Moral and Intellectual Constitution of Man*. 3rd ed. London: W. Pickering, 1834.

Chambers, G. F. "Fitzroy's Weather Forecasts." *Popular Science Review* 6 (1867): 265–66.

———. "How to Study Meteorology." *Popular Science Review* 4 (1867): 140.

———. *The Story of the Weather*. London: George Newnes, 1897.

Chambers, Robert. *Vestiges of Natural History of Creation*, edited by J. Secord. Reprint ed. Chicago: University of Chicago Press, 1994.

Chandler, James, Arnold Davidson, and Harry Harootunian, eds. *Questions of Evidence: Proof, Practice, and Persuasion across the Disciplines*. Chicago: University of Chicago Press, 1994.

Chapman, Allan. "Private Research and Public Duty: George Biddell Airy and the Search for Neptune." *Journal of the History of Astronomy* 19 (1988): 121–39.

———. "Science and the Public Good: George Biddell Airy (1801–92) and the Concept of a Scientific Civil Servant." In *Science, Politics, and the Public Good*, edited by N. A. Rupke, 36–62. London: Macmillan, 1988.

Chesney, George. "Indian Famines." *Nineteenth Century* 2 (1877): 603–20.

———. *Indian Polity: A View of the System of Administration in India*. London: Longman and Green, 1868.

Cheyne, A. C. *The Transforming of the Kirk: Victorian Scotland's Religious Revolution*. Edinburgh: St Andrew's Press, 1983.

Christie, S. Hunter. "The Reply of the President and Council to a Letter Addressed to Them by the Secretary of State for Foreign Affairs, on the Subject of the Cooperation of Different Nations in Meteorological Observations." *Abstracts of the Papers Communicated to the Royal Society of London* 6 (1850–54): 188–92.

Claridge, John. *The Country Calendar or the Shepherd of Banbury's Rules*. London: Sylvan Press, 1946.

Clark, James. *The Influence of Climate in the Prevention and Cure of Chronic Diseases More Particularly of the Chest and Digestive Organs Comprising an Account of the Principal Places Resorted to by Invalids in England and the South of France; A Comparative Estimate of Their Respective Merits in Particular Diseases and General Directions for Invalids While Travelling and Residing Abroad*. London: Thames and George Underwood, 1829.

Clarke, I. F. *The Pattern of Expectation 1644–2001*. New York: Basic Books, 1979.

"The Classification of Clouds and the Cloud Atlas." In *Report of the International Meteorological Conference at Munich 1891*, 18–20. London: HMSO, 1892.

Clayden, Arthur. *Cloud Studies*. London: J. Murray, 1905.

Clements, R. E. "The Study of the Old Testament." In *Nineteenth Century Religious Thought in the West*, edited by J. C. Ninian Smart, Steven Katz, and Patrick Sherry, 3 vols., 3:109–41. Cambridge: Cambridge University Press, 1985.

Clerke, Agnes. *A Popular History of Astronomy during the Nineteenth Century*. 2nd ed. Edinburgh: Adam and Charles Black, 1887.

"Cloud Atlas." *American Meteorological Journal* 7 (1890–91): 424.

Clouston, Charles. *Popular Weather Prognostics of Scotland*. Edinburgh: privately printed, 1867.

Codell, Julie F. *Imperial Co-Histories: National Identities and the British and Colonial Press*. Madison, NJ: Fairleigh Dickinson University Press, 2003.

Cohn, Bernard. "Cloth, Clothes, and Colonialism." In *Colonialism and Its Forms of Knowledge*, 106–62. Princeton, NJ: Princeton University Press, 1996.

———. "Notes on the History of the Study of Indian Society and Culture." In *An Anthropologist among the Historians*, 136–71. Delhi: Oxford University Press, 1987.

———. "Representing Authority in Victorian India." In *An Anthropologist among the Historians*, 632–82. Delhi: Oxford University Press, 1987.

Collini, Stefan. *Public Moralists: Political Thought and Intellectual Life in Britain, 1850–1930*. Oxford: Oxford University Press, 1991.

Cook, Chris, and Brendan Keith. *British Historical Facts 1830–1900*. London: Macmillan, 1984.

Cooke, Christopher. *Curiosities of Occult Literature*. London: A. Hall, Smart and Allen, 1863.

———. "Storm Signals and Forecasts, Their Utility and Importance with Respect to Navigation and Commerce." *Journal of the Royal Society of Arts* 15 (1867): 242–49.

Cooper, Brian P., and Margueritte S. Murphy. "The Death of the Author at the Birth of Social Science: The Case of Harriet Martineau and Adolphe Quetelet." *Studies in the History and Philosophy of Science* 31 (2000): 1–36.

Cooter, Roger, and Steven Pumfrey. "Separate Spheres and Public Places: Reflections on the History of Science Popularization and Science in Popular Culture." *History of Science* 32 (1994): 237–67.

Cornish, Francis Ware. *The English Church: The Nineteenth Century*. 2 vols. New York: AMS Press, 1910.

"Correspondence between the Board of Trade and the Royal Society in Reference to the Meteorological Department." *Proceedings of the Royal Society* 14 (1865): 306–19.

Cory, F. W. *How to Foretell the Weather with the Pocket Spectroscope*. London: Chatto and Windus, 1884.

Cotton, H. *England and India: An Address Delivered at the Positivist School*. London: Kegan Paul, 1883.

———. "Has India Food for Its People?" *Fortnightly Review* 22 (1877): 863–77.

Cox, S. "Sensibility as Argument." In *Sensibility in Transformation: Creative Resistance to Sentiment from the Augustans to the Romantics*, edited by S. M. Conger, 63–82. London: Associated University Presses, 1990.

Craft, Erik D. "Private Weather Organizations and the Founding of the United States Weather Bureau." *Journal of Economic History* 59, no. 4 (1999): 1063–71.

———. "The Value of Weather Information Services for Nineteenth Century Great Lakes Shipping." *American Economic Review* 88, no. 5 (1998): 1059–76.

Crary, Jonathan. *Techniques of the Observer: On Vision and Modernity in the Nineteenth Century*. Cambridge, MA: MIT Press, 1990.

Crawford, Elizabeth, Terry Shinn, and Sverker Sörlin. *Denationalizing Science: The Contexts of International Scientific Practice*. Dordrecht: Kluwer Academic Publishers, 1992.

Crouthamel, James L. *Bennett's New York Herald and the Rise of the Popular Press*. Syracuse: Syracuse University Press, 1989.

Crouzet, Francois. *The Victorian Economy*. Translated by A. Forster. London: Methuen and Co., 1982.

Cruikshank, George. "Al-maniac Day—A Rush for the Murphies." In *Comic Almanack, an Ephemeris in Jest and Earnest: 1835–1843*, 162. London: J. C. Hatton, 1860.

"Current Notes: Atlas of Clouds." *American Meteorological Journal* 6 (1889–90): 383–84.

Curry, Patrick. *Confusion of Prophets: Victorian and Edwardian Astrology*. London: Collins and Brown, 1992.

———. *Prophecy and Power: Astrology in Early Modern England*. Cambridge, UK: Polity, 1989.

Curtin, Philip. *Death by Migration: Europe's Encounter with the Tropical World in the Nineteenth Century*. Cambridge: Cambridge University Press, 1989.

Curzon, George Nathaniel. *Frontiers*. Oxford: Clarendon Press, 1908.

Dale, Peter Allan. *In Pursuit of a Scientific Culture: Science, Art, and Society in the Victorian Age.* Madison: University of Wisconsin Press, 1989.
Dallas, W. L. "The Prediction of Droughts." In *Report of the International Meteorological Congress—Chicago 1893*, edited by Oliver L. Fassig, 13–18. Washington, DC: Weather Bureau, 1894.
Danahay, Martin. "Matter Out of Place: The Politics of Pollution in Ruskin and Turner." *Clio* 21 (1991): 61–77.
Daniell, John F. *Meteorological Essays and Observations.* London: Thomas and George Underwood, 1823.
Daston, Lorraine. *Classical Probability in the Enlightenment.* Princeton, NJ: Princeton University Press, 1988.
————. "Marvellous Facts and Miraculous Evidence in Early Modern Europe." In *Questions of Evidence: Proof, Practice, and Persuasion across the Disciplines*, edited by J. Chandler, A. I. Davidson, and H. Harootunian, 243–74. Chicago: University of Chicago Press, 1991.
Davis, John. "Weather Forecasting and the Development of Meteorological Theory at the Paris Observatory 1853–1878." *Annals of Science* 41 (1984): 380.
Day, James. *Meterology [sic] as Applied to Practical Science.* London: privately printed, [c. 1853].
Deacon, Margaret. *Scientists and the Sea 1650–1900.* 2nd ed. Aldershot: Ashgate, 1997.
De Luna Martins, Luciana. "Mapping Tropical Waters: British Views and Visions of Rio de Janeiro." In *Mappings*, edited by D. Crosgrove, 146–68. London: Reaktion, 1999.
De Morgan, Augustus. *Essay on Probabilities and on Their Application to Life Contingencies and Insurance Offices.* London: Longman, 1838.
Desmond, Adrian. *Archetypes and Ancestors: Palaeontology in Victorian London, 1850–1875.* Chicago: University of Chicago Press, 1984.
————. *Huxley: From Devil's Disciple to Evolution's High Priest.* London: Penguin, 1998.
————. *The Politics of Evolution: Morphology, Medicine, and Reform in Radical London.* Chicago: University of Chicago Press, 1989.
————. "Redefining the X Axis: Professionals, Amateurs and the Making of Mid-Victorian Biology—A Progress Report." *Journal of the History of Biology* 34 (2001): 3–50.
Desmond, Adrian, and James Moore. *Darwin.* London: Penguin, 1992.
Dettelbach, Michael. "The Face of Nature: Precise Measurement, Mapping and Sensibility in the Work of Alexander von Humboldt." *Studies in the History and Philosophy of Biology and Biomedical Sciences* 30 (1999): 473–504.
————. "Global Physics and Aesthetic Empire: Humboldt's Physical Portrait of the Tropics." In *Visions of Empire: Voyages, Botany, and Representations of Nature*, edited by D. P. Miller and P. H. Reill, 258–93. Cambridge: Cambridge University Press, 1996.
————. "Humboldtian Science." In *Cultures of Natural History*, edited by N. Jardine, J. A. Secord, and E. C. Spary, 287–304. Cambridge: Cambridge University Press, 1996.
Dick, Steven J. "Centralizing Navigational Technology in America: The U.S. Navy's Depot of Charts and Instruments, 1830–1842." *Technology and Culture* 33 (1992): 467–509.
Dickens, Charles. "The Shipwreck." *All the Year Round* 11 (1860): 321.
————. *The Uncommercial Traveller.* London: Chapman and Hall, 1868.
Digby, William. *The Famine Campaign in Southern India 1876–1878.* 2 vols. London: Longman and Green, 1878.
Dingley, Robert. "Ruins of the Future: Macaulay's New Zealander and the Spirit of the Age." In *Histories of the Future: Studies in Fact, Fantasy and Science Fiction*, edited by A. Sandison and R. Dingley, 15–33. New York: Palgrave, 2000.
Dolan, Brian. "Pedagogy through Print: James Sowerby, John Mawe and the Problem of Colour in Early Nineteenth-Century Natural History Illustration." *British Journal for the History of Science* 31 (1998): 275–304.

Donnison, Jean. *Midwives and Medical Men: A History of the Struggle for the Control of Childbirth.* 2nd ed. London: Historical Publications, 1988.
Doré, Gustav, and Blanchard Jerrold. *London: A Pilgrimage.* New York: Dover Publications, 1970.
Dove, H. W. *The Law of Storms Considered in Connection with the Ordinary Movements of the Atmosphere.* Translated by R. Scott. 2nd ed. London: Longman, Roberts, Green, 1862.
Drew, John. *Practical Meteorology.* 2nd ed. London: John van Voorst, 1860.
Drewitt, F. Daubrey. *Edward Jenner.* London: Longman, 1933.
Duns, John. *Biblical Natural Science Being an Explanation of All References in Holy Scripture to Geology, Botany, Zoology and Physical Geography.* 2 vols. London: William MacKenzie, 1860–69.
Dutta, Simanti. *Imperial Mappings in Savage Spaces: Baluchistan and British India.* Delhi: B. R. Pub. Corp., 2002.
Eden, Ashley, Edward Robert Bulwer Lytton, and John Strachey. *The Indian Famine of 1877: Being a Statement of the Measures Proposed by the Government of India for the Prevention and Relief of Famines in the Future.* London: C. Kegan, 1878.
[Editorial]. *Journal of Astronomic Meteorology and Record of the Science and Phenomena of the Weather* 1 (April 1862): 1.
Edney, Matthew. *Mapping an Empire: The Geographical Construction of British India, 1765–1843.* Chicago: University of Chicago Press, 1997.
Ellegard, Alvar. *Darwin and the General Reader: The Reception of Darwin's Theory of Evolution in the British Periodical Press, 1859–1872.* Stockholm: Almqvist and Wiksell, 1958.
Ellis, Ieuan. *Seven against Christ: A Study of "Essays and Reviews."* Leiden: E. J. Brill, 1980.
Ellison, Thomas. *The Cotton Trade of Great Britain.* New York: A. Kelley, 1968.
Embree, Ainslie. "Frontiers into Boundaries: The Evolution of the Modern State." In *Imagining India: Essays on Indian History,* edited by Mark Juergensmeyer, 67–84. Delhi: Oxford University Press, 1989.
———. "The Idea of India in Classical Western Political Thought." In *Imagining India: Essays on Indian History,* edited by Mark Juergensmeyer, 28–40. Delhi: Oxford University Press, 1989.
Eyler, John M. *Victorian Social Medicine: The Ideas and Methods of William Farr.* Baltimore: Johns Hopkins University Press, 1979.
Fancher, Raymond E. "Biography and Psychodynamic Theory: Some Lessons from the Life of Francis Galton." *History of Pyschology* 1 (1998): 99–115.
Fara, Patricia. *Sympathetic Attractions: Magnetic Practices, Beliefs, and Symbolism in Eighteenth-Century England.* Princeton, NJ: Princeton University Press, 1996.
Fayrer, Sir Joseph. *Rainfall and Climate in India.* London: E Stanford, 1881.
Feldman, Theodore. "The Ancient Climate in the 18th and Early 19th Century." In *Science and Nature: Essays in the History of the Environmental Sciences,* edited by M. Shortland, 23–40. Stanford in the Vale: British Society for the History of Science, 1993.
———. "Late Enlightenment Meteorology." In *The Quantifying Spirit,* edited by T. Frangsmyr, J. L. Heilbron, and R. E. Rider, 143–77. Berkeley: University of California Press, 1990.
Fitch, Raymond. *The Poison Sky: Myth and Apocalypse in Ruskin.* Columbus: Ohio University Press, 1982.
Fitzroy, Robert. *Barometer and Weather Guide.* 4th ed. London: HMSO, 1861.
———. *Memorandum.* Privately printed, 1852.
———. *Meteorological Papers.* London: HMSO, 1857.
———. *Narrative of the Surveying Voyages of His Majesty's Ships Adventure and Beagle between the Years 1826 and 1836 Describing Their Examination of the Southern Shores of South America, and the Beagle's Circumnavigation of the Globe.* 3 vols. London: Henry Colburn, 1839.
———. *Notes on Meteorology.* London: HMSO, 1858.
———. "On Weather Glasses." *Lifeboat Journal* 4 (1860): 329–30.

———. *Weather Book*. 2nd ed. London: Longman, 1863.
Flammarion, Camille. *The Atmosphere*, edited by J. Glaisher. New York: Harper, 1873.
Flegg, Columba Graham. *Gathered under the Apostles: A Study of the Catholic Apostolic Church*. Oxford: Clarendon Press, 1992.
Fleming, James Rodger, ed. *Historical Essays on Meteorology 1919–1995*. Boston: American Meteorological Society, 1996.
———. *Historical Perspectives on Climate Change*. New York: Oxford University Press, 1998.
———. "Meteorological Observing Systems before 1870 in England, France, Germany, Russia, and the USA: A Review and Comparison." *World Meteorological Organization Bulletin* 46 (1997): 249–58.
———. *Meteorology in America, 1800–1870*. Baltimore: Johns Hopkins University Press, 1990.
Force, James E. "Hume and Johnson on Prophecy and Miracles: Historical Context." *Journal of the History of Ideas* 43 (1982): 463–75.
Forrest, Derek. *Francis Galton: The Life of a Victorian Genius*. London: Paul Elek, 1974.
Forster, Thomas. *Researches about Atmospheric Phenomena*. London: T. Underwood, 1813.
Fowler, William. *Mozley and Tyndall on Miracles*. London: Longman and Green, 1868.
Friedman, Robert M. *Appropriating the Weather: Vilhelm Bjerknes and the Construction of a Modern Meteorology*. Ithaca, NY: Cornell University Press, 1989.
Frisinger, H. Howard. *The History of Meteorology: To 1800*. New York: Science History, 1977.
Fullbrook, Charles. *The Wet and Dry Seasons of England from the Year 1846 to 1860*. London: Henry Renshaw, 1861.
"Further Correspondence between the Board of Trade and the Royal Society in Reference to the Magnetism of Ships and the Meteorological Department." *Proceedings of the Royal Society* 14 (1865): 516–41.
Galison, Peter, Caroline Jones, and Amy Slaton, eds. *Picturing Science, Producing Art*. New York: Routledge, 1998.
Galton, Francis. *English Men of Science: Their Nature and Nurture*. London: Macmillan, 1874.
———. *Meteorographica; or, Methods of Mapping the Weather Illustrated by Upwards of 600 Printed and Lithographed Diagrams Referring to the Weather of a Large Part of Europe, During the Month of December 1861*. London: Macmillan, 1863.
———. "Notes on Modern Geography." In *Cambridge Essays Contributed by Members of the University*, 4 vols., edited by J. Parker, 79–109. London: J. Parker, 1855–58.
———. "On Stereoscopic Maps, Taken from Models of Mountainous Countries." *Journal of the Royal Geographical Society* 35 (1865): 99–104.
———. "Statistical Inquiries into the Efficacy of Prayer." *Fortnightly Review* 12 (1872): 125–35.
Garber, Elizabeth. "Thermodynamics and Meteorology." *Annals of Science* 33 (1976): 51–65.
Gates, Barbara, and Ann B. Shteir. *Natural Eloquence: Women Reinscribe Science*. Madison: University of Wisconsin Press, 1997.
"George James Symons [obituary]." *Quarterly Journal of the Royal Meteorological Society* 26 (1900): 155–59.
Gibbs, W. J. *Origins of Australian Meteorology*. Canberra: Australian Government Publishing Service, 1975.
Gigerenzer, Gerd, Zeno Swijtink, Theodore Porter, Lorraine Daston, John Beatty, Lorenz Kruger. *The Empire of Chance: How Probability Changed Science and Everyday Life*. Cambridge: Cambridge University Press, 1989.
Gilley, Sheridan. "Edward Irving: Prophet of the Millennium." In *Revival and Religion since 1700: Essays for John Walsh*, edited by J. Garnett and C. Matthew, 95–110. London: Hambledon, 1993.
Glaisher, James. "George Biddell Airy Esq. F.R.S." *Leisure Hour* 11 (1862): 648–51.
———. "The Influence of the Moon on Rainfall." *Proceedings of the British Meteorological Society* 4 (1869): 327–50.

———. "The Influence of the Moon on the Direction of the Wind." *Proceedings of the British Meteorological Society* 3 (1867): 359–78.

———. *Philosophical Instruments and Processes Represented in the Great Exhibition*. London: David Bogue, 1852.

———. *Report on the Meteorology of London, and Its Relation to the Epidemic of Cholera*. London: HMSO, 1855.

Goldman, Lawrence, Fred Lewes, Edward Higgs, and Simon Szreter. "The General Register Office of England and Wales and the Public Health Movement, 1837–1914: A Comparative Perspective." *Social History of Medicine* 4 (1991): 401–97.

Golinski, Jan. "Barometers of Change: Meteorological Instruments as Machines of Enlightenment." In *The Sciences in Enlightened Europe*, edited by J. Golinski, W. Clark, and S. Schaffer, 69–93. Chicago: University of Chicago Press, 1999.

———. "The Human Barometer: Weather Instruments and the Body in Eighteenth-Century England." Paper presented at the annual meeting of the American Society for Eighteenth-Century Studies, Notre Dame, IN, September 10, 1998.

Gooday, Graeme. "'Nature' in the Laboratory: Domestication and Discipline with the Microscope in Victorian Life Science." *British Journal for the History of Science* 24 (1991): 307–41.

———. "Precision Measurement and the Genesis of Physics Teaching Laboratories in Victorian Britain." PhD diss., University of Kent at Canterbury, 1989.

———. "Precision Measurement and the Genesis of Physics Teaching Laboratories in Victorian Britain." *British Journal for the History of Science* 23 (1990): 25–51.

———. "Sunspots, Weather and the Unseen Universe: Balfour Stewart's Anti-materialist Representation of Energy." In *Science Serialized: Representations of the Sciences in Nineteenth-Century Periodicals*, edited by S. Shuttleworth and G. Cantor. Cambridge, MA: MIT Press, forthcoming.

Goodison, Nicholas. *English Barometers 1680–1860: A History of Domestic Barometers and Their Makers and Retailers*. Woodbury: Antique Collector's Club, 1977.

Gopal, Sarvepalli. *British Policy in India, 1858–1905*. Cambridge: Cambridge University Press, 1965.

Gore, George. *The Scientific Basis of National Progress, Including That of Morality*. London: Williams and Norgate, 1882.

Graham, J. H. *The Destruction of Daylight: A Study in the Smoke Problem*. London: G. Allen, 1907.

Grant, Robert. *History of Physical Astronomy*. London: Robert Baldwin, 1852.

"The Great Trigonometrical Survey." *Calcutta Review* 38 (1863): 26–62.

Greene, Mott T. *Geology in the Nineteenth Century: Changing Views of a Changing World*. Ithaca, NY: Cornell University Press, 1982.

Greenough, Paul. "Hunter's Drowned Land: An Environmental Fantasy of the Victorian Sunderbans." In *Nature and the Orient: The Environmental History of South and Southeast Asia*, edited by Richard Grove, V. Damodaran, and S. Sangwan, 237–72. Oxford: Oxford University Press, 1998.

Grogg, Ann Hofstra. "The Illustrated London News 1842–1852." PhD diss., Indiana University, 1977.

Grove, Richard H. "The East India Company, the Raj and El Niño: The Critical Role Played by Colonial Scientists in Establishing the Mechanisms of Global Climate Teleconnections, 1770–1930." In *Nature and the Orient: The Environmental History of South and Southeast Asia*, edited by Richard Grove, V. Damodaran. and S. Sangwan, 301–23. Oxford: Oxford University Press, 1998.

Grove, Richard H., Vinita Damodaran, and Satpal Sangwan, eds. *Nature and the Orient: The Environmental History of South and Southeast Asia*. Oxford: Oxford University Press, 1998.

Gucken, W. *Nineteenth Century Spectroscopy: Development of the Understanding of the Spectra, 1802–1897*. Baltimore: Johns Hopkins University Press, 1969.

Guillemin, Amédée. *The Heavens: An Illustrated Handbook of Popular Astronomy*, translated by M. N. Lockyer, edited by N. Lockyer and R. Proctor. 2nd ed. London: Richard Bentley, 1866.
Guy, William. "On the Claims of Science to Public Recognition and Support; With Special Reference to the So-Called 'Social Sciences.'" *Journal of the Royal Statistical Society* 33 (1870): 433–51.
Hacking, Ian. *The Taming of Chance*. Cambridge: Cambridge University Press, 1990.
Hackmann, W. D. "Scientific Instruments: Models of Brass and Aids to Discovery." In *The Uses of Experiment*, edited by David Gooding and Simon Schaffer, 31–65. Cambridge: Cambridge University Press, 1989.
Hall, Marie Boas. *All Scientists Now: The Royal Society in the Nineteenth Century*. New York: Cambridge University Press, 1984
———. "Public Science in Britain: The Role of the Royal Society." *Isis* 72 (1981): 627–29.
Hamblyn, Richard. *The Invention of Clouds: How an Amateur Meteorologist Forged the Language of the Skies*. New York: Farrar, Straus, and Giroux, 2001.
Hamlin, Christopher. "James Geikie, James Croll, and the Eventful Ice Age." *Annals of Science* 39 (1982): 565–83.
Hancher, Michael. "An Imagined World: The Imperial Gazetteer." In *Imperial Co-Histories: National Identities and the British and Colonial Press*, edited by Julie F. Codell, 45–67. Madison, NJ: Fairleigh Dickinson University Press, 2003.
Hankins, Thomas, and Robert J. Silverman. *Instruments and the Imagination*. Princeton, NJ: Princeton University Press, 1995.
Hankins, Thomas, and Albert Van Helden. "Instruments in the History of Science." *Osiris* 9 (1994): 1–6.
Hardy, Thomas. *Far from the Madding Crowd*. Oxford: Oxford University Press, 1993.
———. *The Mayor of Casterbridge*. Edited by D. Kramer. Oxford: Oxford University Press, 1987.
Harris, Michael. "Astrology, Almanacs and Booksellers." *Publishing History* 8 (1980): 92–109.
Harris, Michael, and Robin Myers, eds. *The Stationers' Company and the Book Trade, 1550–1990*. Winchester, UK: St. Paul's Bibliographies, 1998.
Harrison, J. F. C. *The Second Coming: Popular Millenarianism 1790–1850*. New Brunswick, NJ: Rutgers University Press, 1979.
Harrison, J. Park. "Lunar Influence on Temperature." *Proceedings of the Royal Society* 14 (1865): 223–31.
Harrison, Mark. *Climates and Constitutions: Health, Race, Environment and British Imperialism in India, 1600–1850*. New Delhi: Oxford University Press, 1999.
Harrison, Peter. "Prophecy, Early Modern Apologetics and Hume's Arguments against Miracles." *Journal for the History of Ideas* 60 (1999): 241–56.
Hartley, Beryl. "The Living Academies of Nature: Scientific Experiment in Learning and Communicating the New Skills of Early Nineteenth Century Landscape Painting." *Studies in the History and Philosophy of Science* 27 (1996): 149–80.
Hartwig, G. *The Aerial World: A Popular Account of the Phenomena and Life of the Atmosphere*. London: Longman and Green, 1886.
Hays, J. N. "The Rise and Fall of Dionysius Lardner." *Annals of Science* 38 (1981): 527–42.
Headrick, Daniel. *The Invisible Weapon: Telecommunications and International Politics, 1851–1945*. Oxford: Oxford University Press, 1991.
Hellmann, Gustav. *Meteorologische karten 1688, 1817, 1846, 1863, 1864. Sechs tafeln in lichtdruck mit einer einleitung*. Berlin: Asher, 1897.
———. "Die Organisation des meteorologischen Dienstes in den Haupstaaten Europa's." *Zeitschrift des Königlichen Pruessischen Statistischen Bureaus* 20 (1880): 1–52, 427–52.
Henson, Louise, Geoffrey Cantor, Gowan Dawson, Richard Noakes, Sally Shuttleworth, and John R. Topham, eds. *Culture and Science in the Nineteenth Century Media*. London: Ashgate, 2004.

Hentschel, Klaus. *Mapping the Spectrum: Techniques of Visual Representation in Research and Teaching*. Oxford: Oxford University Press, 2002.
Herschel, John. F. W. *A Manual of Scientific Enquiry Prepared for the Use of Her Majesty's Navy: And Adapted for Travellers in General*. 2nd ed. London: John Murray, 1851.
———. *Meteorology*. 2nd ed. Edinburgh: Adam and Charles Black, 1862.
———. *Preliminary Discourse on the Study of Natural Philosophy*. Reprint ed. Chicago: University of Chicago Press, 1987.
———. "[Review of Quetelet's] Letters on Probability." *Edinburgh Review* 92 (1850): 1–57.
———. "The Weather and Weather Prophets." *Good Words* 5 (1864): 57–64.
Heyck, Thomas William. *The Transformation of Intellectual Life in Victorian England*. London: Croom Helm, 1982.
Hildebrandsson, Hugo. "Remarks Concerning the Nomenclature of Clouds for Ordinary Use." *Quarterly Journal of the Royal Meteorological Society* 13 (1887): 148–66.
———. *Sur la classification des nuages employés a l'observatoire meteorologique d'Uppsala*. Uppsala: Berling for the University of Uppsala, 1879.
Hildebrandsson, Hugo, and Gustav Hellmann. *Codex of Resolutions Adopted at International Meteorological Meetings 1872–1907*. London: HMSO, 1909.
Hildebrandsson, Hugo, Albert Riggenbach, L. Teisserenc de Bort, eds. *International Cloud Atlas*. Paris: Gauthier-Villars, 1896.
Hildebrandsson, Hugo, and L. Teisserenc de Bort. *Les bases de la météorologie dynamique: Historique - état de nos connaissances*. Paris: Gauthier Villars, 1898.
Hilton, Boyd. *The Age of Atonement: The Influence of Evangelicalism on Social and Economic Thought 1785–1865*. Oxford: Clarendon Press, 1988.
———. "Chalmers as Political Economist." In *The Practical and the Pious: Essays on Thomas Chalmers*, edited by A. C. Cheyne, 141–56. Edinburgh: St. Andrew's Press, 1985.
Hilton, Timothy. *John Ruskin: The Early Years 1819–59*. New Haven, CT: Yale University Press, 1985.
———. *John Ruskin: The Later Years*. New Haven, CT: Yale University Press, 2000.
Hilts, Victor. "Aliis Extenderum; or, The Origins of the Statistical Society of London." *Isis* 69 (1978): 21–43.
Hood, Thomas. "A Flying Visit." *Comic Annual* 10 (1839): 133–48.
Hooker, Joseph. *The Flora of British India*. 7 vols. London: J. Reeve, 1875–97.
———. *Himalayan Journal: Notes of a Naturalist in Bengal, the Sikkim and Nepal Himalayas, the Khasia Mountains*. 2 vols. London: John Murray, 1855.
Houghton, Walter, ed. *The Wellesley Index to Victorian Periodicals*. 5 vols. Toronto: University of Toronto Press, 1966–89.
Howard, Luke. *Barometrographica: Twenty Years' Variation of the Barometer in the Climate of Britain, Exhibited in Atmospheric Curves with the Attendant Winds and Weather*. London: Richard and John E. Taylor, 1847.
———. *The Climate of London Deduced from Meteorological Observations Made in the Metropolis and at Various Places Around It*. 2nd ed. 3 vols. London: Harvey and Darton, 1833.
———. *A Cycle of 18 Years*. London: J. Ridgway, 1842.
———. *Essay on the Modifications of Clouds*. 3rd ed. London: John Churchill, 1865.
———. *Papers on Meteorology, Relating Especially to the Climate of Britain and to the Variation of the Barometer*. London: Taylor and Francis, 1850.
Howe, Ellic. *Urania's Children: The Strange World of the Astrologers*. London: Kimber, 1967.
Hufbauer, Karl. *Exploring the Sun: Solar Science Since Galileo*. Baltimore: Johns Hopkins University Press, 1991.
Hughes, David, and Carole Stott. "Two Piazzi Smyth Comet Paintings." *Annals of Science* 46 (1989): 165–72.

Hulme, Edward, et al. *Art-Studies from Nature, as Applied to Design for the Use of Architects, Designers and Manufacturers*. London: Virtue, 1872.
Humboldt, Alexander von. *Cosmos: A Sketch of the Physical Description of the Universe*. Translated by E. C. Otte. 2 vols. Baltimore: Johns Hopkins University Press, 1997.
Hunt, Bruce. "Doing Science in a Global Empire: Cable Telegraphy and Electrical Physics in Victorian Britain." In *Victorian Science in Context*, edited by Bernard Lightman, 312–33. Chicago: University of Chicago Press, 1997.
Hunt, J. L. "James Glaisher, FRS (1809–1903)." *Weather* 33 (1978): 242–49.
Hunt, John Dixon. *The Wider Sea: A Life of John Ruskin*. New York: Viking, 1982.
Hunter, William W. *Annals of Rural Bengal*. 7th ed. London: Smith and Elder, 1897.
———. *Famine Aspects of Bengal Districts*. London: Trübner, 1874.
[———]. "India." In *Encyclopaedia Britannica*, 9th ed. 1890. 16:731–813.
———. "Orissa and Orissa Tributary States." *Imperial Gazetteer of India*, 10:426–68. 2nd ed. London: Smith and Elder, 1881.
Hunter, William W., and Norman Lockyer. "Sunspots and Famines." *Nineteenth Century* 2 (1877): 583–602.
Huxley, Leonard. *Life and Letters of Sir Joseph Dalton Hooker . . . Based on Materials Collected and Arranged by Lady Hooker*. New York: Appleton, 1918.
Huxley, Thomas H. "On the Method of Zadig: Retrospective Prophecy as a Function of Science." In *Science and Culture and Other Essays*, 135–55. New York: Appleton, 1893.
Jacobs, L. "A Short History of Former Homes of the Meteorological Office." *Meteorological Magazine* 102 (1973): 48–50.
———. "The 200-Years' Story of Kew Observatory." *Meteorological Magazine* 98 (1969): 162–71.
James, Frank A. J. L. "The Establishment of Spectro-Chemical Analysis as a Practical Method of Qualitative Analysis, 1854–1861." *Ambix* 30 (1983): 30–53.
James, Henry. *Instructions for Taking Meteorological Observations at the Principal Foreign Stations of the Royal Engineers*. London: HMSO, 1851.
James, Louis. *Print and the People 1819–1851*. London: Allen Lane, 1976.
Jankovic, Vladimir. "Ideological Crests versus Empirical Troughs: John Herschel's and William Radcliffe Birt's Research on Atmospheric Waves, 1843–50." *British Journal for the History of Science* 31 (1998): 21–40.
———. "The Place of Nature and the Nature of Place: The Chorographic Challenge to the History of British Provincial Science." *History of Science* 38 (2000): 79–113.
———. *Reading the Skies: A Cultural History of English Weather 1650–1820*. Chicago: University of Chicago Press, 2001.
Jensen, J. Vernon. "The X Club: Fraternity of Victorian Scientists." *British Journal for the History of Science* 5 (1970): 63–72.
Jevons, Harriet. *Letters and Journal of W. Stanley Jevons*. London: Macmillan, 1886.
Jevons, William Stanley. *The Coal Question: An Enquiry Concerning the Progress of the Nation, and the Probable Exhaustion of Our Coal-Mines*. 2nd ed. London: Macmillan, 1866.
———. *Investigations in Currency and Finance*. New York: A. M. Kelley, 1964.
———. "On Clouds, Their Various Forms and Producing the Causes." *Sydney Magazine of Science and Art* 1 (1858): 163–76.
———. *The Principles of Science: A Treatise on Logic and Scientific Method*. 2nd ed. London: Macmillan, 1905.
Jobé, Joseph, ed. *The Great Age of Sail*. Translated by M. Kelly. 2nd ed. New York: Viking, 1971.
Johnson, James. *Influences of Tropical Climates on European Constitutions*. 3rd ed. New York: E. Duyckinck, 1826.
Johnston, Alexander Keith. *The Physical Atlas: A Series of Maps and Illustrations of the Geographic*

Distribution of Natural Phenomena, Embracing I. Geology II. Hydrography III. Meteorology IV. Natural History. Edingburgh: Blackwood, 1850.

Jones, George Chetwynd Griffith. *The Great Weather Syndicate*. London: F. V. White and Co., 1906.

Juergensmeyer, Mark, ed. *Imagining India: Essays on Indian History*. Delhi: Oxford University Press, 1989.

Kämtz, Ludwig Friedrich. *A Complete Course of Meteorology*. Translated by C. V. Walker. London: H. Nailliere, 1845.

Keith, Alexander. *Evidence of the Truth of the Christian Religion Derived from the Literal Fulfillment of Prophecy Particularly as Illustrated by the History of the Jews and by the Discoveries of Recent Travellers*. 38th ed. London: T. Nelson, 1861.

Kennedy, Dane. "The Perils of the Midday Sun: Climatic Anxieties in the Colonial Tropics." In *Imperialism and the Natural World*, edited by J. MacKenzie, 118–40. Manchester: Manchester University Press, 1900.

Kennedy, E. A. *The Hourly Weather Guide or Cloud Atlas*. London: privately printed, 1893.

Kevan, Simon M. "Quests for Cures: A History of Tourism for Climate and Health." *International Journal of Biometeorology* 37 (1993): 113–24.

Khrgian, A. K. *Meteorology: A Historical Survey*. Translated by R. Hardin. 2nd ed. Jerusalem: Israel Program for Science Translations Ltd., 1970.

King, George. "On the Construction of Mortality Tables from Census Returns and Records of Deaths." *Journal of the Institute of Actuaries* 42 (1908): 225–77.

King, Henry. *Madras Manual of Hygiene*. Madras: Keys, 1875.

Kingsley, Charles. "How to Study Natural History." In *Scientific Lectures and Essays*, 289–312. London: Macmillan, 1885.

Kingsley, Frances. *Charles Kingsley: His Letters and Memories of His Life*. 2 vols. London: H. S. King, 1877.

Kington, J. A. "A Century of Cloud Classification." *Weather* 24 (1969): 84–89.

————. "A Historical Review of Cloud Study." *Weather* 23 (1968): 349–56.

Kington, John. *The Weather of the 1780s over Europe*. Cambridge: Cambridge University Press, 1988.

Kirchoff, F. "A Science against Sciences: Ruskin's Floral Mythology." In *Nature and the Victorian Imagination*, edited by U. C. Knoepflmacher and G. B. Tennyson, 246–58. Berkeley: University of California Press, 1977.

Klein, Judy L. *Statistical Visions in Time: A History of Time Series Analysis 1662–1938*. Cambridge: Cambridge University Press, 1997.

Klonk, Charlotte. "Science and the Perception of Nature: British Landscape Art in the Late Eighteenth and Early Nineteenth Centuries." PhD diss., Cambridge University, 1992.

Knight, Charles. *Passages of a Working Life during Half a Century*. 3 vols. Shannon: Irish University Press, 1971.

Knight, William. "The Function of Prayer in the Economy of the Universe." *Contemporary Review* 21 (1873): 183–98.

Kristof, Ladis K. D. "The Nature of Frontiers and Boundaries." *Annals of the Association of American Geographers* 49 (1959): 269–82.

Kruger, Lorenz, Lorraine Daston, and Michael Heidelberger, eds. *The Probabilistic Revolution*. 2 vols. Cambridge, MA: MIT Press, 1987.

Kuhn, Thomas. "The Function of Measurement in Modern Physical Science." *Isis* 52 (1961): 161–93.

Kula, Witold. *Measures and Men*. Princeton, NJ: Princeton University Press, 1986.

Kumar, Deepak. "Calcutta: The Emergence of a Science City 1784–1856." *Indian Journal of the History of Science* 29 (1994): 1–8.

Kutzbach, Gisela. *The Thermal Theory of Cyclones: A History of Meteorological Thought in the Nineteenth Century*. Boston: American Meteorological Society, 1979.

Lamb, Herbert. *Historic Storms of the North Sea, British Isles and Northwest Europe*. Cambridge: Cambridge University Press, 1991.

Lamotte, Francoise, and Maurice Lantier. *Urbain LeVerrier: Savant universel, gloire nationale, personalité cotentine*. Coutances: OCEP, 1977.

Landow, George. *The Aesthetic and Critical Theories of John Ruskin*. Princeton, NJ: Princeton University Press, 1971.

————. "The Rainbow: A Problematic Image." In *Nature and the Victorian Imagination*, edited by U. C. Knoepflmacher and G. B. Tennyson, 341–69. Berkeley: University of California Press, 1977.

Laplace, Pierre Simon. *Essai philosophique sur les probabilités*. Paris: n.p., 1814.

Lawrence, Christopher. "Incommunicable Knowledge: Science, Technology and the Clinical Art in Britain 1850–1914." *Journal of Contemporary History* 20 (1985): 503–20.

Leach, Nicholas, ed. *British Life-boat Stations, Based on the Work of Graham Farr*. 4th ed. Welwyn Garden City: Life-boat Enthusiasts Society, 1993.

"Leading Principles of Meteorology." *British Quarterly* 3 (1845): 34–51.

Leighton, John. *Pictorial Beauties of Nature; or, Sketches in Various Departments of Natural History: With Coloured Illustrations by Famous Artists*. London: Ward, Lock, and Tyler, 1872.

Lenoir, Timothy. "Helmholtz and the Materialities of Communication." *Osiris* 9 (1994): 185-207.

Le Roy Ladurie, Emmanuel. *Times of Feast, Times of Famine: A History of Climate since the Year 1000*. London: George Allen and Unwin, 1972.

Lewin, Thomas. *Invasion of Britain by Julius Caesar*. London: Longman, Green, Longman and Roberts, 1859.

————. *Invasion of Britain by Julius Caesar*. 2nd ed. London: Longman, 1862.

Lewis, Charles Lee. *Matthew Fontaine Maury: The Pathfinder of the Seas*. Annapolis, MD: United States Naval Institute, 1927.

Ley, Clement. *Cloudland: A Study on the Structure and Character of Clouds*. London: E. Stanford, 1894.

————. "Cloud Photography." *Symons's Monthly Meteorological Magazine* 15 (1880): 54-55.

[Liefchild, John]. "Weather Forecasts and Storm Warnings." *Edinburgh Review* 74 (1866): 51–85.

Lightman, Bernard. "Astronomy for the People: R. A. Proctor and the Popularization of the Victorian Universe." In *Facets of Faith and Science*, edited by J. van de Meer, 31–45. Lanham, MD.: Pascal Center for Advanced Studies in Faith and Science, University Press of America, 1996.

————, ed. *Victorian Science in Context*. Chicago: University of Chicago Press, 1997.

————. "'The Voices of Nature': Popularizing Victorian Science." In *Victorian Science in Context*, 187–211. Chicago: University of Chicago Press, 1997.

Livingstone, David. *The Geographical Tradition: Episodes in the History of a Contested Enterprise*. Oxford: Oxford University Press, 1992.

————. "Tropical Climate and Moral Hygiene: The Anatomy of a Victorian Debate." *British Journal for the History of Science* 32 (1999): 93–110.

Livingstone, David N., D. G. Hart, and Mark A. Noll, eds. *Evangelicals and Science in Historical Perspective*. New York: Oxford University Press, 1999.

Loomis, Elias. "On Two Storms Which Were Experienced Throughout the United States in the Month of February, 1842." *Transactions of the American Philosophical Society* 9 (1846): 161–84.

Loudon, J. C. *Encyclopedia of Gardening, Comprising the Theory and Practice of Horticulture*. 3rd ed. London: A. Spottiswoode, 1835.

Lowe, E. J. "A Calendar of Nature." *Recreative Science* 1 (1859): 150–54.

Lubbock, John. *On the Senses, Instincts and Intelligence of Animals with Special Reference to Insects*. 2nd rev. ed. London: K. Paul, Trench, Trubner, 1899.

Lupton, Arthur. *Modern Prophecies; or, a Collection and Examination of Some of the Most Important Predictions Which Are Now Current*. London: Whitfield, Green and Sons, 1865.

————. *Modern Prophecies*. Re-issued by Harry Lupton. London: Whitfield, Green and Sons, 1887.

Lyon, Eileen Groth. *Politicians in the Pulpit: Christian Radicalism in Britain from the Fall of the Bastille to the Disintegration of Chartism*. Aldershot: Ashgate, 1999.

Maas, Harro. "An Instrument Can Make a Science: Jevons's Balancing Acts in Economics." *History of Political Economy* 33 (2001): 277–302.

————. "Of Clouds and Statistics: Inferring Causal Structures from the Data." *Measurements in Physics and Economics Discussion Paper Series*. DP MEAS 7/99. London: London School of Economics, 1999.

Macaulay, Thomas Babington. *Essays and Lays of Ancient Rome*. London: Longman, 1889.

————. "Review of von Ranke's *The Ecclesiastical and Political History of the Popes of Rome, During the Sixteenth and Seventeenth Centuries*." *Edinburgh Review* 72 (1840): 227–58.

Macgregor, David R. *The Tea Clippers: Their History and Development, 1833–1875*. 2nd ed. London: Conway Maritime Press, 1983.

MacKenzie, Donald. *Statistics in Britain: Social Construction of Scientific Knowledge*. Edinburgh: Edinburgh University Press, 1990.

MacKenzie, J., ed. *Imperialism and the Natural World*. Manchester: Manchester University Press, 1990.

MacLeod, Alexander. *A Man's Gift and Other Sermons with a Memorial Sketch by the Rev. A. G. Fleming and Personal Reminiscences by the Rev. Principal Fairbairn*. London: J. Nisbet, 1895.

————. *Scripture, Meteorology, and Modern Science*. Glasgow: George Baillie, 1867.

MacLeod, Roy. "Evolutionism, Internationalism, and Commercial Enterprise in Science: The International Scientific Series 1871–1910." In *Development of Science Publishing in Europe*, edited by A. J. Meadows, 63–93. Amsterdam: Elsevier Science Publishers, 1980.

————, ed. *Government and Expertise: Specialists, Administrators, and Professionals, 1860–1919*. Cambridge: Cambridge University Press, 1988.

————, ed. *Nature and Empire: Science and the Colonial Enterprise*. *Osiris* 15 (2001).

————. "The 'Practical Man': Myth and Metaphor in Anglo-Australian Science." *Australian Cultural History* 8 (1989): 24–49.

————. *Public Science and Public Policy in Victorian England*. Aldershot, UK: Variorum, 1996.

MacLeod, Roy, and Peter Collins, eds. *The Parliament of Science: The British Association for the Advancement of Science, 1831–1981*. Northwood, UK: Science Reviews, 1981.

Manchester, A. H. *A Modern Legal History of England and Wales 1750–1950*. London: Butterworths, 1980.

Marriott, William. "An Account of the Bequest of George James Symons to the Royal Meteorological Society." *Quarterly Journal of the Royal Meteorological Society* 27 (1901): 241–60.

————. "The Application of Photography to Meteorological Phenomena." *Quarterly Journal of the Royal Meteorological Society* 16 (1890): 146–52.

————. "The Earliest Telegraphic Daily Meteorological Reports and Weather Maps." *Quarterly Journal of the Royal Meteorological Society* 28 (1902): 123–31.

[Marshall, Frederic]. "Weather." *Blackwood's Edinburgh Magazine* 118 (November 1875): 611–28.

Martin, Robert M. *The Indian Empire: History, Topography, Geology, Climate, Population, etc.* 3 vols. London: London Printing and Publishing Co., 1858–61.

Martineau, Harriet. *Suggestions towards the Future Government of India*. London: Smith Elder, 1858.

Marvin, Carolyn. *When Old Technologies Were New: Thinking about Electric Communication in the Late Nineteenth Century*. New York: Oxford University Press, 1988.

Mathe-Shires, L. "Imperial Nightmares: The British Image of 'the Deadly Climate' of West Africa, c. 1840–74." *European Review of History* 8 (2001): 137–56.

Maury, Matthew Fontaine. *Physical Geography of the Sea*. New York: Harper, 1855.
McAlpin, Michelle B. *Subject to Famine: Food Crises and Economic Change in Western India 1860–1920*. Princeton, NJ: Princeton University Press, 1983.
McKee, Alexander. *The Golden Wreck*. New York: Morrow, 1862.
McLean, Ruari. *Victorian Book Design and Colour Printing*. 2nd ed. Berkeley: University of California Press, 1972.
Meadows, A. J. *Science and Controversy: A Biography of Sir Norman Lockyer*. Cambridge, MA: MIT Press, 1972.
Meldrum, Charles. "On Synoptic Charts of the Indian Ocean." *Symons's Monthly Meteorological Magazine* 3 (1868): 143–45.
Mellersh, H. E. L. *FitzRoy of the Beagle*. London: Rupert Hart Davis, 1968.
Merryweather, George. *An Essay Explanatory of the Tempest Progosticator in the Building of the Great Exhibition for the Works of Industry of All Nations*. London: John Churchill, 1851.
"Meteorology in India." *Calcutta Review* 52 (1871): 270–93.
"Meteorology: Its Progress and Practical Applications." *London Quarterly* 2 (1854): 127–50
Middleton, W. Knowles. *The History of the Barometer*. Baltimore: Johns Hopkins Press, 1968.
———. *A History of the Theories of Rain and Other Forms of Precipitation*. New York: F. Watts, 1966.
———. *Invention of the Meteorological Instruments*. Baltimore: Johns Hopkins Press, 1969.
Mill, Hugh Robert. "George James Symons." *Meteorological Magazine* 73 (1938): 164–69.
Mill, John Stuart. *Collected Works of John Stuart Mill*. Edited by John M. Robson. 33 vols. Toronto: University of Toronto Press, 1973.
Milne, Graeme J. *Trade and Traders in Mid-Victorian Liverpool: Mercantile Business and the Making of a World Port*. Liverpool: Liverpool University Press, 2000.
Mirowksi, Philip. "Macroeconomic Instability and 'Natural' Processes in Early Neoclassical Economics." *Journal of Economic History* 44 (1984): 345–54.
Monmonier, Mark. *Air Apparent: How Meteorologists Learned to Map, Predict and Dramatize Weather*. Chicago: University of Chicago Press, 1999.
Moon, Paul. *FitzRoy: Governor in Crisis 1843–45*. Auckland, NZ: David Ling, 2000.
Moore, R. J. "India and the British Empire." In *British Imperialism in the Nineteenth Century*, edited by C. C. Eldridge, 64–84. New York: St Martin's Press, 1984.
Morrell, Jack, and Arnold Thackray. *Gentlemen of Science: Early Correspondence of the British Association for the Advancement of Science*. London: Royal Historical Society, University College London, 1984.
———. *Gentlemen of Science: Early Years of the British Association for the Advancement of Science*. Oxford: Oxford University Press, Clarendon Press, 1981.
Morris, Albert T. *A Treatise on Meteorology*. Edinburgh: R. Grant, 1866.
Morus, Iwan. *Frankenstein's Children: Electricity, Exhibition, and Experiment in Early-Nineteenth-Century London*. Princeton, NJ: Princeton University Press, 1998.
Morus, Iwan, Simon Schaffer, and James A. Secord. "Scientific London." In *London: World City*, edited by C. Fox, 129–42. New Haven, CT: Yale University Press in association with the Museum of London, 1992.
Moyer, Albert. *Joseph Henry: Rise of an American Scientist*. Washington, DC: Smithsonian Institution Press, 1997.
Mozley, J. B. *Eight Lectures on Miracles Preached before the University of Oxford in the Year 1865*. 3rd ed. London: Rivington, 1872.
Muir, Percy. *Victorian Illustrated Books*. London: B. T. Batsford, 1971.
Mullin, Robert Bruce. "Science, Miracles, and the Prayer-Gauge Debate." In *When Science and Christianity Meet*, edited by David C. Lindberg and Ronald L. Numbers, 203–24. Chicago: University of Chicago Press, 2003.

Murphy, Patrick. *The Anatomy of Seasons: Weather Guide Book, and Perpetual Companion to the Almanac*. London: J. B. Bailliere, 1834.
————. *An Inquiry into the Nature and Causes of Miasmata*. London: n.p., 1825.
————. *Meteorology, Considered in Its Connexion with Astronomy, Climate and the Geographical Distribution of Animals and Plants, Equally as with the Seasons and Changes of the Weather*. London: J. B. Balliere, 1836.
————. *Rudiments of the Primary Forces of Gravity, Magnetism and Electricity, in Their Agency on the Heavenly Bodies*. London: Whittaker, Treacher, 1830.
Myers, Robin. *The Stationers' Company Archives: An Account of the Records, 1554–1984*. Winchester, UK: St. Paul's Bibliographies, 1990.
Nalbach, Alex. "'The Software of Empire': Telegraphic News Agencies and Imperial Publicity, 1865–1914." In *Imperial Co-Histories: National Identities and the British and Colonial Press*, edited by Julie F. Codell, 46–67. Madison, NJ: Fairleigh Dickinson University Press, 2003.
Nebecker, Frederik. *Calculating the Weather: Meteorology in the 20th Century*. San Diego: Academic Press, 1995.
Negretti, Henry, and Joseph Warren Zambra. *Negretti and Zambra's Encyclopaedic Illustrated and Descriptive Catalogue of Optical, Mathematical, Philosophical, Photographic and Standard Meteorological Instruments*. London: Hayman and Lilly, 1870–79.
————. *A Treatise on Meteorological Instruments: Explanatory of Their Scientific Principles, Method of Construction and Practical Utility*. London: Negretti and Zambra, 1864.
"The New Code of Signals." *Mercantile Magazine* 4 (1857): 148–50.
Newman, Francis. *Controversy about Prayer*. London: Thomas Scott, 1873.
Noakes, Richard. "Cranks and Visionaries: Science, Spiritualism and Transgression in Victorian Britain." PhD diss., Cambridge University, 1998.
"Norman Robert Pogson." *Monthly Notices of the Royal Astronomical Society* 52 (1892): 235–38.
"Notes on British Storms." *Leisure Hour* 13 (1864): 181–85.
"Obituary: Frederic Chambers." *Monthly Notices of the Royal Astronomical Society* 76 (1915–16): 258–59.
"Obituary: Henry Francis Blanford." *Proceedings of the Royal Society of London* 54 (1893): xii–xix.
Olesko, Kathryn M. "The Meaning of Precision: The Exact Sensibility in Early Nineteenth Century Germany." In *Values of Precision*, edited by M. N. Wise, 103–44. Princeton, NJ: Princeton University Press, 1995.
Oliphant, Margaret. *The Life of Edward Irving, Minister of the National Scotch Shurch, Illustrated by His Journals and Correspondence*. 2 vols. London: Hurst and Blackett, 1862.
Oliver, J. Westwood. *Sunspottery*. London: n.p., 1883.
Oliver, W. H. *Prophets and Millenialists: The Uses of Biblical Prophecy in England from the 1790s to the 1840s*. Auckland: Auckland University Press, 1978.
"On Divination and on Prophecy, as Distinct from Divination and Superior in Its Nature, Origin and End." *The Yorkshireman* 4 (1836): 158–60, 267–72.
Ophir, Adi, and Steven Shapin. "The Place of Knowledge: A Methodological Survey." *Science in Context* 4 (1991): 3–21.
"Opinion de M. Biot sur les observatories météorologiques permanents que l'on propose d'etablir en divers point de l'Algerie." *Compte rendu des séances de l'academie des sciences* 61 (1855): 1153–90.
Oppenheim, Janet. *"Shattered Nerves": Patients, Doctors and Depression in Victorian England*. New York: Oxford University Press, 1991.
Orange, A. D. "The Idols of the Theatre: The British Association and Its Early Critics." *Annals of Science* 32 (1975): 277–94.
Ormerod, Eleanor A., ed. *The Cobham Journals, Abstracts and Summaries of Meteorological and*

Phrenological Observations Made by Miss Caroline Molesworth at Coham Surrey in the Years 1825 to 1850. London: Edward Stanford, 1880.
Ospovat, Don. "Lyell's Theory of Climate." *Journal of the History of Biology* 10 (1977): 317–39.
Outram, Dorinda. "New Spaces in Natural History." In *Cultures of Natural History*, edited by J. A. Secord, N. Jardine, and E. Spary, 249–65. Cambridge: Cambridge University Press, 1996.
Pang, Alex Soojung-Kim. "Social Event of the Season: Solar Eclipse Expeditions and Victorian Culture." *Isis* 84 (1993): 252–77.
———. "Victorian Observing Practices: Printing Technology and Representations of the Solar Corona (2) The Age of Photomechanical Reproduction." *Journal of the History of Astronomy* 26 (1995): 63–75.
———. "Visual Representation and Post-constructivist History of Science." *Historical Studies in the Physical and Biological Sciences* 28 (1997): 139–71.
Parsons, William (Earl of Rosse). "Address Delivered before the Royal Society." *Abstracts of the Papers Communicated to the Royal Society of London* 5 (1843–50): 857–95.
———. "Address Delivered before the Royal Society." *Abstracts of the Papers Communicated to the Royal Society of London* 6 (1850–54): 343–72.
Pattison, Mark, ed. *Essays on the Endowment of Research*. London: H. S. King, 1876.
Payne Smith, Robert. *Prophecy: A Preparation for Christ*. London: Macmillan, 1869.
Pearce, Alfred J. *Textbook of Astrology*. London: Cousins, 1879.
———. *The Weather Guide Book: A Concise Exposition of Astronomic-Meteorology*. London: Simpkin and Marshall, 1864.
Pearson, Karl. *The Life, Letters and Labours of Francis Galton*. 3 vols. in 4. Cambridge: Cambridge University Press, 1914–30.
Peart, Sandra J. "Sunspots and Expectations: W. S. Jevons' Theory of Economic Fluctuations." *Journal of the History of Economic Thought* 13 (1991): 243–65.
Perkins, Maureen. *Visions of the Future: Almanacs, Time and Cultural Change: 1775–1870*. Oxford: Clarendon Press, 1996.
Peterson, Thomas F., Jr. "The Zealous Marketing of Rain-Band Spectroscopes." *Rittenhouse: Journal of the American Scientific Instrument Enterprise* 7 (1993): 91–96.
Philindus. "Famines and Floods in India." *Macmillan's Magazine* 37 (1878): 236–56.
"The Philosophical Limits of Prediction." *Dublin University Magazine* 3 (1879): 404–11, 526–38.
Piddington, Henry. *Horn-book of Storms for the India and China Seas*. Calcutta: Bishop's College Press, 1844.
Pinsel, Marc. "The Wind and Current Chart Series Produced by Matthew Fontaine Maury." *Navigation* 28 (1981): 123–36.
Plant, Thomas. *Meteorology: Its Study Important for Our Good; for the Prevention of Loss of Life and Property from Storms and Floods*. Birmingham: privately printed, 1862.
Poey y Aguirre, Andres. *Comment on observe les nuages pour prevoir le temps*. 3rd ed. Paris: Gauthier-Villars, 1879.
Poovey, Mary. *A History of the Modern Fact: Problems of Knowledge in the Sciences of Wealth and Society*. Chicago: University of Chicago Press, 1998.
Popper, Karl. "Of Clouds and Clocks." In *Objective Knowledge: An Evolutionary Approach*, 206–56. Rev. ed. New York: Oxford University Press, 1979.
Porter, Theodore. "Precision and Trust: Early Victorian Insurance and the Politics of Calculation." In *Values of Precision*, edited by M. Norton Wise, 173–97. Princeton, NJ: Princeton University Press, 1994.
———. *The Rise of Statistical Thinking*. Princeton, NJ: Princeton University Press, 1986.
———. *Trust in Numbers: The Pursuit of Objectivity in Science and Public Life*. Princeton, NJ: Princeton University Press, 1995.

Pounds, Norman. "The Origin of the Idea of Natural Frontiers in France." *Annals of the Association of American Geographers* 41 (1951): 146–57.
Powell, Baden. *The Order of Nature Considered in Reference to the Claims of Revelation*. London: Longman, 1859.
Poynting, John Henry. "A Comparison of the Fluctuations of the Price of Wheat and in the Cotton and Silk Imports into Great Britain." *Journal of the Royal Statistical Society* 47 (1884): 34–74.
Prakash, Gyan. *Another Reason: Science and the Imagination of Modern India*. Princeton, NJ: Princeton University Press, 1999.
Pratt, Frederic. "Modern Meteorology." *Meteorology Magazine* 1 (1864): 31–35
Prescott, J. R. V. *The Geography of Frontiers and Boundaries*. Chicago: Aldine, 1965.
Proctor, Richard. *Wages and Wants of Science-Workers*. London: Smith & Elder, 1876.
Prout, William. *Chemistry, Meteorology, and the Function of Digestion*. London: William Pickering, 1834.
Prouty, Roger W. *The Transformation of the Board of Trade, 1830–1855: A Study of Administrative Reorganization in the Heyday of Laissez Faire*. London: Heinemann, 1957.
Pyenson, Lewis. *Civilizing Mission: Exact Sciences and French Overseas Expansion, 1830–1940*. Baltimore: Johns Hopkins University Press, 1993.
Quetelet, Adolphe. *Letters Addressed to the Grand Duke of Saxe Coburg and Gotha on the Theory of Probability as Applied to the Moral and Political Sciences*. Translated by O. G. Downes. London: Charles and Edwin Layton, 1849.
———. *Sciences mathematiques et physiques chez les Belges au commencement du XIXe siecle*. Brussels: H. Thiry-Van Buggenhoudt, 1866.
Raper, Henry. *The Practice of Navigation and Nautical Astronomy*. London: R. B. Bate, 1840.
———. *The Practice of Navigation and Nautical Astronomy*. 2nd ed. London: R. B. Bate, 1841.
Reardon, B. M. G. *Religious Thought in the Victorian Age*. 2nd ed. London: Longman, 1995.
Reed, Arden. *Romantic Weather: The Climates of Coleridge and Baudelaire*. Hanover, NH: University Press of New England, 1983.
Rees, Graham L. *Britain's Commodity Markets*. London: Paul Elek, 1972.
"Report of the Meteorological Committee to the President and Council of the Royal Society on the Work Done in the Meteorological Office since Their Appointment in 1866 to December 31, 1875." *Proceedings of the Royal Society* 24 (1876): 189–210
Report of the Proceedings of the Meteorological Congress at Vienna: Protocols and Appendices. London: HMSO, 1873.
Report on the Administration of the Meteorological Department in Western India [for the years 1879–80, 1880–81]. Bombay, 1879–81.
Report on the Administration of the Meteorological Department of the Government of India [for the years 1875–76 through 1890–91]. Calcutta, 1876–91.
Report on Weather Telegraphy and Storm Warnings, Presented to the Meteorological Congress at Vienna by a Committee Appointed at the Leipzig Conference. London: E. Stanford for HMSO, 1873.
Richards, Joan. "The Probable and the Possible in Early Victorian England." In *Victorian Science in Context*, edited by Bernard Lightman, 51–71. Chicago: University of Chicago Press, 1997.
Richardson, Benjamin Ward. *Thomas Sopwith, with Excerpts from His Diary of Fifty-Seven Years*. London: Longman and Green, 1891.
Richardson, Lewis Fry. *Weather Prediction by Numerical Process*. New York: Dover Publications, 1965.
Ridder, A. de. *150 ans de météorologie en Belgique: 1833–1983*. Brussels: Institut Royal Météorologique de Belgique, 1984.
Ritchie, G. S. *The Admiralty Chart: British Naval Hydrography in the Nineteenth Century*. New York: American Elsevier Publishing, 1967.

Robinson, Arthur H. *Early Thematic Cartography in the History of Cartography*. Chicago: University of Chicago Press, 1982.

———. *Nature of Maps: Essays towards Understanding Maps and Mapping*. Chicago: University of Chicago Press, 1976.

Robinson, Arthur H., and Helen Wallis. *Cartographical Innovations: An International Handbook of Mapping Terms to 1900*. Tring, UK: Map Collector Publications in association with the International Cartographic Association, 1987.

———. "Humboldt's Map of Isothermal Lines: A Milestone in the History of Thematic Cartography." *Cartographic Journal* 2 (1967): 119–23.

Robson, John M. "Textual Introduction [to *A System of Logic*]." In *Collected Works of John Stuart Mill*, edited by John M. Robson, 7:xlvix–cviii. 33 vols. Toronto: University of Toronto Press, 1973.

Rodgers, Frederic. "The Integrity of the British Empire." In *Empire and Imperialism*, edited by P. J. Cain, 121–33. South Bend, IN: St. Augustine Press, 1999.

Rogers, William. *Murphy's Weather Almanac, a Farce in One Act*. London: W. Strange, Turner and Fisher, 1838.

Rogerson, J. W. *The Bible and Criticism in Victorian Britain: Profiles of F. D. Maurice and William Robertson Smith*. Sheffield: Sheffield Academic Press, 1995.

———. *Old Testament Criticism in the Nineteenth Century: England and Germany*. London: Society for the Promotion of Christian Knowledge, 1984.

Romanes, G. *Animal Intelligence*. London: Kegan Paul, 1882.

Rose, H. J. *The Question Why Should We Pray for Fair Weather Answered*. Oxford and London: J. H. and James Parker, 1860.

Roswadowski, Helen. "Fathoming the Ocean: Discovery and Exploration of the Deep Sea, 1840–1880." PhD diss., University of Pennsylvania, 1996.

Rothermel, Holly. "Images of the Sun: De la Rue, Airy and Celestial Photography." *British Journal for the History of Science* 26 (1993): 137–69.

Rouse, Joseph. "What Are Cultural Studies of Knowledge?" *Configurations* 1 (1993): 1–22.

Rudwick, Martin. "The Emergence of a Visual Language for Geological Science 1760–1840." *History of Science* 14 (1976): 149–95.

———. *The Meaning of Fossils: Episodes in the History of Paleontology*. New York: Science History Publications, 1976.

Rupke, Nicolaas A. "Introduction." In *Cosmos: A Sketch of a Physical Description of the Universe*. Vol. 1, edited by N. A. Rupke, 3–27. Baltimore: Johns Hopkins University Press, 1997.

———. *Science, Politics, and the Public Good: Essays in Honour of Margaret Gowing*. London: Macmillan Press, 1988.

Ruskin, John. "A Caution to Snakes." In *The Library Edition of the Works of John Ruskin*, edited by E. Cook and A. Wedderburn, 26:295–332. London: Longman, 1903–12.

———. *Eagle's Nest*. In *The Library Edition of the Works of John Ruskin*, edited by E. Cook and A. Wedderburn, 22:113–287. London: Longman, 1903–12.

———. *The Library Edition of the Works of John Ruskin*. Edited by E. Cook and A. Wedderburn. 34 vols. London: Longman, 1903–12.

———. "The Nature and Authority of Miracle." *Contemporary Review* 21 (1872–73): 627–42.

———. *Queen of the Air*. In *The Library Edition of the Works of John Ruskin*, edited by E. Cook and A. Wedderburn, 19:282–426. London: Longman, 1903–12.

———. *The Storm-Cloud of the Nineteenth Century*. In *The Library Edition of the Works of John Ruskin*, edited by E. Cook and A. Wedderburn, 34:xxi–80. London: Longman, 1903–12.

Rusnock, Andrea. "Correspondence Networks and the Royal Society 1700–1750." *British Journal for the History of Science* 32 (1999): 155–69.

Ryle, J. C. *Coming Events and Present Duties, Being Miscellaneous Sermons on Prophetical Subjects.* London: William Hunt and Co., 1867.

Sabine, Edward. "Note on a Correspondence." *Proceedings of the Royal Society of London* 15 (1866): 29–38.

———. "Results of the First Year's Performance of the Photographical Self-Recording Meteorological Instruments at the Central Observatory of the British System of Meteorological Observations." *Proceedings of the Royal Society of London* 18 (1869–70): 3–12.

Said, Edward W. *Orientalism.* New York: Random House, 1994.

Sangwan, Satpal. "From Gentlemen Amateurs to Professionals: Reassessing the Natural Science Tradition in Colonial India 1780–1840." In *Nature and the Orient*, edited by R. Grove, V. Damodran, and S. Satpal, 210–36. Delhi: Oxford University Press, 1998.

———. "Reordering the Earth: The Emergence of Geology as Scientific Discipline in Colonial India." *Earth Sciences History* 12 (1993): 224–33.

Sargent, Frederick. *Hippocratic Heritage: A History of Ideas about Weather and Human Health.* New York: Pergammon Press, 1982.

Saxby, Stephen M. "The Coming Winter and the Weather." *Nautical Magazine* 31 (1861): 664–71.

———. *Foretelling the Weather: Being a Description of a Newly Discovered Lunar Weather-System.* London: Longman, 1862.

———. "Lunar Equinoctials—The Barometer at Fault." *Nautical Magazine* 29 (1860): 482–84.

———. *Saxby's Weather System; or, Lunar Influence on Weather.* London: Longman, 1864.

———. "Weather Warnings and a Great Day Auroral Storm." *Nautical Magazine* 32 (1863): 146–55.

Schaaf, Larry. *Out of the Shadows: Herschel, Talbot and the Invention of Photography.* New Haven, CT: Yale University Press, 1992.

———. "Piazzi Smyth at Teneriffe." *History of Photography* 4 (1980): 289–309; 5 (1981): 27–50.

Schabas, Margaret. *A World Ruled by Number: William Stanley Jevons and the Rise of Mathematical Economics.* Princeton, NJ: Princeton University Press, 1990.

Schaffer, Simon. "Astronomers Mark Time: Discipline and the Personal Equation." *Science in Context* 2 (1988): 115–45.

———. "Late Victorian Metrology and Its Instrumentation." In *Invisible Connections: Instruments, Institutions, and Science*, edited by R. Bud and S. Cozzens, 23–50. Bellingham, WA: SPIE Press, 1992.

———. "Metrology, Metrication, and Victorian Values." In *Victorian Science in Context*, 438–74. Chicago: University of Chicago Press, 1997.

———. "Natural Philosophy and Public Spectacle in the Eighteenth Century." *History of Science* 21 (1983): 1–43.

———. "On Astronomical Drawing." In *Picturing Science, Producing Art*, edited by P. Galison and C. Jones, 441–74. New York: Routledge, 1998.

Schlagintweit, Hermann de. "Numerical Elements of Indian Meteorology [Series 1]." *Philosophical Transactions of the Royal Society* 153 (1863): 525–42.

———. "Numerical Elements of Indian Meteorology: 2. Insolation and Its Connexion with Atmospheric Moisture." *Proceedings of the Royal Society of London* 14 (1865): 111–22.

———. "Numerical Elements of Indian Meteorology: 3. Temperatures of the Atmosphere and Isothermal Profiles of High Asia." *Proceedings of the Royal Society of London* 14 (1865): 547–48.

Schramm, Jan-Melissa. *Testimony and Advocacy in Victorian Law, Literature and Theology.* Cambridge: Cambridge University Press, 2000.

Schuster, J. A., and Richard Yeo, eds. *The Politics and Rhetoric of Scientific Method.* Dordrecht: Reidel, 1986.

The Science of the Weather in a Series of Letters and Essays. Glasgow: Laidlow, 1867

Scott, Douglas, ed. *Luke Howard: His Correspondence with Goethe and His Continental Journey of 1816*. York: William Sessions, 1976.
Scott, Robert H. "Forecasting the Weather." *Good Words* 22 (1881): 451–56, 565–70.
————. "The History of the Kew Observatory." *Proceedings of the Royal Society* 39 (1885): 37–86.
————. "On the Connexion between Oppositely Disposed Currents of Air and the Weather Subsequently Experienced in the British Islands." *Proceedings of the Royal Society of London* 18 (1869–70): 12–15.
[————]. "The Weather and Its Prediction." *Quarterly Review* 148 (1879): 489–500.
————. "Weather Charts in the Newspapers." *Journal of the Royal Society of Arts* 23 (1875): 776–82.
————. "A Word about the Weather." *Chambers' Edinburgh Journal* 51 (1874): 511.
Secord, Anne. "'Be What You Would Seem to Be': Samuel Smiles, Thomas Edward and the Making of a Working Class Scientific Hero." *Science in Context* 16 (2003): 147–73.
————. "Botany on a Plate: Pleasure and the Power of Pictures in Promoting Early Nineteenth-Century Scientific Knowledge." *Isis* 93 (2002): 28–57.
————. "Corresponding Interests: Artisans and Gentlemen in Nineteenth-Century Natural History." *British Journal for the History of Science* 27 (1994): 383–408.
————. "Science in the Pub: Artisan Botanists in Early Nineteenth-Century Lancashire." *History of Science* 32 (1994): 269–315.
Secord, James. "Extraordinary Experiment: Electricity and the Creation of Life in Victorian England." In *Uses of Experiment: Studies in the Natural Sciences*, edited by David Gooding et al., 337–83. Cambridge: Cambridge University Press, 1989.
————. *Victorian Sensation: The Extraordinary Publication, Reception, and Secret Authorship of Vestiges of the Natural History of Creation*. Chicago: University of Chicago Press, 2000.
Seiberling, Grace, and Carolyn Bloore. *Amateurs, Photography, and the Mid-Victorian Imagination*. Chicago: University of Chicago Press, 1986.
Shapiro, Barbara. *"Beyond Reasonable Doubt" and "Probable Cause": Historical Perspectives on the Anglo-American Law of Evidence*. Berkeley: University of California Press, 1991.
Shattock, Joanne, and Michael Wolff. *Victorian Periodical Press: Samplings and Soundings*. Toronto: University of Toronto Press, 1982.
Shaw, W. Napier. *Manual of Meteorology*. Vol. 1: *Meteorology in History*. Cambridge: Cambridge University Press, 1926.
Sheets-Pyenson, Susan. "Popular Science Periodicals in Paris and London: The Emergence of a Low Scientific Culture 1820–1875." *Annals of Science* 42 (1985): 549–72.
Sheynin, O. B. "On the History of the Statistical Method in Meteorology." *Archive for History of Exact Sciences* 31 (1984): 53–95.
Shortland, Michael, ed. *Science and Nature: Essays in the History of the Environmental Sciences*. Stanford in the Vale, UK: British Society for the History of Science, 1990.
Shortland, Michael, and Bruce Somerville. "Thomas Henry Huxley, H. G. Wells and the Method of Zadig." In *Thomas Henry Huxley's Place in Science*, edited by Alan P. Barr, 296–322. Athens: University of Georgia Press, 1997.
Sivasundaram, Sujit. "The Periodical as Barometer: Spiritual Measurement and the Evangelical Magazine." In *Science and Culture in the Nineteenth Century Media*, edited by L. Henson and J. Topham, 43–56. London: Ashgate, 2004.
Skrine, Francis Henry. *Life of Sir William Wilson Hunter*. London: Longman, 1901.
Slamon, Wesley C. "John Venn's Logic of Chance." In *Probabilistic Thinking, Thermodynamics and the Interaction of the History and Philosophy of Science*. Vol 2: *Ideas in History*, edited by J. Hintikka, D. Gruender, and E. Agazzi, 125–38. Dordrecht, Netherlands: Reidel, 1978.
Smith, Crosbie, and M. Norton Wise. *Energy and Empire: A Biographical Study of Lord Kelvin*. Cambridge: Cambridge University Press, 1989.

Smith, H. T. "Marine Meteorology: Its History and Progress." *Marine Observer* 2 (1925): 33–35, 90–92, 173–75.

Smith, Lindsay. *Victorian Photography, Painting, and Poetry: The Enigma of Visibility in Ruskin, Morris and the Pre-Raphaelites*. Cambridge: Cambridge University Press, 1995.

Smith, Robert W. "A National Observatory Transformed: Greenwich in the Nineteenth Century." *Journal for the History of Astronomy* 22 (1991): 5–20.

Smith, William Robertson. *Lectures and Essays*. Edited by J. S. Black and G. W. Chrystal. London: A & C Black, 1912.

————. "On Prophecy." In *Lectures and Essays*, edited by J. S. Black and G. W. Chrystal, 341–66. London: A & C Black, 1912.

————. "Prophet." *Encyclopaedia Britannica*, 9th ed. 1885. 9:814–21.

Smyth, Charles Piazzi. *Cloud Forms That Have Been; to the Glory of God Their Creator and the Wonderment of Learned Men as Recorded by Instant Photographs Taken at Clova, [and] Ripon in 1894 and 1895*. 3 vols. N.p., 1895.

————. "Colour in Practical Astronomy, Spectroscopically Examined." *Transactions of the Royal Society of Edinburgh* 28 (1879): 779–804.

————. "Gaseous Spectra in Vacuum Tubes under Small Dispersion and at Low Temperatures." *Transactions of the Royal Society of Edinburgh* 30 (1883): 93–149.

————. "Hyperborean Storm of the 2nd and 3rd October 1860." *Edinburgh Astronomical Observations* 13 (1871): T83–T141.

————. *Madeira Meteorologic: Being a Paper on the Above Subject Read before the Royal Society, Edinburgh on 1st of May 1882*. Edinburgh: D. Douglas, 1883.

————. "A Memoir on Astronomical Drawing." *Monthly Notices of the Royal Astronomical Society* 5 (1843): 277–79.

————. "Meteorological Spectroscopy in the Small and Rough." *Edinburgh Astronomical Observations* 14 (1870–77): 27–42.

————. "Rainband Spectroscopy." *Journal of the Scottish Meteorological Society* 5 (1878): 84–97.

————. "Spectroscopy, Phenomenal, Meteorological and Solar." *Edinburgh Astronomical Observations* 14 (1877): 27–42.

————. *Teneriffe: An Astronomer's Experiment; or, Specialities of a Residence above the Clouds*. London: L. Reeve, 1858.

Smyth, William H. "Taormina, Sicily." In *The Sea Chart: An Historical Survey*, edited by Michael Sanderson and Derek Howse, 120–21. New York: McGraw Hill, 1973.

"Spirit of the Annual Periodicals." *Mirror of Literature* 4 (1824): 402–10.

Srivastara, Hari Shanker. *The History of Indian Famines and Development of Famine Policy 1853–1918*. Agra, India: Sri Ram Mehra and Co., 1968.

Stafford, Barbara. *Voyage into Substance: Art, Science, Nature and the Illustrated Travel Account 1760–1840*. Cambridge, MA: MIT Press, 1984.

Stafford, Robert A. "Annexing the Landscapes of the Past: British Imperial Geology in the Nineteenth Century." In *Imperialism and the Natural World*, edited by J. MacKenzie, 67–89. Manchester: Manchester University Press, 1990.

————. *Scientist of Empire: Sir Roderick Murchison, Scientific Exploration and Victorian Imperialism*. Cambridge: Cambridge University Press, 1989.

Standage, Tom. "Everyone Complains about the Weather . . . Piers Corbyn Is Doing Something about It." *Wired*, December 4, 2001. http://www.wired.com/wired/archive/7.02/weather.html

————. *The Neptune File: Planet Detectives and the Discovery of Worlds Unseen*. London: Penguin, 2000.

Steinmetz, Andrew. *A Manual of Weathercasts and Storm Prognostics on Land and Sea; or, The Signs Whereby to Judge the Coming Weather Adapted for All Countries*. London: Routledge, 1866.

———. *Sunshine and Showers: A Compendium of Popular Meteorology*. London: Reeve and Co., 1867.

Stenton, Michael, ed. *Who's Who of British Members of Parliament*. Vol. 1: *1832–1882*. Atlantic Highlands, NJ: Humanities Press, 1976.

Stewart, Balfour. "An Account of Certain Experiments on Aneroid Barometers, Made at Kew Observatory." *Proceedings of the Royal Society of London* 16 (1867–68): 472–81.

Stewart, Balfour, and Norman Lockyer. "The Place of Life in a Universe of Energy." *Macmillan's Magazine* 18 (1868): 319–27.

———. "The Sun as Type of the Material Universe." *Macmillan's Magazine* 18 (1868): 246–57.

Stewart, Balfour, and Peter G. Tait. *The Unseen Universe; or, Physical Speculations on a Future State*. London: Macmillan, 1875.

Stewart, Larry. *The Rise of Public Science: Rhetoric, Technology and Natural Philosophy in Newtonian Britain, 1660–1750*. Cambridge: Cambridge University Press, 1992.

Stigler, Steven M. *The History of Statistics: The Measurement of Uncertainty before 1900*. Cambridge, MA: Belknap Press, 1986.

Storr, Vernon F. *The Development of English Theology in the Nineteenth Century 1800–1860*. London: Longman, 1913.

Strachey, Richard. "Introductory Lecture on Scientific Geography." *Proceedings of the Royal Geographical Society* 20 (1875–76): 179–203.

Strong, John V. "The Infinite Ballot Box of Nature." *Proceedings of the Philosophy of Science Association* 1 (1976): 197–211.

Sulivan, H. Norton. *Life and Letters of the Late Admiral Sir Bartholomew James Sulivan*. London: John Murray, 1896.

Sutton, Michael. "Herschel and the Development of Spectroscopy." *British Journal for the History of Science* 7 (1974): 42–60.

Sykes, William H. "Discussion of Meteorological Observations Taken in India, at Various Heights, Embracing Those at Dodabetta on the Neelgherry Mountains, at 8640 Feet Above the Level of the Sea." *Philosophical Transactions of the Royal Society of London* 140 (1850): 297–378.

Symons, George J. "British Association 1868." *Symons's Monthly Meteorology Magazine* 3 (1868): 146–47.

———. "Cloud Atlases." *Symons's Monthly Meteorology Magazine* 24 (1889): 160.

———. "Daily Atlantic Weather Maps." *Symons's Monthly Meteorological Magazine* 14 (1879): 48–53.

———. "History of the English Meteorological Societies." *Quarterly Journal of the Royal Meteorological Society* 7 (1881): 66–107.

———. "Indian Meteorology." *Symons's Monthly Meteorology Magazine* 13 (1878): 161–67.

———. "International Cloud Atlas." *Symons's Monthly Meteorological Magazine* 24 (1889): 160.

———. "Meteorology at Dundee." *Symons's Monthly Meteorological Magazine* 2 (1866): 99–105, 114–18.

———. "On the Climates of the Various British Colonies." *Symons's Monthly Meteorological Magazine* 12 (1877): 55–60.

———. *Rain: How When Where and Why It Is Measured: Being a Popular Account of Rainfall Investigations*. London: Edward Standford, 1867.

———. "Reviews: *Cloudland*." *Symons's Monthly Meteorological Magazine* 31 (1896): 81–82.

Szreter, Simon. "The General Register Office and the Public Health Movement in Britain, 1837–1914." *Social History of Medicine* 4 (1991): 435–63.

Taub, Liba. *Ancient Meteorology*. London: Routledge, 2003.

Taylor, Isaac. *Natural History of Enthusiasm*. New York: J. Leavitt, 1831.

"Telegraph." *Encyclopaedia Britannica*. 11th ed. 1910–11. 26:510–29.

Temple, Frederick. *The Relations between Religion and Science*. London: Macmillan, 1884.

Temple, Frederick, et al. *Essays and Reviews*. London: J. W. Parker, 1860.
Thom, Alexander. *An Inquiry into the Nature and Course of Storms in the Indian Ocean South of the Equator*. London: Smith, Elder and Co., 1845.
Thomas, Morley. *The Beginnings of Canadian Meteorology*. Toronto: ECW Press, 1991.
Thompson, C. Halford. "American Storm-Warnings." *Gentleman's Magazine* 23 (1879): 597–615.
[Thompson, Henry]. "The 'Prayer for the Sick.'" *Contemporary Review* 20 (1872): 205–10.
Thomson, William. *A Change of Air: Climate and Health*. London: A and C Black, 1979.
Thrower, Norman J. *Maps and Civilization: Cartography in Culture and Society*. Chicago: University of Chicago Press, 1996.
Todhunter, Isaac. *William Whewell . . . An Account of His Writings with Selections from His Literary and Scientific Correspondence*. 2 vols. London: Macmillan, 1876.
Topham, Jonathan. "Beyond the 'Common Context': The Production and Reading of the Bridgewater Treatises." *Isis* 89 (1998): 233–62.
———. "Science and Popular Education in the 1830s: The Role of the Bridgewater Treatises." *British Journal for the History of Science* 15 (1992): 397–430.
Topper, David. "Natural Science and Visual Art: Reflections on the Interface." In *Beyond History of Science: Essays in Honor of Robert E. Schofield*, edited by E. Garber, 296–310. Cranbury, NJ: Associated University Presses, 1990.
Tucker, Jennifer. "Photography as Witness, Detective, and Impostor: Visual Representation in Victorian Science." In *Victorian Science in Context*, edited by B. Lightman, 378–408. Chicago: University of Chicago Press, 1997.
———. "Voyages of Discovery on Oceans of Air: Scientific Observation and the Image of Science in an Age of 'Balloonacy.'" *Osiris* 11 (1996): 144–76.
Turner, Frank. *Between Science and Religion: The Reaction to Scientific Naturalisn in Later Victorian England*. New Haven, CT: Yale University Press, 1974.
———. "Public Science in Britain, 1880–1919." *Isis* 71 (1980): 589–608.
———. "Rainfall, Plagues, and the Prince of Wales." In *Contesting Authority: Essays in Victorian Intellectual Life*, 151–70. Cambridge: Cambridge University Press, 1993.
———. "The Victorian Conflict between Science and Religion: A Professional Development." *Isis* 69 (1978): 356–76.
Turner, Gerard L. *Nineteenth Century Scientific Instruments*. Berkeley: Sotheby Publications/ University of California, 1983.
———. *Scientific Instruments and Experimental Philosophy, 1550–1850*. Aldershot, UK: Variorum, 1990.
———. *Scientific Instruments, 1500–1900: An Introduction*. London: Philip Wilson; Berkeley: University of California Press, 1998.
Tweyman, Stanley, ed. *Hume on Miracles*. Bristol: Thoemmes Press, 1996.
Twining, William. *Theories of Evidence: Bentham and Wigmore*. London: Weidenfeld and Nicolson, 1985.
Tyndall, John. "Action of Free Molecules on Radiant Heat and to Conversion Thereby into Sound." *Philosophical Transactions of the Royal Society of London* 173 (1882): 291–354.
———. "Reflections on Prayer and Natural Law." In *Fragments of Science*, 1–7. New York: Appleton, 1898.
Venn, John. *Logic of Chance*. 4th ed. New York: Chelsea, 1962.
Vincent, David. *Literacy and Popular Culture: England 1750–1914*. Cambridge: Cambridge University Press, 1989.
Vogel, Julius. "Greater or Lesser Britain." In *Empire and Imperialism: The Debate of the 1870s*, edited by P. Cain, 77–104. Bristol: Thoemmes Press, 1999.
Walker, J. Malcolm. "The Meteorological Societies of London." *Weather* 48 (1993): 364–72.
———. "Pen Portraits of Presidents: Richard James Morrison." *Weather* 49 (1994): 71–72.

Waller, John C. "Gentlemanly Men of Science: Sir Francis Galton and the Professionalization of the British Life-Sciences." *Journal of the History of Biology* 34 (2001): 83–114.

Wallis, Roy, ed. *On the Margins of Science: The Social Construction of Rejected Knowledge*. Keele: University of Keele, 1979.

Walton, Elijah. *Clouds, Their Forms and Combinations*. London: Longman, 1868.

Ward, Wilfred. *William George Ward and the Catholic Revival*. London: Macmillan, 1893.

[Ward, William G.]. "Science, Prayer, Free Will, and Miracles." *Dublin Review* 8 (1867): 255–98.

Warner, Brian. *Charles Piazzi Smyth, Astronomer Artist: His Cape Years*. Cape Town: Balkema, 1988.

Warner, Deborah J. "What Is a Scientific Instrument?" *British Journal for the History of Science* 23 (1990): 83–93.

Watt, Andrew. "Early Days of the Society." *Journal of the Scottish Meteorological Society* 36 (1910): 304–12.

Watts, W. Marshall. *An Introduction to the Study of Spectrum Analysis*. London: Longman and Green, 1904.

Weale, John, ed. *London and Its Vicinity Exhibited in 1851*. London: n.p., 1851.

"The Weather of the United States." *Chambers' Journal of Popular Literature, Science and Art* 57 (1880): 721–23.

"Weather Watchers and Their Work." *Strand Magazine* 3 (1892): 182–89.

Welby, Horace. *Predictions Realized in Modern Times*. London: Kent and Co., 1862.

Wells, H. G. *Tono-bungay*. Edited by B. Cheyette. New York: Oxford University Press, 1996.

Welsh, Alexander. *Strong Representations: Narrative and Circumstantial Evidence in England*. Baltimore: Johns Hopkins University Press, 1992.

Westherall, David. "The Growth of Archeaological Societies." In *The Study of the Past in the Victorian Age*, edited by V. Brand, 21–34. Oxford: Oxbow Books, 1998.

Wheeler, Michael, ed. *Ruskin and Environment: The Storm Cloud of the Nineteenth Century*. Manchester: Manchester University Press, 1995.

Whewell, William. *Astronomy and General Physics Considered with Reference to Natural Theology*. London: W. Pickering, 1833.

———. *The Philosophy of the Inductive Sciences Founded upon Their History*. 2 vols. London: J. W. Parker, 1840.

[———]. "Preliminary Discourse on the Study of Natural Philosophy [A Review]." *Quarterly Review* 45 (1831): 374–407.

Whitbread, Samuel Charles. "President's Address." *Proceedings of the British Meteorological Society* 1 (1850): 2–4.

White, Gilbert. *Natural History of Selbourne*. Harmondsworth, UK: Penguin, 1977.

White, W. H., ed. *The Journal of the Astronomic Meteorology and Record of the Science and Phenomena of the Weather*. London: privately printed, 1862.

———. *A Series of Letters on Weather Changes and Their Approximate Causes*. Glasgow: R. H. Melville, 1865.

Wilkins, W. J. *Daily Life and Work in India*. London: T. Fisher Unwin, 1888.

Wilks, M., ed. *Prophecy and Eschatology*. Oxford: Blackwell for the Ecclesiastical History Association, 1994.

Williams, Frances Leigh. *Matthew Fontaine Maury*. New Brunswick, NJ: Rutgers University Press, 1963.

Williams, Rowland. *Lampeter Theology*. London: Bell and Daldy, 1856.

Winter, Alison. "'Compasses All Awry': The Iron Ship and the Ambiguities of Cultural Authority in Victorian Britain." *Victorian Studies* 38 (1994): 69–98.

———. "The Construction of Orthodoxies and Heterodoxies in Early Victorian Life Sciences." In *Victorian Science in Context*, edited by B. Lightman, 24–50. Chicago: University of Chicago Press, 1997.

———. *Mesmerized: Powers of Mind in Victorian Britain*. Chicago: University of Chicago Press, 1998.
Wise, M. Norton, ed. *Values of Precision*. Princeton, NJ: Princeton University Press, 1995.
Woodward, Christopher. *In Ruins*. London: Chatto and Windus, 2001.
Wynter, Andrew. "The Clerk of the Weather." In *Subtle Brains and Lissome Fingers, Being Some of the Chisel Marks of Our Industrial and Scientific Progress*, 4–17. London: R. Hardwicke, 1863.
Yavetz, Ido. "A Victorian Thunderstorm: Lightning Protection and Technological Pessimism in the Nineteenth Century." In *Technology, Pessimism and Postmodernism*, edited by Yaron Ezrahi, Everett Mendelsohn, and Howard Segal, 53–76. Amherst: University of Massachusetts Press, 1994.
Yeo, Richard. "An Idol of the Market Place: Baconianism in the Nineteenth Century." *History of Science* 23 (1985): 251–98.
Zaniello, Thomas. "The Spectacular English Sunsets of the 1880s." In *Victorian Science and Victorian Values: Literary Perspectives*, edited by J. Paradis and T. Postlewait, 247–67. New York: New York Academy of Sciences, 1981.

Index